Intermediate 2

BIOLOGY

2ND EDITION

Team Co-ordinator: James Torrance

Writing Team:

James Torrance

James Fullarton

Clare Marsh

James Simms

Caroline Stevenson

Diagrams by James Torrance

Cover Photograph: the cover photograph shows a happy, well fed and inquisitive Common Seal. Common Seals can frequently be seen basking in the sea water lochs and coastal waters of Scotland. In September and October they come ashore at sheltered and undisturbed beaches to give birth and care for their pups.

Hodder Gibson

A MEMBER OF THE HODDER HEADLINE GROUP

The Publishers would like to thank the following for permission to reproduce copyright material:

Photo credits

Figure 01.18 © Biomedical Imaging Unit, Southampton General Hospital/SPL; Figure 02.03 © Andrew Syred/SPL; Figure 02.15 © Dr David Patterson/SPL; Figure 03.05 © Prof. K. Seddon & Dr T. Evans, Queen's University Belfast/SPL; Figure 04.10 © Tony Marshall/EMPICS; Figure 06.03 © David R. Frazier Photolibrary, Inc./Alamy; Figure 06.03 © Huw Evans/Alamy; Figure 06.05 © Holt Studios International Ltd/Alamy; Figure 06.07 © Juniors Bildarchiv/Alamy; Figure 07.04 © Terry Mead/SPL; Figure 07.05 © Richard R Hansen/SPL; Figure 07.06 © Mike Powles/Still Pictures; Figure 07.07 © Roger Wilmshurst/SPL; Figure 07.14 © Chinch Gryniewicz; Ecoscene/Corbis; Figure 07.25 © St Bartholemew's Hospital/SPL; Figure 07.28 © Peter Parks/NHPA; Figure 07.39 © M. H. Sharp/SPL; Figure 07.40 © The Photolibrary Wales/Alamy; Figure 07.41 © WoodyStock/Alamy; Figure 08.12 © Biophoto Associates/SPL; Figure 10.02 © David Aubrey/SPL; Figure 10.05 © A. B. Dowsett/SPL; Figure 10.08 © SCIMAT/SPL; Figure 11.07 © Biophoto Associates/SPL; Figure 11.15 © Anatomical Travelogue/SPL; Figure 11.23 © Biophoto Associates/SPL; Figure 12.05 © Manfred Kage/SPL; Figure 13.02 © Nathan Benn/Corbis; Figure 13.04 © LUNAGRAFIX/SPL; Figure 13.15 © John Burbidge/SPL; Figure 14.01 © Steve Gschmeissner/SPL; Figure 14.06 © Dr Kari Lounatmaa/SPL; Figure 15.01 © Dr Colin Chumbley/SPL; Figure 15.09 © Anatomical Travelogue/SPL; Figure 15.11 © CNRI/SPL; Figure 15.14 © Adam Hart-Davis/SPL.

All other photos by the author.

Acknowledgements

All artworks by James Torrance

Every effort has been made to trace all copyright holders, but if any have been inadvertently overlooked the Publishers will be pleased to make the necessary arrangements at the first opportunity.

Although every effort has been made to ensure that website addresses are correct at time of going to press, Hodder Gibson cannot be held responsible for the content of any website mentioned in this book. It is sometimes possible to find a relocated web page by typing in the address of the home page for a website in the URL window of your browser.

Whilst every effort has been made to check the instructions of the practical work in this book, it is still the duty and legal obligation of the school to carry out their own risk assessments

Orders: please contact Bookpoint Ltd, 130 Milton Park, Abingdon, Oxon OX14 4SB. Telephone: (44) 01235 827720. Fax: (44) 01235 400454. Lines are open from 9.00 – 5.00, Monday to Saturday, with a 24-hour message answering service. Visit our website at www.hoddereducation.co.uk. Hodder Gibson can be contacted direct on: Tel: 0141 848 1609; Fax: 0141 889 6315; email: hoddergibson@hodder.co.uk

Contents

Preface

This book has been written to articulate closely with Standard Grade Biology. It is intended to act as a valuable resource for pupils studying Biology Intermediate 2 as a one-year bridging course on the way to Higher Grade Biology or Higher Grade Human Biology the following session.

The book provides a concise set of notes that adheres to the SQA syllabus for Biology Intermediate 2. Each section of the book matches a unit of the syllabus; each chapter corresponds to a content area. The text is interspersed with a variety of special features:

- *Testing Your Knowledge*: key questions designed to continuously assess *knowledge and understanding*, especially useful as homework and as instruments of diagnostic assessment to check that full understanding of course content has been achieved
- *Activities*: pieces of structured course work directly related to the syllabus and designed to foster the development of specified *problem-solving skills* including selection of relevant information, presentation of information, processing of information, planning experimental procedures, drawing valid conclusions and making predictions
- *Practical Activities*: assignments designed to give students extensive day-to-day experience of syllabus-related practical work and allow them to gain confidence in *practical abilities*.
- *Practical Activities and Reports*: assignments designed to exactly match the required performance criteria and provide students with further opportunities to develop *practical abilities* and to practise writing *scientific reports* that include description of procedure, recording of results, presentation of results, drawing of conclusions and evaluation of procedure.

Each chapter is followed by a feature entitled *Applying Your Knowledge*, which consists of a variety of questions designed to give students practice in exam questions and further consolidate *knowledge and understanding* and *problem-solving skills* by acting as extensions to classwork and as homework.

At regular intervals throughout the book *What You Should Know* summaries of key facts and concepts are given as 'cloze' tests accompanied by appropriate word banks.

Living Cells

Plants, animals and micro-organisms are made of cells. Cells are functioning living units and nothing smaller than one cell can lead an independent life in the biosphere

1 Structure and function of cells

Cellular structure

Every living organism is made up of one or more **cells**. Cells are the basic units of life. Nothing smaller than a cell can lead an independent life and show all the characteristics of a living thing.

Comparison of a typical animal cell (e.g. a cell from human cheek epithelium) and a typical plant cell (e.g. a cell from a leaf of an *Elodea* plant) reveals that the two have some structural similarities and some differences. These are highlighted in Figure 1.1. The functions of the various cellular structures are summarised in Table 1.1.

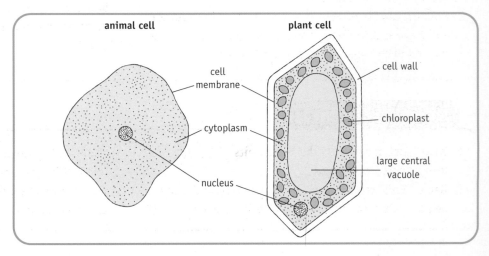

Figure 1.1 *Comparison of animal and plant cell*

	cell structure	description	function
found in plant and animal cells	nucleus	large, normally spherical structure containing genetic materials	controls cell's activities and passes information on from generation to generation
	cytoplasm	fluid, jelly-like material	acts as site of cell's biological processes and biochemical reactions (e.g. respiration)
	cell membrane	thin layer surrounding cytoplasm	controls the passage of substances into and out of the cell
found in plant cells only	chloroplast	discus-shaped structure containing green chlorophyll	absorbs light energy needed for manufacture of carbohydrates during photosynthesis
	cell wall	outer layer made of basket-like mesh of cellulose fibres	supports the cell and contributes to plant's overall semi-rigid structure
	large central vacuole	fluid-filled sac-like structure in cytoplasm	stores water and solutes as cell sap and regulates water content by osmosis

Table 1.1 *Functions of cell structures*

Practical Activity and Report

Examining cells

YOU NEED

1 microscope and light source
1 pair of forceps
1 safety razor
5 glass slides
5 cover slips
1 dropping bottle of iodine solution
1 piece of onion leaf
1 piece of rhubarb stalk
1 specimen of *Elodea*
1 sample of live yeast cells suspended in water

WHAT TO DO

1 Set up a light microscope.
2 Mount a sample of your cheek cells in iodine solution. Add a cover slip. Concentrate on one cell and look for the cell membrane, nucleus and cytoplasm.
3 Mount a sample of onion leaf epidermis in iodine solution. Add a cover slip. Concentrate on one cell and look for the cell wall, nucleus and cytoplasm.
4 Mount a sample of rhubarb stalk epidermis in water. Add a cover slip. Concentrate on one cell and look for the cell wall and the vacuole full of red cell sap.
5 Mount a leaf from an *Elodea* plant in water. Add a cover slip. Concentrate on one cell and look for the cell wall, chloroplasts and evidence of movement of cytoplasm (cytoplasmic streaming).
6 Place a drop of water containing yeast cells on a slide. Add a cover slip and view under the highest magnification possible. Concentrate on one cell and look for the cell wall and cytoplasm.
7 Attempt question 2 on page 17 at the end of this chapter.

Testing Your Knowledge

1 What name is given to the basic units of life that can lead an independent existence? (1)
2 Name THREE structural features that a typical plant cell and a typical animal cell have in common. (3)
3 Name THREE structural features present in an *Elodea* leaf cell but absent from a cheek epithelial cell. (3)
4 Give the function of each of the following structures: cell membrane, cell wall, nucleus. (3)

Commercial and industrial uses of cells

Yeast

Yeast is a single-celled **fungus**. Its structure is shown in Figure 1.2. It does not contain chlorophyll and cannot make its own food by photosynthesis.

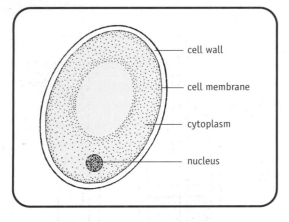

Figure 1.2 *Structure of yeast cell*

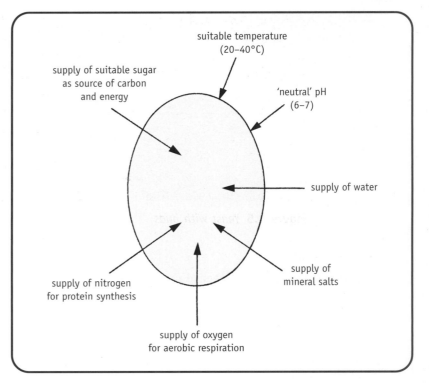

Figure 1.3 *Conditions for growth of yeast*

Conditions for growth

Figure 1.3 indicates the conditions that yeast requires for growth. This growth takes the form of an increase in cell number by asexual reproduction. Yeast multiplies by **budding** (Figures 1.4 and 1.5)

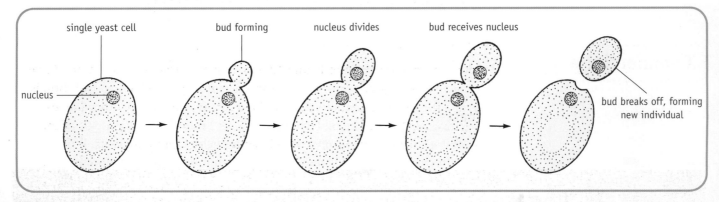

Figure 1.4 *Budding in yeast*

Figure 1.5 *Yeast with buds*

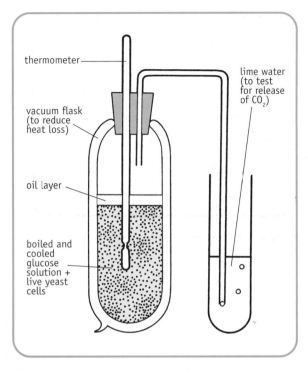

Figure 1.6 *Fermentation experiment*

Aerobic and anaerobic respiration

Respiration is the process by which a living organism releases energy from its food. When the process uses oxygen it is called **aerobic** respiration; when it occurs in the absence of oxygen it is called **anaerobic** respiration.

Anaerobic respiration releases less energy per unit of sugar than aerobic respiration. For this reason the vast majority of living things thrive in oxygen and respire aerobically whenever possible. They only resort to anaerobic respiration when oxygen is unavailable. Yeast is no exception to this rule and growth and budding are promoted by aerobic conditions.

Alcoholic fermentation

The experiment shown in Figure 1.6 is set up to investigate the action of live yeast cells on sugar (glucose) in the absence of oxygen. The sugar solution is boiled before use to remove dissolved oxygen and to kill any other micro-organisms present. The oil layer keeps air, and therefore oxygen, out of the yeast and sugar mixture. A control is set up using dead yeast cells.

	results after 2 days	
	experiment	control
thermometer readings	temperature rises from 20 to 23°C	temperature remains unchanged
lime water test for carbon dioxide	changes from clear to cloudy (milky)	remains unchanged
distillation at 80°C of liquid remaining in flask	ethanol (alcohol) is collected	no ethanol is collected

Table 1.2 *Fermentation results*

Table 1.2 shows a typical set of results. Since the control remains unchanged, it is concluded that living yeast cells (in the absence of oxygen) are able to use sugar as their source of energy. Some of this energy is lost as heat. Carbon dioxide and alcohol are released as waste products. This process is called **alcoholic fermentation**. It is summarised in the following word equation:

$$\text{sugar} \rightarrow \text{alcohol} + \text{carbon dioxide} + \text{energy}$$

Baking

Alcoholic fermentation is an essential stage in **bread-making**. When the kneaded dough (see Figures 1.7 and 1.8) is left in a warm place for a few hours, the small quantity of sugar present in the mixture is rapidly fermented to alcohol and carbon dioxide by **yeast**. The bubbles of CO_2 become trapped in the dough, making it rise. The alcohol is driven off during the baking process.

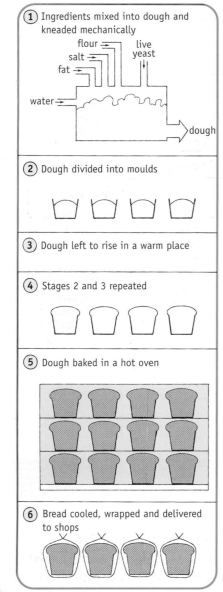

① Ingredients mixed into dough and kneaded mechanically

flour → live yeast
salt →
fat →
water →
→ dough

② Dough divided into moulds

③ Dough left to rise in a warm place

④ Stages 2 and 3 repeated

⑤ Dough baked in a hot oven

⑥ Bread cooled, wrapped and delivered to shops

Figure 1.7 *Dough before and after rising*

Brewing

The production of **beer** depends on alcoholic fermentation of the sugar maltose by yeast. The maltose is obtained from barley grains, which are allowed to germinate under controlled conditions. During this process, amylase enzyme in the grains digests starch to **maltose**. After a few days the malting process is brought to a halt by heat treatment. This destroys the enzyme and prevents the plant seedlings from using the maltose for growth. Instead the maltose sugar is used to make beer.

Figure 1.8 *Bread-making*

Activity

Selecting information on the commercial use of a micro-organism

WHAT TO DO

Study the diagram of the commercial brewing of beer in Figure 1.9 and then answer the following questions.

a) Which micro-organism is being put to commercial use in this process?
b) (i) Name TWO conditions required for growth by the microbe during alcoholic fermentation.
   (ii) For each of these describe the steps taken by a commercial brewer to try to maintain them at optimum level.
c) Why are hops used during beer brewing?
d) Name THREE by-products of the process.
e) Why does the fermentation process normally stop when the alcohol concentration reaches 12–15%?
f) By what means are dead yeast cells removed from the final product before it is packaged?

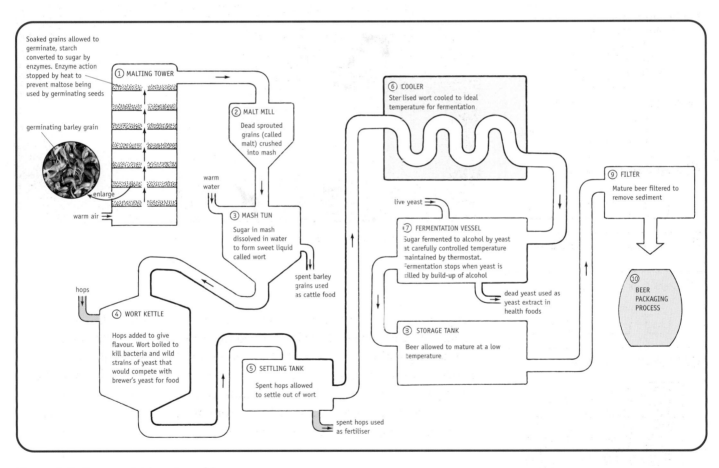

Figure 1.9 *Commercial brewing of beer*

Wine-making

The production of **wine** also depends on alcoholic fermentation. Wine-making is one of the oldest biotechnological processes known to humankind. In ancient times, ripe grapes were crushed and any wild yeast cells that just happened to be present on the grape skins fermented the fruit juice to wine.

Modern technology has refined this process by employing specific strains of yeast known to produce desired types of wine. First the ripe grapes are crushed to release the fruit juice rich in sugar. Then a small quantity of sulphur dioxide is added to kill wild yeasts and other microbes that are present on the grape skins. The sulphur dioxide is soon lost to the atmosphere by gaseous diffusion.

The wine **yeast** is now added and, in the absence of oxygen in the fermentation tank, it converts the sugar in the fruit juice to **alcohol** and carbon dioxide. As the concentration of alcohol builds up to around 12–15%, the yeast cells become poisoned, die and drop to the bottom of the fermentation tank, forming a sediment. Once the wine has cleared, it is filtered and bottled.

Commercial wine-making is carried out on a huge scale. Figure 1.10 shows a simple version of the process on a small scale using a wine-making kit.

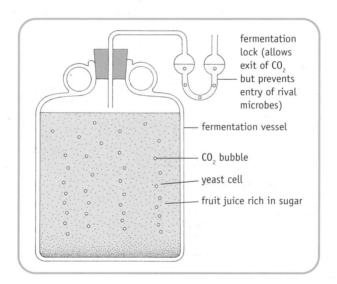

fermentation lock (allows exit of CO_2 but prevents entry of rival microbes)

fermentation vessel

CO_2 bubble

yeast cell

fruit juice rich in sugar

Figure 1.10 *Home wine-making set-up*

Alternative fuel production

Fossil fuels such as coal and oil are **non-renewable** resources that will eventually run out. For this reason scientists continue to search for alternative energy sources that are **renewable** and can be developed in a cost-effective way.

Gasohol

Alcohol formed by fermentation of plant material is rich in energy and can be used as a fuel. About half of Brazil's million-ton sugar cane crop is converted to alcohol every year for use as fuel for cars with specially adapted engines. The product on sale at the pumps is called **gasohol** (see Figure 1.11). It consists of alcohol mixed with a little petrol. Provided that the sun continues to shine, sugar made by photosynthesis will be available to produce this **alternative fuel**.

Biogas

Methane is the chemical name of the gas widely used as a source of fuel. Underground reserves of methane are non-renewable and will eventually run out. Fortunately microbes can be used to produce an alternative supply of this fuel. During anaerobic respiration some types of bacteria break down organic waste materials and convert them to methane (often called **biogas**) as in the equation:

$$\text{organic waste} \xrightarrow{\text{anaerobic bacteria}} \text{biogas} + CO_2$$

These anaerobic bacteria can be put to use on the farm to convert manure and leafy crop remains to methane. Figure 1.12 shows a biogas generator in action. The biogas is often used for cooking, heating and lighting. It can also be compressed into cylinders and used to fuel tractors and farm machinery. In China more than seven million homes already have their own small family-sized biogas generator.

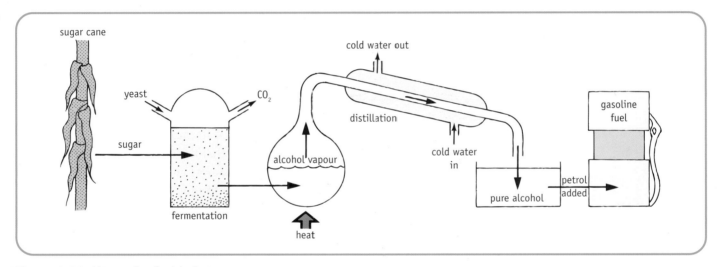

Figure 1.11 *Alternative fuel industry*

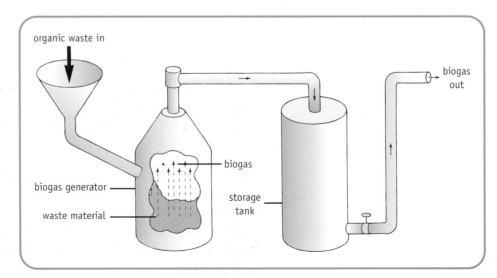

Figure 1.12 *Biogas generator*

Testing Your Knowledge

1. a) Identify FOUR structural features that a yeast cell has in common with an *Elodea* leaf cell. (4)
   b) Name ONE structure absent from a yeast cell but present in *Elodea*. (1)
2. a) Give the word equations of aerobic and anaerobic respiration in yeast. (2)
   b) On which of these chemical reactions do the baking and brewing industries depend? (1)
3. a) Describe the role of carbon dioxide in bread-making. (1)
   b) Does bread normally contain alcohol? Explain your answer. (1)
4. a) Write a simple word equation to summarise the process of malting in barley grains about to be used in the brewing of beer. (1)
   b) Why is malting not carried out during wine-making? (2)
   c) What procedure is carried out to prevent rival microbes competing for the wine yeast's food supply? (1)
   d) Explain the purpose of the fermentation lock in Figure 1.10. (2)
   e) Why is cloudy wine not ready for consumption? (1)
5. Brazil has a warm sunny climate suitable for the rapid growth of sugar cane. Describe how this South American country has overcome its lack of the fossil fuels needed to run cars. (2)

Antibiotics

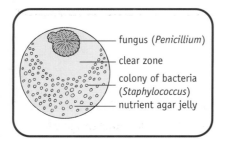

- fungus (*Penicillium*)
- clear zone
- colony of bacteria (*Staphylococcus*)
- nutrient agar jelly

Figure 1.13 *Fleming's famous plate*

In 1928 Alexander Fleming found a fungal contaminant growing on one of his plates of bacteria (Figure 1.13). He noticed that the area around the fungal colony, instead of being cloudy with bacteria, was **clear**. He therefore concluded that some substance made by the fungus *(Penicillium)* and secreted out into the nutrient agar was inhibiting the growth of nearby bacteria. This substance, an **antibiotic**, was later isolated and called **penicillin**.

Tests showed that penicillin was effective against several species of disease-causing bacteria inside the human body without being toxic to the human patient. Penicillin has been used successfully against numerous bacterial infections such as pneumonia, gonorrhoea and meningitis.

Other antibiotics

Many other antibiotics have now been discovered. An **antibiotic** is a naturally occurring chemical produced by one type of micro-organism (e.g. a fungus) and passed out into the surrounding environment where it kills or inhibits the growth of other types of micro-organisms (e.g. several species of bacteria).

In a natural ecosystem such as soil, a microbe uses an antibiotic to kill its rivals and allow it to make use of the available food in the absence of competition.

Specificity of action

If a micro-organism's growth is prevented by an antibiotic, the microbe is said to be **sensitive** to the antibiotic. If the antibiotic has no effect, the microbe is said to be **resistant**.

From the experiment shown in Figure 1.14, it can be concluded that bacterial species 1 is sensitive to streptomycin and resistant to penicillin whereas bacterial species 2 is sensitive to penicillin and resistant to streptomycin.

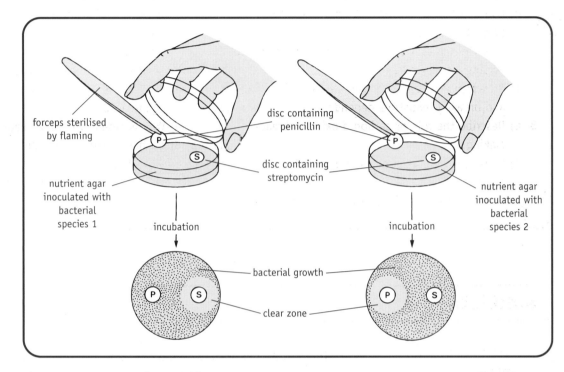

Figure 1.14 *Action of antibiotics*

There is no one antibiotic that is effective against all species of bacteria (see Table 1.3). However, some are found to be effective over a wider range of bacterial strains than others (Figure 1.15).

disease-causing bacterium	disease caused	antibiotic			
		penicillin	**streptomycin**	**tetracycline**	**chloramphenicol**
Corynebacterium diphtheriae	diphtheria	+++	−	++	++
Mycobacterium tuberculosis	tuberculosis	−	+++	−	−
Salmonella typhii	typhoid	−	−	+	+++
Streptococcus pneumoniae	pneumonia	+++	−	+++	+++

+++ = very effective, ++ = effective, + = slightly effective, − = ineffective

Table 1.3 *Effectiveness of different antibiotics*

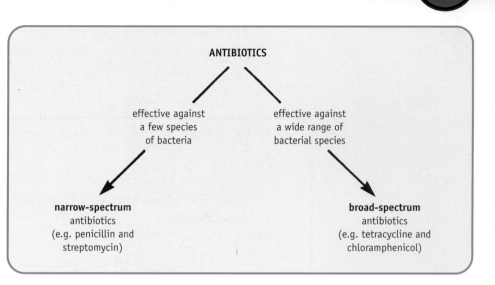

Figure 1.15 *Narrow- and broad-spectrum antibiotics*

Effects of antibiotics on sensitive bacteria

Different antibiotics act on bacteria in different ways. Some of these are shown in Figure 1.16, which features a generalised version of a bacterial cell. Over the last 50 years, many antibiotics have been isolated and mass-produced for use against disease-causing bacteria. As a result, the incidence of deadly diseases such as diphtheria, meningitis, pneumonia, tuberculosis and typhoid has declined to almost zero in many countries of the world.

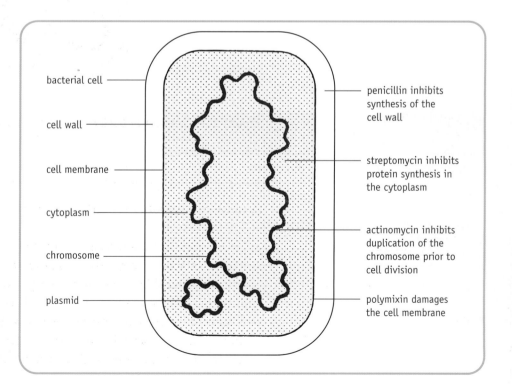

Figure 1.16 *How antibiotics work*

Resistant strains

However, the battle against bacteria has not been won. Among almost all types of bacteria there occur naturally a few individuals that are **resistant** to antibiotics. These bacteria just happen to have the genetic material that makes them resistant to antibiotics. Despite the presence of the antibiotic, they survive and multiply while their competitors, which are sensitive to the antibiotic, die.

Mechanisms of resistance

Figure 1.17 shows some of the ways in which bacteria may successfully resist an antibiotic.

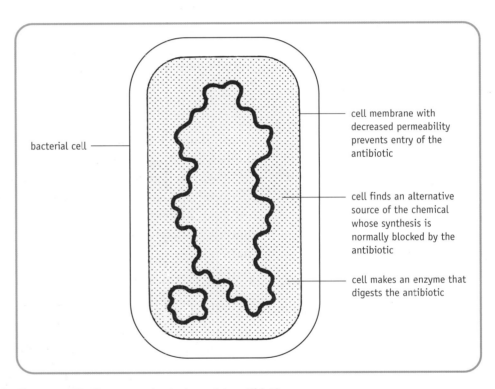

Figure 1.17 *How some bacteria resist antibiotics*

Need for a range of antibiotics

A range of antibiotics is needed in the treatment of bacterial diseases for the following reasons:
- No one antibiotic is effective against all species of bacteria.
- People vary in their genetic make-up and therefore a minority of people tend to suffer side effects or be allergic to a certain antibiotic. If several antibiotics are available then a suitable one can be used.
- Strains of bacteria are constantly emerging that are resistant to one or more antibiotics. If several different antibiotics are available then there is a good chance that at least one of them will be effective against the bacterial strain.

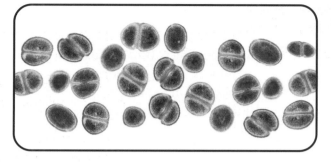

Figure 1.18 *MRSA (Methicillin-resistant Staphylococcus aureus)*

MRSA

Staphylococcus aureus is a species of bacterium commonly found on the surface of the human skin. It causes one in five of the infections acquired by hospital patients during treatment. In the past it was successfully controlled by a range of antibiotics. However, there now exists a strain of *S. aureus* known as MRSA (see Figure 1.18) which is **resistant** to almost all antibiotics. Experts claim that it is only a matter of time before a strain resistant to all known antibiotics appears. For this reason the search for new antibiotic-producing fungi and their products remains as important as ever.

Figure 1.19 *Preparing a slide of yoghurt bacteria*

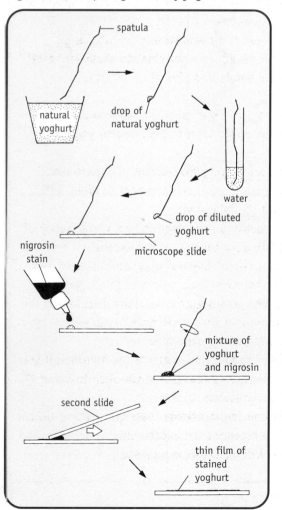

Practical Activity

Examining yoghurt bacteria

YOU NEED

1 carton of natural yoghurt
1 spatula
1 test tube
2 microscope slides
1 dropping bottle of nigrosin stain
1 microscope and light source

WHAT TO DO (see Figure 1.19)

1 Using a spatula, add a drop of natural yoghurt to a little water in a test tube and stir.
2 Add a drop of diluted yoghurt to a glass slide.
3 Add a drop of nigrosin stain.
4 Using the spatula, mix the yoghurt and stain.
5 Using a second glass slide, drag the mixture along the slide leaving a thin layer.
6 Allow the specimen to dry for 5 minutes.
7 Set up the microscope.
8 Examine the specimen under the highest magnification and look for yoghurt bacteria.
9 Try to decide the number of types present, their shape(s) and whether they absorbed or repelled the black stain.
10 Compare your findings with the bacteria shown in Figure 1.20 on page 16.

Practical Activity and Report

Demonstration of lactic acid production by yoghurt bacteria

YOU NEED

2 test tubes
2 rubber stoppers
1 test tube stand
2 metal spatulas
natural yoghurt
boiled and cooled natural yoghurt
UHT milk
1 marker pen or 2 labels
1 dropping bottle of universal indicator
1 pH reference scale
access to incubator at 30°C
1 roll of sellotape
1 pair of scissors

WHAT TO DO

1 Read all the instructions in this section and prepare your table of results before carrying out the investigation.

2 Label two test tubes A and B and add your initials.

3 Half-fill each tube with sterile (UHT) milk.

4 Using a spatula, add a large scoop of natural yoghurt (rich in live yoghurt bacteria) to tube A.

5 Using the second spatula, add an equal scoop of boiled and cooled natural yoghurt to tube B.

6 Put 20 drops of universal indicator in each tube.

7 Seal each tube with a rubber stopper and sellotape and then shake each tube thoroughly.

8 Draw up a table which refers to the pH of tubes A and B both before and after incubation.

9 Note the pH of the contents of each tube in your table and then place the tubes in an incubator at 30°C.

10 After 24 hours examine the two tubes without opening them, note the pH of each and complete your table.

11 If other students have carried out the same experiment, pool the results.

12 Return your sealed tubes to the teacher for safe disposal.

REPORTING

Write up your report by doing the following:

1 Copy the title given at the start of this activity.

2 Put the subheading '**Aim**' and state the aim of your experiment.

3 a) Put the subheading '**Method**'.

b) Draw a labelled diagram of your apparatus set up and ready to go into the incubator.

c) Briefly describe the experimental procedure that you followed using the **impersonal passive voice**. **Note:** The impersonal passive voice avoids the use of 'I' and 'we'. Instead it makes the apparatus the subject of the sentence. In this experiment, for example, you could begin your report by saying 'Two test tubes labelled A and B were half-filled with milk...etc. (**not** 'I half-filled two test tubes'...etc.)

d) Continuing in the impersonal passive voice, state how your results were obtained.

4 Put a subheading '**Results**' and draw a final version of your table of results.

5 Put a subheading '**Presentation of results**' and present your results as a bar chart to compare the pH in each tube before and after the time in the incubator.

6 Put a subheading '**Conclusion**' and draw a conclusion about the type of substance produced by yoghurt bacteria.

7 Put a final subheading '**Evaluation of Experimental Procedure**', consider your experiment carefully and then answer the following:

a) Strictly speaking, why does adding a large scoop of yoghurt to each tube leave the experiment open to a source of error? Suggest how this source of error could be overcome in a repeat of the experiment.

b) Explain why it is better to pool the class results when drawing a conclusion rather than depend on the results of only one group.

c) Feel free to comment on any of the following if you have an additional point that you wish to make:

(i) further sources of error

(ii) further improvements that you would include in a repeat of the experiment

(iii) limitations of the equipment.

Lactic acid formation

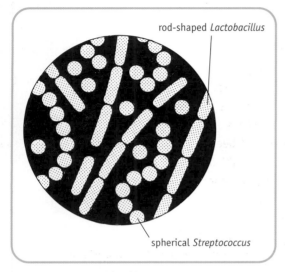

rod-shaped *Lactobacillus*

spherical *Streptococcus*

Figure 1.20 *Lactic acid bacteria*

Milk is rich in sugar, protein and fat and is therefore an excellent source of food for many types of living thing, including bacteria. During the **souring** of milk, certain strains of naturally occurring bacteria growing in the milk respire anaerobically. They feed on a sugar in the milk called **lactose** and convert it to **lactic acid**, as shown in the following equation:

$$\text{lactose} \xrightarrow{\text{enzyme action}} \text{lactic acid}$$

This process is called lactic acid fermentation and the bacteria responsible are referred to as **lactic acid bacteria**. Figure 1.20 shows a sample of these bacteria against a background of nigrosin stain.

Effect of lactic acid

The lactic acid formed by the bacteria results in a drop in pH, which makes molecules of milk protein coagulate (clump together) and produce the semi-solid food known as **yoghurt**. Since the growth of most micro-organisms is prevented by conditions of low pH, the lactic acid also acts as a **natural preservative**, protecting the yoghurt against the action of decomposers.

Testing Your Knowledge

1 a) What is meant by the term antibiotic? (2)
  b) (i) Identify the first antibiotic to be discovered and isolated by humans.
     (ii) Name the micro-organism that produces this antibiotic. (2)
  c) In what way does the production of an antibiotic help a fungus to survive under normal conditions in its natural habitat? (2)
  d) State TWO features that an antibiotic must possess before it can be put to medical use by doctors. (2)

2 a) With reference to the experiment shown in Figure 1.14, copy the following sentences choosing the correct word from the choice given in brackets:
     (i)   Bacterial species 1 is (sensitive/resistant) to penicillin.
     (ii) Bacterial species 1 is (sensitive/resistant) to streptomycin.
     (iii) Bacterial species 2 is (sensitive/resistant) to penicillin.
     (iv) Bacterial species 2 is (sensitive/resistant) to streptomycin. (4)

  b) Why is it not possible to successfully treat all bacterial diseases with penicillin? (1)
  c) Tetracycline is described as a 'broad-spectrum' antibiotic. Explain why. (1)

3 a) Give TWO ways in which antibiotics can inhibit the growth of or kill sensitive bacteria. (2)
  b) Give TWO ways in which resistant bacteria can prevent antibiotics having an adverse effect on them. (2)

4 a) Give TWO advantages of having a choice of several antibiotics with which to treat a particular disease. (2)
  b) Some experts claim that humans are heading towards a postantibiotic era where they will no longer be able to depend on antibiotics to cure diseases. Briefly explain how this situation could arise. (2)

5 Copy the following sentences choosing the correct answer from each choice in brackets:
  During the production of yoghurt, (bacteria/yeast) respire (aerobically/anaerobically) and convert (maltose/lactose) to (lactic acid/alcohol). This chemical brings about the coagulation of milk (proteins/sugars) and acts as (an antibiotic/a preservative). (3)

Applying Your Knowledge

1 Match the terms in list X with their descriptions in list Y.

List X
1) alcohol
2) antibiotic
3) bacteria
4) carbon dioxide
5) cell membrane
6) cell wall
7) chloroplast
8) cytoplasm
9) fermentation
10) lactic acid
11) lactose
12) maltose
13) nucleus
14) penicillin
15) resistant
16) sensitive
17) vacuole

18) yeast

List Y
a) gas that makes dough rise during baking
b) unicellular fungus used in baking and brewing
c) mesh of cellulose fibres that surrounds and supports a plant cell
d) product of anaerobic respiration used as an alternative fuel
e) spherical structure that controls cellular activities
f) sugar from cereal grains used in the brewing of beer
g) the first antibiotic to be isolated and used to fight disease
h) thin layer that controls entry and exit of materials into and out of a cell
i) unicellular organisms needed for yoghurt-making
j) process on which wine- and beer-making depend
k) describes a microbe that is inhibited by a certain antibiotic
l) describes a microbe that is neither inhibited nor killed by a certain antibiotic
m) jelly-like material in which biochemical reactions occur in all living cells
n) sugar in milk acted on by yoghurt-forming bacteria
o) large sac-like structure in a plant cell that regulates the water content of the cell
p) chemical substance produced by bacterial action on lactose
q) general name for a substance formed by one micro-organism that inhibits growth of other micro-organisms
r) discus-shaped structure containing chlorophyll for photosynthesis

2 Figure 1.21 shows the four types of cell referred to in the Practical Activity on page 3.

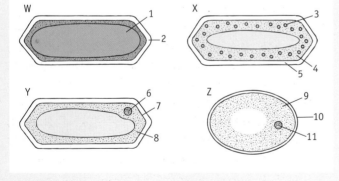

Figure 1.21

a) Identify cell types W, X, Y and Z. (4)
b) Name cell structures 1 to 11 (11)
c) (i) State the functions of parts 1, 3 and 9.
    (ii) Which of these is present in all living cells? (4)

3 Figure 1.22 shows a three-dimensional version of a cell with most of the front half cut away. Imagine that the remaining small front part is now also cut off. Make a simple, labelled diagram to show the possible appearance of the inside of the cell. (5)

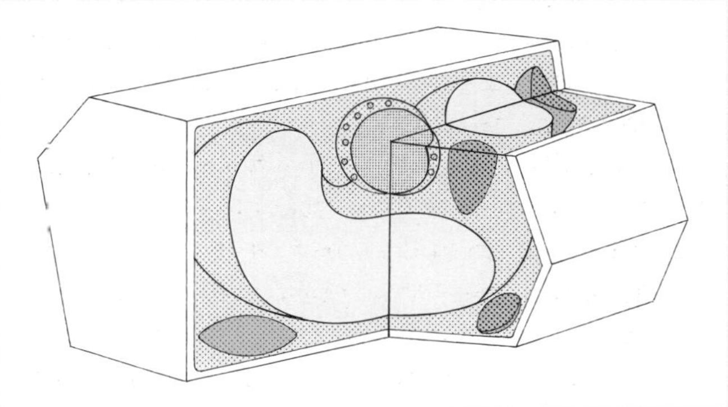

Figure 1.22

4 1 g of live yeast was added to glucose solution at pH 4, as shown in Figure 1.23, and the time taken to collect 5 cm³ of carbon dioxide was recorded. The experiment was repeated at six different pH values.

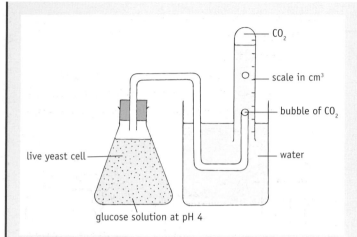

glucose solution at pH 4

live yeast cell

CO_2

scale in cm^3

bubble of CO_2

water

Figure 1.23

The results are given in Table 1.4

pH	time to collect 5 cm^3 of carbon dioxide (min)
4	294
5	188
6	104
7	116
8	149
9	214
10	273

Table 1.4

a) Present the results as a line graph by plotting the data given in Table 1.4 and joining up the points with a ruler. (2)
b) From your graph state the pH value at which yeast showed:
   (i) most activity   (ii) least activity. (2)
c) From your graph estimate the time that would have been required by yeast to release 5 cm^3 of CO_2 at pH 9. (1)
d) (i) Of the pH values used in this experiment, which is the optimum for the activity of yeast cells?
   (ii) How could you obtain an even more accurate measurement of the pH at which yeast is most active? (2)
e) The temperature of the laboratory in which the experiment was carried out varied between 18 and 23°C. How could this source of error be eliminated in a repeat of the experiment?

Describe any changes that would need to be made to the experimental set-up. (2)
5 The flow chart in Figure 1.24 represents the brewing of beer. Give the letter that indicates the stage at which:
   a) boiling is carried out to kill rival microbes
   b) starch is being converted to maltose inside soaked barley grains
   c) live yeast is added to sterilised wort
   d) hops are added to give flavour
   e) dead yeast cells are removed as yeast extract for manufacture of health foods. (5)

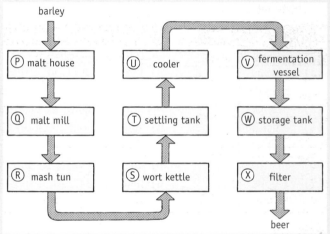

barley

(P) malt house   (U) cooler   (V) fermentation vessel

(Q) malt mill   (T) settling tank   (W) storage tank

(R) mash tun   (S) wort kettle   (X) filter

beer

Figure 1.24

6 The graph in Figure 1.25 shows some of the changes that took place during the fermentation of a closed batch of beer.

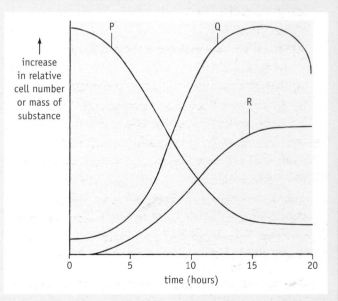

increase in relative cell number or mass of substance

P   Q

R

time (hours)

Figure 1.25

a) Match P, Q and R with live yeast cells, glucose and alcohol. (3)

b) Suggest why R is absent during the first two hours of the process. (1)

c) Explain the shape of curve Q. (3)

d) Explain the shape of curve P. (2)

e) (i) Identify the factor that you think limited the growth of the yeast cells.

   (ii) Explain your answer. (2)

7 Table 1.5 refers to the costs incurred in the manufacture of ethanol from three different raw materials.

expense	relative cost to manufacturer (units)		
	straw	sugar cane	timber
purchase of raw material	8.7	14.6	40.1
chemical treatment	8.7	2.3	12.9
fuel to run plant	412.4	30.3	139.8
use of water	1.3	0.7	0.7

Table 1.5

a) Which raw material is the least expensive to purchase? (1)

b) Which raw material requires the most costly chemical treatment? (1)

c) Which expense is equal for two of the raw materials? (1)

d) Which raw material needs most fuel to run its treatment plant? (1)

e) Which raw material is cheapest overall? (1)

f) By how many times is the most expensive raw material dearer than the cheapest one overall? (1)

8 A sterile inoculating loop was used to apply a sample of fungus species A to the nutrient agar in a Petri dish as shown in the left part of Figure 1.26 at line A. The loop was flamed and used to apply a sample of fungus species B at line B. The procedure was repeated in order to apply bacterial strains W, X, Y and Z at the lines indicated by these letters.

The right-hand side of the diagram shows the growth of fungi A and B and the bacterium W after three days.

a) To which fungus is bacterium W

   (i) sensitive?

   (ii) resistant?

   (iii) Explain how you arrived at your answers.

   (iv) Suggest the means by which the fungus you gave as your answer to (i) exerts its effect over bacterial strain W. (4)

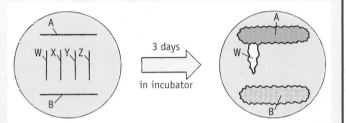

Figure 1.26

b) Copy the right-hand part of the diagram and draw in the following bacterial colonies:

   (i) X, which is resistant to both fungi

   (ii) Y, which is sensitive to both fungi

   (iii) Z, which is sensitive to fungus A but resistant to fungus B. (3)

9 A multidisc bearing antibiotics 1 to 6 on its side arms was placed on a colony of bacterial strain P growing on nutrient agar in a Petri dish. The procedure was repeated for bacterial strains Q, R and S and the four Petri dishes were incubated for two days at 30°C. Figure 1.27 shows the results.

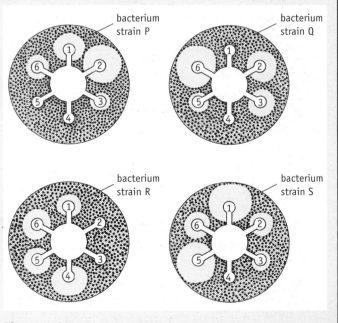

Figure 1.27

a) How many strains of bacteria were sensitive to antibiotic 1? (1)

b) Which strains of bacteria were sensitive to antibiotic 2? (1)

c) How many strains of bacteria were resistant to antibiotic 4? (1)

d) Which bacterium was most sensitive to antibiotic 5? (1)

e) How many strains of bacteria were resistant to antibiotic 6? (1)

f) How many antibiotics are effective against bacterium P? (1)

g) Which antibiotic is most effective against bacterium Q? (1)

h) How many antibiotics had no effect against bacterium R? (1)

i) Which antibiotic was least effective against bacterium S? (1)

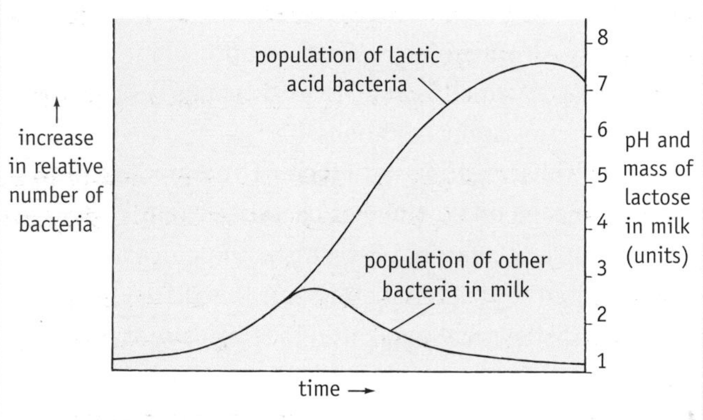

10 The graph in Figure 1.28 shows changes in two populations of bacteria in fresh milk left in a warm room for several days.

a) (i) Describe the relationship that exists between the number of lactic acid bacteria and the other strains of milk bacteria.

(ii) Account for this relationship. (2)

b) Make a copy of (or trace) the diagram to include both y axes.

(i) Draw a line to represent the change in lactose concentration that would occur over this timescale.

(ii) Draw a second line on your graph to represent the change in pH that would occur. (3)

Figure 1.28

Structure and function of cells

Word bank

alcohol
antibiotic
carbon dioxide
cells
chlorophyll
chloroplasts
compete
cytoplasm
fuel
lactic acid
membrane
nucleus
preservative
range
resistant
sensitive
vacuole
wall

Table 1.6 *Word bank for Chapter 1*

What You Should Know

(Chapter 1) (See Table 1.6 for word bank)

1 All living things are composed of one or more _____, the basic units of life.

2 Plant and animal cells have a nucleus, _____ and a cell membrane. Only plant cells have a cell wall, a large central vacuole and, sometimes, _____.

3 The cell's activities are controlled by the _____; the cell's biological processes occur in the cytoplasm; the movement of substances in and out of the cell is controlled by the cell _____. A plant cell is supported by the cell _____; its water content is regulated by the large _____; _____ for photosynthesis is contained in chloroplasts.

4 Yeast is a single-celled organism that produces _____ and carbon dioxide during anaerobic respiration. The release of _____ during baking makes the dough rise.

5 Yeast is used to produce alcohol during brewing, wine-making and the manufacture of alternative _____.

6 An _____ is a substance made by one micro-organism that inhibits the growth of or kills other micro-organisms.

7 Many fungi make antibiotics that help them _____ and survive in their natural habitats.

8 When bacteria are inhibited or killed by an antibiotic, they are said to be _____ to it; when bacteria remain unaffected by an antibiotic, they are said to be _____ to it.

9 A wide _____ of antibiotics is needed because no one antibiotic is effective against all bacteria and many naturally resistant strains of bacteria exist.

10 Yoghurt bacteria produce _____, which coagulates soluble proteins in milk, forming yoghurt. By lowering the food's pH and thereby inhibiting the growth of other microbes, lactic acid acts as a _____.

21

2 Diffusion and osmosis

Diffusion

The molecules of a liquid (and a gas) move about freely all the time. In the experiment shown in Figure 2.1, a crystal of purple potassium permanganate is dropped into a beaker of water. The diagram illustrates the events that follow. The purple particles move from a region of **high concentration** (the dissolving crystal) to a region of **low concentration** (the surrounding water) until the concentration of purple particles (and water) is uniform throughout.

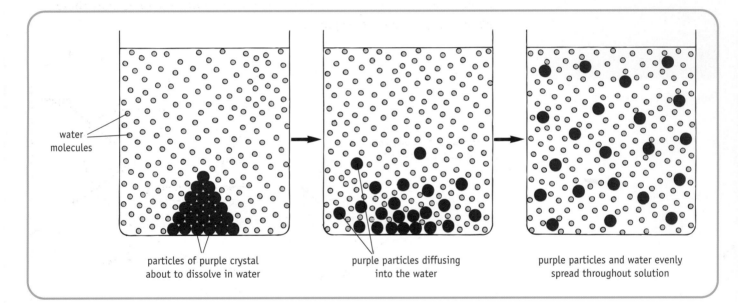

water molecules

particles of purple crystal about to dissolve in water

purple particles diffusing into the water

purple particles and water evenly spread throughout solution

Figure 2.1 *Diffusion*

Diffusion is the name given to the movement of the molecules of a substance from a region of high concentration of that substance to a region of low concentration until the concentration becomes equal.

Concentration gradient

The difference in concentration that exists between a region of high concentration and a region of low concentration is called the **concentration gradient**. During diffusion movement of molecules always occurs down a concentration gradient from high to low concentration.

Importance of diffusion to organisms

Unicellular organisms

Diffusion of oxygen and carbon dioxide

In a unicellular animal such as *Paramecium* oxygen is constantly being used up by the cell contents during respiration. This results in the concentration of oxygen molecules inside the cell being lower than in the surrounding water. The cell membrane is freely permeable to the tiny oxygen molecules. Oxygen therefore **diffuses into** the cell from a higher concentration to a lower concentration (see Figure 2.2).

At the same time, the living cell contents are constantly making carbon dioxide (CO_2) by respiration. This results in the concentration of carbon dioxide inside the cell being higher than in the surrounding water. Since the cell membrane is also freely permeable to tiny carbon dioxide molecules, these **diffuse out**, as shown in Figure 2.2.

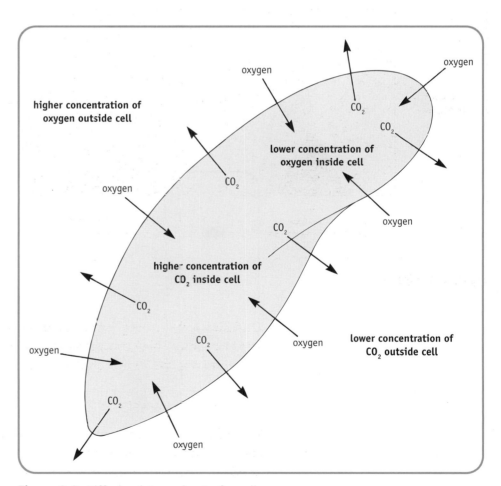

higher concentration of
oxygen outside cell

lower concentration of
oxygen inside cell

higher concentration of
CO_2 inside cell

lower concentration of
CO_2 outside cell

Figure 2.2 *Diffusion into and out of a cell*

Figure 2.3 *Phagocytosis in* Amoeba

Diffusion is important to a unicellular organism since it is the means by which useful substances such as oxygen enter it and waste materials such as carbon dioxide leave.

Diffusion of dissolved food

Amoeba engulfs its food by phagocytosis (see Figure 2.3). This means that the food becomes enclosed in a food vacuole and digested by enzymes to form soluble end products. These dissolved substances pass from a region of higher concentration (the food vacuole) to a region of lower concentration (the cytoplasm) by **diffusing down** a concentration gradient into the cytoplasm. This process provides all parts of the animal with nourishment (see Figure 2.4).

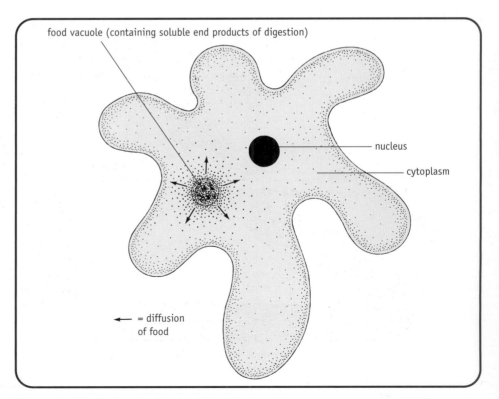

Figure 2.4 *Diffusion of dissolved food in Amoeba*

Multicellular animals

In a multicellular animal such as a human being, diffusion also plays an important role in the exchange of respiratory gases. Blood returning to the lungs from respiring cells (see Figure 2.5) contains a higher concentration of carbon dioxide and a lower concentration of oxygen than the air in the air sac. Carbon dioxide therefore **diffuses out** of the blood and oxygen **diffuses in**. When the blood reaches living body cells, the reverse process occurs and the cells gain oxygen from the blood and lose carbon dioxide by diffusion.

Similarly, diffusion is essential for the movement down a concentration gradient of molecules of

- dissolved food (e.g. glucose and amino acids) from the animal's bloodstream to respiring cells
- waste products (e.g. urea) from the cells to the bloodstream.

Green plants

Green plants depend on the process of diffusion to obtain the raw materials that they need for photosynthesis. Diffusion allows the movement down a concentration gradient of molecules of

- carbon dioxide from the air outside the leaves via the stomata and moist air spaces to the green leaf cells
- water from the soil solution via the root hairs, xylem vessels and moist air spaces to the green leaf cells (also see Figure 5.2 on page 83).

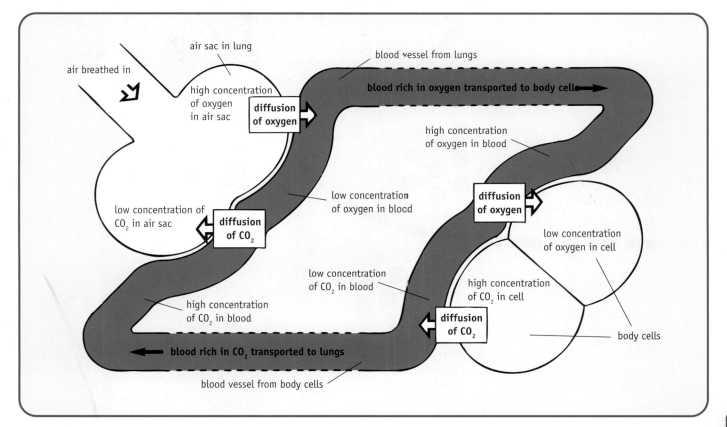

Figure 2.5 *Diffusion in a human being*

Role of cell membrane

Although the membrane of a cell is freely permeable to small molecules such as oxygen, carbon dioxide and water, it is not equally permeable to all substances. Larger molecules such as dissolved food can pass through the membrane slowly. Molecules which are even larger, such as starch, are unable to pass through. Thus the cell membrane controls the passage of substances into and out of a cell.

The exact means by which the cell membrane exerts this control is not yet fully understood. It is known that most cell membranes possess **tiny pores**. It is thought that many small molecules enter or leave by these pores (sometimes with the help of special molecules in the membrane). Other molecules are kept inside or outside the cell because they are too big to pass through the pores (see Figure 2.6).

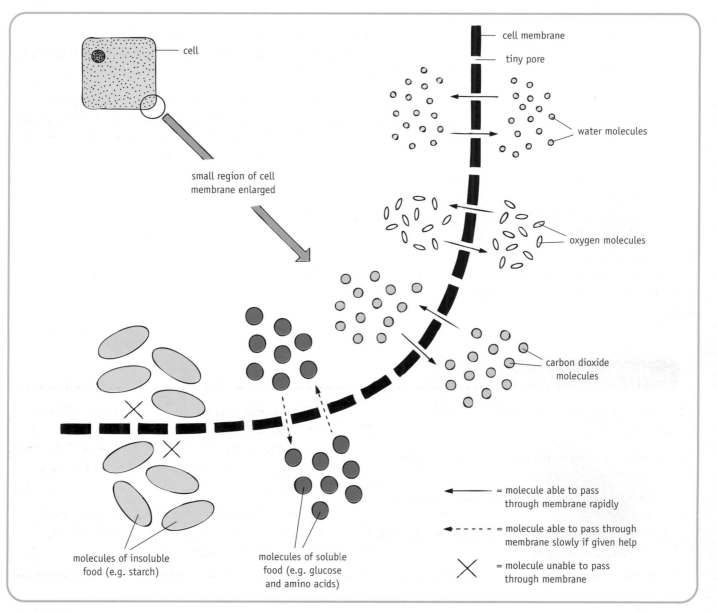

Figure 2.6 *Simplified model of the role of the cell membrane*

Relative water concentrations

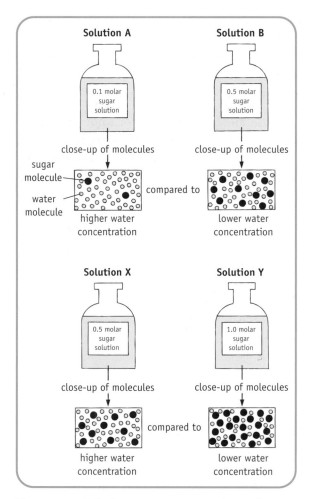

Figure 2.7 *Relative water concentrations of solutions*

When two solutions differ in water concentration, the solution with the **higher** water concentration is said to be **hypotonic** and the solution with the **lower** water concentration is said to be **hypertonic**. Since pure water has the highest possible water concentration, it is hypotonic to all other solutions.

Comparison of the solutions shown in Figure 2.7 shows that solution A is hypotonic to B and that B is hypertonic to A. Solution A is also hypotonic to solutions X and Y. Further comparison shows that solution X is hypotonic to Y and that Y is hypertonic to X. Solution Y is also hypertonic to solutions A and B.

When two solutions are found to be **equal** in water concentration, they are said to be **isotonic**. Solutions B and X in Figure 2.7 are isotonic.

(**Note**: The concentration of a chemical solute (e.g. sugar) dissolved in a solution can also be called its molarity. The higher a solution's molarity, the higher its concentration of solute and therefore the lower its water concentration.)

Testing Your Knowledge

1 Define the terms *diffusion* and *concentration gradient*. (2)

2 a) Name TWO essential substances that enter an animal cell by diffusion. (2)
  b) Name a waste material that diffuses out of an animal cell. (1)
  c) What structure controls the passage of substances into and out of a cell? (1)

3 a) With reference to Figure 2.5 explain why diffusion is important to human beings. (2)
  b) Predict what would happen to the rate of diffusion of CO_2 if a person exercised vigorously. Explain your answer. (2)

4 Explain the difference between the terms *hypotonic* and *hypertonic* with reference to 1% and 10% salt solutions and their water concentrations. (2)

Demonstration of osmosis using potato tissue

INFORMATION

- 0.1 molar (M) sucrose (sugar) solution has a higher water concentration than the contents of potato cells.
- 1.0 molar (M) sucrose solution has a lower water concentration than the contents of potato cells.
- 0.1 M sucrose is hypotonic and 1.0 M sucrose is hypertonic to potato cell sap.

YOU NEED

1 potato tuber
1 cork borer
1 ceramic tile
1 safety razor
paper towels
2 boiling tubes
2 labels or 1 marker pen
1 boiling tube stand
0.1 and 1.0 M sucrose solutions
access to an electronic balance

WHAT TO DO

1. Read all the instructions in this section and prepare your results table before carrying out the experiment.
2. Label two boiling tubes A and B and add your initials.
3. Half-fill tube A with the hypotonic solution (0.1 M sucrose) and half-fill tube B with the hypertonic solution (1.0 M sucrose).
4. Using a cork borer, cut out two cylinders from a potato tuber.
5. Trim the cylinders to equal length using a safety razor.
6. Dry the cylinders on a paper towel to remove liquid from their outer surfaces.
7. Gently squeeze each cylinder between thumb and forefinger and decide if the potato tissue feels firm or soft.
8. Using an electronic balance, weigh the first cylinder, note its initial mass and immerse it in tube A.
9. Repeat step 8 for the second cylinder and immerse it in tube B.
10. After 24 hours, dry each cylinder, reweigh it and note its final mass.
11. Gently squeeze each cylinder and decide if the potato tissue feels very firm (turgid), firm or soft (flaccid) compared with before.

12. If other students have done the same experiment, pool your results.

REPORTING

Write up your report by doing the following:

1. Copy the title given at the start of this activity.
2. Put the subheading '**Aim**' and state the aim of your experiment.
3. a) Put the subheading '**Method**'.
   b) Draw a diagram of your apparatus set up at the start of the experiment with the regions of higher and lower water concentration clearly labelled in both boiling tubes.
   c) Using the impersonal passive voice, briefly describe the experimental procedure that you followed and state how you obtained your results.
4. Put the subheading '**Results**' and draw a final version of your table of results, recording initial texture, final texture, initial mass, final mass, change in mass and percentage (%) change in mass of each cylinder.
5. Put a subheading '**Presentation of Results**' and present your results as a bar graph to compare the percentage change in mass of the two cylinders.
6. Put a subheading '**Conclusions**' and write a short paragraph to state what you have found out from a study of your results. This should deal with each cylinder in turn and include reference to:
   a) the change in texture and mass shown by the cylinder
   b) the molecules that were gained or lost by the cylinder
   c) the name of the process that occurred and the direction in which overall molecular movement took place
   d) the terms turgid, flaccid, higher water concentration, lower water concentration, hypotonic, hypertonic and selectively permeable.
7. Put a final subheading '**Evaluation of Experimental Procedure**' and then answer the following:
   a) Why would the results be invalid if the cylinders were not rolled on a paper towel each time before being weighed?
   b) Why is it better to pool class results if possible rather than depend on the results of one group of pupils?

c) Why is it better to convert the results to a percentage change in mass before comparing the changes in the two cylinders?
d) State how you could check the reliability of the results.
e) Feel free to comment on either of the following if you have an additional point that you wish to make:
   (i) further improvements that you would include in a repeat of the experiment
   (ii) limitations of the equipment.

Effect of hypotonic and hypertonic solutions on fresh beetroot cells

In the experiment shown in Figure 2.8, the very dilute sugar solution outside beetroot cylinder X has a higher water concentration (HWC) than the contents of the beetroot cells which have a lower water concentration (LWC).

The contents of the beetroot cells in cylinder Y have a higher water concentration (HWC) than the surrounding very concentrated sugar solution which has a lower water concentration (LWC).

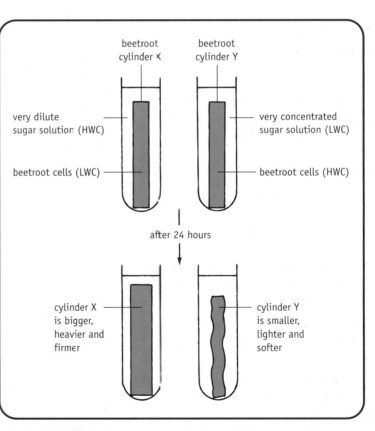

Figure 2.8 *Effect of two sugar solutions on beetroot cells*

Figure 2.9 *Beetroot cylinders after osmosis*

After 24 hours, cylinder X in the hypotonic solution is found to have increased in volume and mass and to have become firmer (**turgid**) in texture. Cylinder Y in the hypertonic solution has decreased in volume and mass and has become softer (**flaccid**) in texture (see Figure 2.9). It is therefore concluded that water molecules have diffused into cylinder X and out of cylinder Y through the cell membrane down the water concentration gradient.

Osmosis

In your investigation with potato cylinders and in the beetroot experiment in Figure 2.8, overall movement of tiny water molecules occurs from a region of higher water concentration (hypotonic solution) to a region of lower water concentration (hypertonic solution) through the **cell membranes**.

The larger sugar molecules are unable to diffuse through the cell membrane. A membrane containing tiny pores that allow the rapid passage through it of water molecules but not large molecules is said to be **selectively permeable**. The movement of water molecules through a selectively permeable membrane is a special case of diffusion called **osmosis**.

Practical Activity and Report

Demonstration of osmosis using Visking tubing model cells

INFORMATION
Visking tubing is a synthetic material. Over a short time-scale (e.g. 2 hours), it acts like a selectively permeable membrane surrounding a cell. Lengths of Visking tubing tied at both ends can therefore be used to act as model cells.

YOU NEED
3 boiling tubes
1 boiling tube stand
3 labels or 1 marker pen
paper towels
3 200 mm lengths of Visking tubing
access to an electronic balance
0.1, 0.5 and 1.0 M sucrose solutions
beaker of warm water

WHAT TO DO
1 Read all the instructions in this section and prepare your results table before carrying out the experiment.
2 Label the three boiling tubes A, B and C and add your initials.
3 Half-fill tube A with 0.1 M sucrose solution, half-fill

tube B with 0.5 M sucrose solution and half-fill tube C with 1.0 M sucrose solution.
4 Soak the three lengths of Visking tubing in warm water for two minutes.
5 Knot one end of one of the Visking tubing lengths and two-thirds fill it with 0.5 M sucrose solution.
6 Knot the other end of the Visking tubing and rinse it under the cold tap to remove any sucrose solution on its outer surface.
7 Roll the Visking tubing 'sausage' (now called cell model A) on a paper towel and dry its outer surface thoroughly.
8 Using the electronic balance, weigh cell model A and note its initial mass in a table of results.
9 Immerse cell model A in the sucrose solution in tube A.
10 Repeat steps 5–9 with the other two lengths of Visking tubing and immerse cell models B and C in tubes B and C after weighing them.
11 After ten minutes roll dry and reweigh each cell model in turn.
12 Return the cell models to their tubes of sucrose.
13 Repeat steps 11 and 12 after 20 and 30 minutes,

keeping a careful note of the results in your table by recording the mass of each cell model at each of the time intervals.

14 If other students have carried out the same experiment, pool the results.

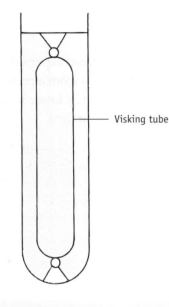

— Visking tube

Figure 2.10 *Visking tubing model cell*

REPORTING

Write up your report by doing the following:

1 Copy the title given at the start of this activity.
2 Put the subheading '**Aim**' and state the aim of your experiment.
3 a) Put the subheading '**Method**'.
   b) Using Figure 2.10 as a guide, draw a diagram of the three cell models set up in their boiling tubes. Label the molarity of all of the sucrose solutions and state whether each is hypotonic, isotonic or hypertonic to the other in the same boiling tube.

c) Using the impersonal passive voice, briefly describe the experimental procedure that you followed and state how you obtained your results.
4 Put the subheading '**Results**' and draw a final version of your table of results.
5 Put a subheading '**Presentation of Results**' and present your results as three line graphs with common axes on the same sheet of graph paper.
6 Put a subheading '**Conclusions**' and write a short paragraph to state what you have found out from a study of your results. This should deal with each cell model in turn and include reference where appropriate to:
   a) the overall pattern of change in mass shown by the cell model
   b) the molecules that were gained or lost by the cell model
   c) the name of the process that occurred and the direction in which overall molecular movement took place
   d) the terms *higher water concentration*, *lower water concentration*, *hypotonic*, *hypertonic*, *isotonic* and *selectively permeable*.
7 Put a final subheading '**Evaluation of Experimental Procedure**' and then answer the following:
   a) Why could the results be invalid if the cell models were not rinsed at the start of the experiment?
   b) Why would the results be invalid if the cell models were not dried before weighing?
   c) State how you could check the reliability of the results.
   d) Feel free to comment on either of the following if you have an additional point that you wish to make:
      (i) further improvements that you would include in a repeat of the experiment
      (ii) limitations of the equipment.

Molecular model of osmosis

Figure 2.11 shows a molecular model of the investigation that you carried out using the three cell models. Cell model A is found to gain mass since water has passed from a region of higher water concentration (**hypotonic** solution) to a region of lower water concentration (**hypertonic** solution) by osmosis.

Cell model B neither gains nor loses water since the solutions inside and outside are equal in water concentration (**isotonic**).

Cell model C is found to lose mass since water has passed out from a region of higher water concentration (**hypotonic** solution) to a region of lower water concentration (**hypertonic** solution) by osmosis.

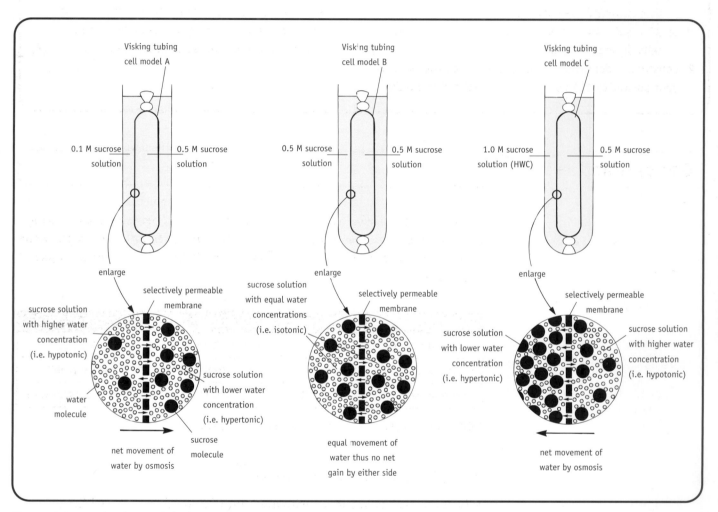

Figure 2.11 *Molecular model of osmosis*

Water concentration gradient

The difference in water concentration that exists between two regions is called the **water concentration gradient.** At the start of the experiment shown in Figure 2.11, a water concentration gradient exists between the regions on either side of the Visking tubing membrane in cell models A and C.

During osmosis, water molecules move down a water concentration gradient from high to low water concentration through the selectively permeable membrane.

Testing Your Knowledge

1 a) Describe TWO changes undergone by a thin cylinder of fresh beetroot when it is immersed for 24 hours in
     (i) hypertonic sucrose solution;
     (ii) hypotonic sucrose solution. (2)
   b) What name is given to the special case of diffusion that causes these changes? (1)
   c) What substance is gained or lost by the beetroot cells during this process? (1)

2 Construct a definition of the process that you gave as your answer to 1b) using all of the following words

and phrases: *lower water concentration, net movement, membrane, water molecules, higher water concentration.* (3)

3 a) When a membrane is described as being *selectively permeable*, what does this mean? (1)
   b) With reference to the terms selectively *permeable membrane* and *concentration gradient of water*, explain how the process of osmosis occurs. (2)

Osmosis and cells

Movement of water by osmosis occurs in living things between neighbouring cells and their immediate environment. In the human body, for example, water passes from the gut cells into the bloodstream. Table 2.1 lists some examples of osmosis in a plant.

direction of water movement	significance
soil solution → root hairs	water absorbed by plant from soil
xylem vessels → stem cells	water makes cells turgid, giving support
xylem vessels → green leaf cells	water used as raw material for photosynthesis

Table 2.1 *Movement of water by osmosis in plant cells*

Role of cell membrane in osmosis

Whenever a cell is in contact with a solution (or another cell) of differing water concentration, osmosis occurs. This is made possible by the fact that the **cell membrane** is selectively permeable. It allows the rapid movement of water molecules through it but only allows larger molecules to move across slowly or not at all.

The direction in which net movement of water molecules occurs depends on the water concentration of the liquid in which the cell is immersed compared with that of the cell contents.

Red blood cells

Since pure water is hypotonic to the contents of a red blood cell, water enters by osmosis until the cell **bursts** (see Figure 2.12). Since 0.85% salt solution is isotonic to the contents of a red blood cell, no net flow of water by osmosis into or out of the cell occurs (also see Figure 2.13). Since 1.7% salt solution is hypertonic to the contents of a red blood cell, water passes out by osmosis and the cell **shrinks** (also see Figure 2.14).

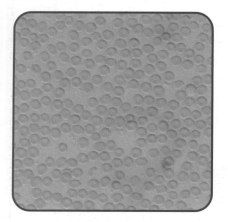

Figure 2.13 *Red blood cells in isotonic salt solution*

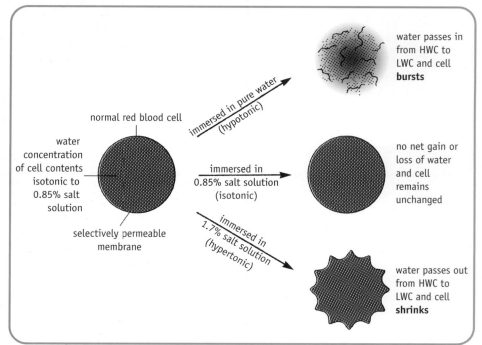

water passes in from HWC to LWC and cell **bursts**

normal red blood cell

immersed in pure water (hypotonic)

water concentration of cell contents isotonic to 0.85% salt solution

immersed in 0.85% salt solution (isotonic)

no net gain or loss of water and cell remains unchanged

selectively permeable membrane

immersed in 1.7% salt solution (hypertonic)

water passes out from HWC to LWC and cell **shrinks**

Figure 2.12 *Osmosis in a red blood cell*

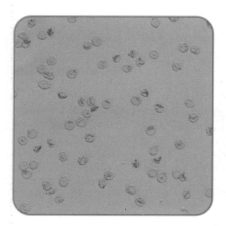

Figure 2.14 *Red blood cells in hypertonic salt solution*

Unicellular animals

Amoeba and *Paramecium* are examples of unicellular animals that live in fresh water. Each animal is bounded by a selectively permeable cell membrane. Inside the membrane, the cell contents act as a region of lower water concentration compared with the surrounding pond water which is a region of higher water concentration. This situation results in the organism **gaining water** continuously by osmosis. Bursting of the cell is prevented by the activities of one or more **contractile vacuoles**.

Paramecium

In *Paramecium* (see Figure 2.15) two contractile vacuoles are present in each cell. They occupy fixed positions fairly close to the cell membrane, one at the front and one at the rear of the organism (see Figure 2.16). Several canals radiate out into the cytoplasm from each vacuole. The canals collect excess water and pass it to their vacuole, making it expand. When fully swollen, a vacuole contracts and discharges its contents to the outside through a pore in the membrane.

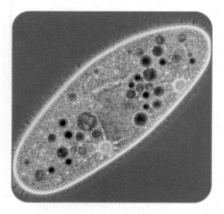

Figure 2.15 *Paramecium*

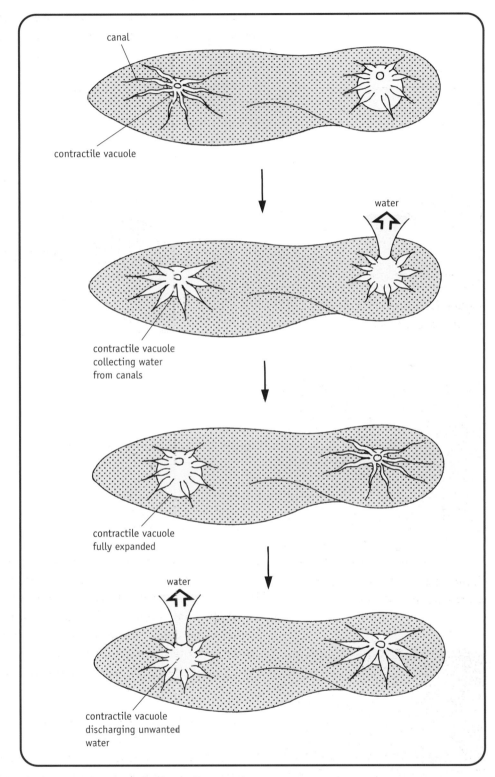

Figure 2.16 *Action of contractile vacuoles*

Since the two contractile vacuoles discharge their contents alternately, one vacuole is about half-full when the other is discharging unwanted water. This process of water regulation is also known as **osmoregulation**.

Plant cells

Since pure water has a higher water concentration than the contents of a normal plant cell (see Figure 2.17), water enters the cell by osmosis. The vacuole swells up and presses the cytoplasm against the cell wall which stretches slightly and presses back, preventing the cell from busting. Cells in this swollen condition are said to be **turgid**. (A young plant depends on the turgor of its cells for support.)

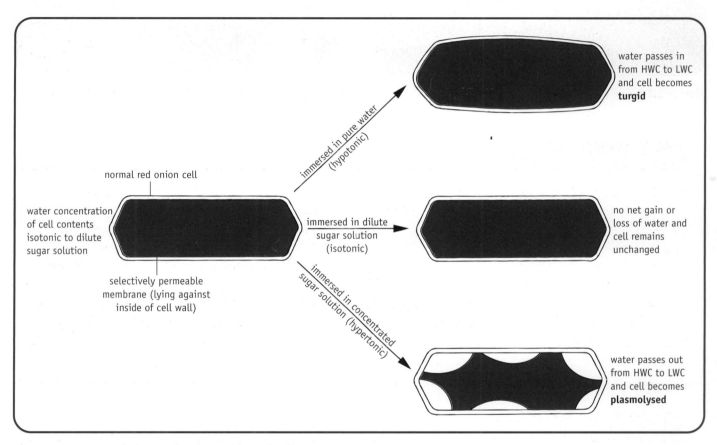

Figure 2.17 *Osmosis in a red onion epidermal cell*

Since the dilute sucrose (sugar) solution (Figures 2.17 and 2.18) has the same water concentration as the cell contents, no net flow of water occurs.

Since the concentrated sucrose solution has a lower water concentration than the cell contents, water passes out of the cell by osmosis. The living contents shrink and pull away from the fairly rigid cell wall. Cells in this state are said to be **plasmolysed** (Figures 2.17 and 2.19). However, they are not dead. When immersed in water, plasmolysed cells regain turgor by taking in water by osmosis.

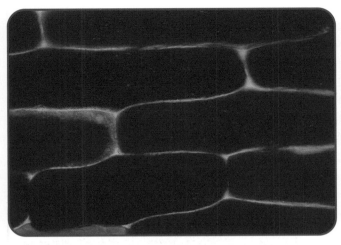

Figure 2.18 *Red onion cells in isotonic sugar solution*

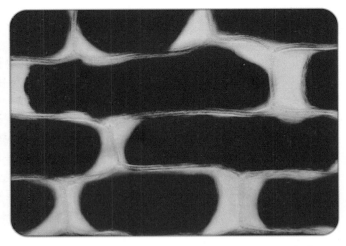

Figure 2.19 *Red onion cells in hypertonic sugar solution*

Testing Your Knowledge

1. a) With reference to the water concentrations involved, explain why red blood cells burst when placed in water yet onion epidermal cells do not. (2)
   b) Why do red blood cells shrink when placed in concentrated salt solution? (1)
   c) Predict the percentage concentration of salt present in normal blood plasma in the human body. Explain how you arrived at your answer. (2)

2. a) Why does *Paramecium* not burst when placed in a solution that is hypotonic to its cell contents? (2)
   b) As the water concentration of *Paramecium's* surroundings increases so does the quantity of oxygen consumed by the animal. Suggest why. (2)

3. a) Describe and explain the osmotic effect of very concentrated sugar solution on onion epidermal cells. (2)
   b) What name is given to cells in this state? (1)
   c) How could such cells be restored to their turgid condition? (1)

Applying Your Knowledge

1. Match the terms in list X with their descriptions in list Y.

List X
1) concentration gradient
2) contractile vacuole
3) diffusion
4) flaccid
5) freely permeable
6) hypertonic

List Y
a) structure that allows rapid passage through it of small molecules (e.g. water) but not larger molecules
b) term used to describe a solution with a higher water concentration than a comparable solution
c) term used to describe a solution with a lower water concentration than a comparable solution
d) term used to describe two solutions that are equal in water concentration
e) the difference in concentration that exists between two regions, resulting in diffusion
f) structure used by a unicellular animal to remove excess water gained by osmosis

7) hypotonic

g) shrinkage of plant cell contents away from cell walls as a result of excessive water loss

8) isotonic

h) term used to describe plant tissue that has lost water by osmosis and become soft

9) osmosis

i) movement of molecules of a substance from high to low concentration

10) plasmolysis

j) term used to describe a structure that allows rapid passage through it of all molecules in solution

11) selectively permeable membrane

k) term used to describe a plant cell or tissue swollen with water taken in by osmosis

12) turgid

l) process of increased movement of water molecules through a selectively permeable membrane to a more concentrated solution

2 Figure 2.20 represents a sugar lump dissolving in a beaker of water. Figure 2.21 represents molecular close-ups of the liquids.

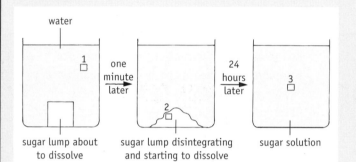

Figure 2.20

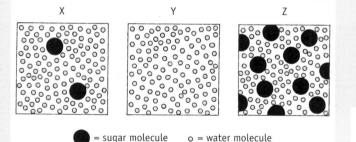

= sugar molecule   o = water molecule

Figure 2.21

a) Match boxes 1, 2 and 3 in Figure 2.20 with their molecular close-ups in Figure 2.21.   (2)

b) Compared with the liquid represented by box 3 in Figure 2.20, which molecular close-up in Figure 2.21 represents a liquid which is
   (i) hypertonic to it;
   (ii) hypotonic to it?   (2)

c) Answer true or false to the following statements:

(i) Water molecules would pass down a water concentration gradient from 2 to 3.

(ii) Water molecules would pass down a water concentration gradient from 1 to 3.

(iii) Sugar molecules would pass down a sugar concentration gradient from 2 to 3.

(iv) Sugar molecules would pass down a sugar concentration gradient from 1 to 2.   (4)

3 a) Name a chemical molecule (other than water) that would be found diffusing out of a mesophyll cell in a green leaf in
   (i) light;
   (ii) darkness.   (2)

   b) Name a carbohydrate stored by plants whose molecules are too large to diffuse out of plant cells.   (1)

4 Three identical cylinders of fresh turnip were immersed in the liquids shown in Figure 2.22 for 24 hours. Each was then removed and held between forefinger and thumb as shown in Figure 2.23.

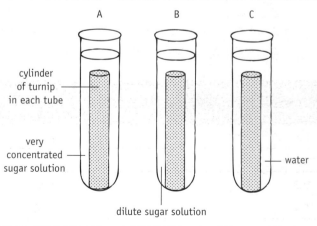

Figure 2.22

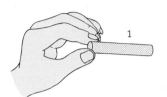

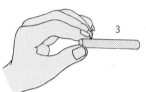

Figure 2.23

a) Match numbers 1, 2 and 3 with letters A, B and C. (3)
b) Justify your choice in each case. (3)

5 Five thin cylinders were cut out of a potato using a cork borer. The cylinders were trimmed to a length of 50 mm and immersed in sucrose sugar solutions of different concentration. The cylinders' lengths were remeasured after 24 hours and the results recorded in Table 2.2.

a) Copy and complete the last two columns in Table 2.2. (2)

b) Draw a line graph of percentage change in length of cylinder against molar concentration of sugar solution (showing gains above and losses below the x axis). (4)

c) Was 0.1 M sugar solution hypotonic or hypertonic to potato cell sap? Explain your answer. (2)

d) Would 0.6 M sugar solution be hypotonic or hypertonic to potato cell sap? Explain your answer. (2)

e) From your graph, estimate the molarity of sugar solution to which potato cell sap is equal in water concentration. (1)

f) Why must the same diameter of cork borer be used to cut out all the cylinders? (1)

g) Why is there no need to roll the cylinders on a paper towel in this investigation? (1)

6 Figure 2.24 shows two osmometers set up at the start of an experiment. Visking tubing acts as a selectively permeable membrane over a short period of time (e.g. 2 hours).

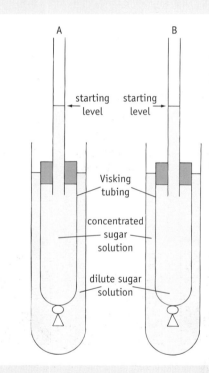

Figure 2.24

concentration of sugar solution (M)	initial length of cylinder (mm)	final length of cylinder (mm)	change in length (mm)	percentage change in length
0.1	50	58	+8	+16
0.2	50	53		
0.3	50	48		
0.4	50	43		
0.5	50	38		

Table 2.2

a) Identify the region of higher and lower water concentration in each osmometer. (2)

b) (i) Predict the direction in which the level will move in each osmometer within the next 30 minutes.

    (ii) Explain your answer to (i) with reference in each case to the movement of molecules down a concentration gradient. (3)

c) What will happen to the *rate* of movement of the level in tube A if the experiment is repeated using water instead of dilute sugar solution? (1)

7 A sample of epidermal cells from a red onion was mounted in a solution isotonic to the cell sap and examined to establish the normal appearance of the cells (see Figure 2.25). The cells were then immersed for 10 minutes in solution 1 and re-examined. This procedure was repeated using different concentrations of sugar solution. The appearance of the cells at each stage is shown in the diagram.

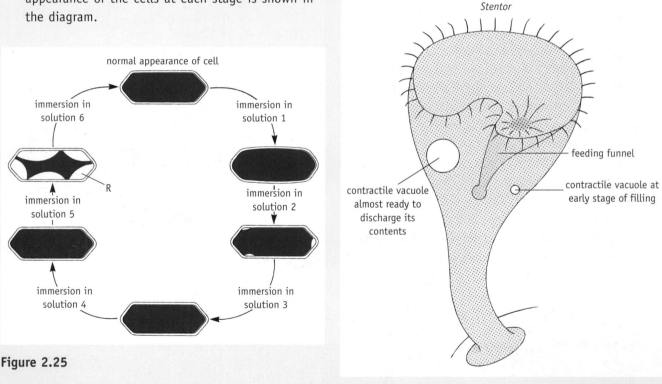

Figure 2.25

a) Which of the five solutions was
    (i) most hypertonic relative to the others?
    (ii) most hypotonic relative to the others? (2)

b) Identify the process occurring at each numbered immersion in Figure 2.25 using the appropriate letter from Table 2.3. (6)

process	overall movement of water molecules
X	movement out of cell exceeds movement in
Y	movement into cell exceeds movment out
Z	movement into cell equals movement out

Table 2.3

c) What substance would be present in region R in Figure 2.18 after the cells had been immersed for 10 minutes? (1)

8 *Stentor* (see Figure 2.26) is a free-moving freshwater unicellular animal that becomes attached by its lower end to a stationary object during feeding. In its natural habitat, *Stentor* gains water by osmosis. Unwanted water is removed by two contractile vacuoles. Each filling and emptying of a contractile vacuole is called a pulsation.

Figure 2.26

In an experiment, specimens of *Stentor* were placed in bathing solutions of different concentrations of salt and viewed under a microscope. Table 2.4 gives average results from the observation of five animals.

concentration of salt in bathing solution (%)	average time for one pulsation (s)
0.1	95
0.3	156
0.5	201
0.7	378

Table 2.4

a) What relationship exists between the salt concentration of the bathing solutions and their relative water concentrations? (1)

b) (i) What relationship exists between the salt concentration of the bathing solutions and the time taken for one pulsation?

(ii) Explain why. (2)

c) The experiment was repeated using pure water and 1% salt solution. *Stentor's* contractile vacuoles were found to stop working in one of these liquids.

(i) Identify the liquid in which this happened.

(ii) Predict the response of the animal's contractile vacuoles to the other liquid.

(iii) Predict the effect that this second liquid would have on mammalian liver cells. (3)

d) (i) Identify the variable factor in this experiment involving *Stentor*.

(ii) Name TWO other factors that must be kept constant to make the procedure valid. (3)

e) Why was an average time calculated for each salt concentration? (1)

f) Suggest why hungry specimens of *Stentor* were chosen for use in this experiment. (1)

9 Part of a plant's root in the ground is shown in Figure 2.27.

a) (i) Name structures 1 and 2.

(ii) Which of these is freely permeable and which is selectively permeable? (3)

b) (i) Which of the four lettered regions has the highest water concentration?

(ii) Which of the four lettered regions has the lowest water concentration?

(iii) What term is used to refer to the difference in water concentration that exists between the two regions? (3)

c) Give the route that water will take by listing the plant cells involved in the correct order. (1)

10 If the plant cells shown in Figure 2.28 remained in contact as shown, then water would pass by osmosis from BOTH:

A R to Q and Q to P.

B Q to S and R to Q.

C P to Q and R to S.

D Q to P and Q to R.

(Choose ONE answer only.) (1)

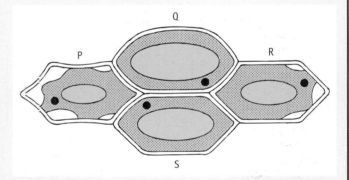

Figure 2.28

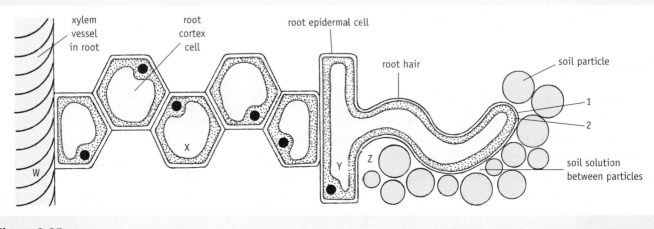

Figure 2.27

3 Enzyme action

Effect of heat on the breakdown of hydrogen peroxide

Hydrogen peroxide is a chemical that breaks down into water and oxygen as shown in the following equation:

$$\text{hydrogen peroxide} \longrightarrow \text{water} + \text{oxygen}$$
$$(2H_2O_2) \qquad\qquad (2H_2O) \quad (O_2)$$

In the experiment shown in Figure 3.1, test tubes containing hydrogen peroxide and drops of detergent are placed in five water baths at different temperatures. The detergent is used to sustain any oxygen bubbles that are released, as a froth.

After 30 minutes the tubes are inspected for the presence of a froth of oxygen bubbles, which indicates the breakdown of hydrogen peroxide. The diagram shows a typical set of results where the volume of froth is found to increase with increase in temperature.

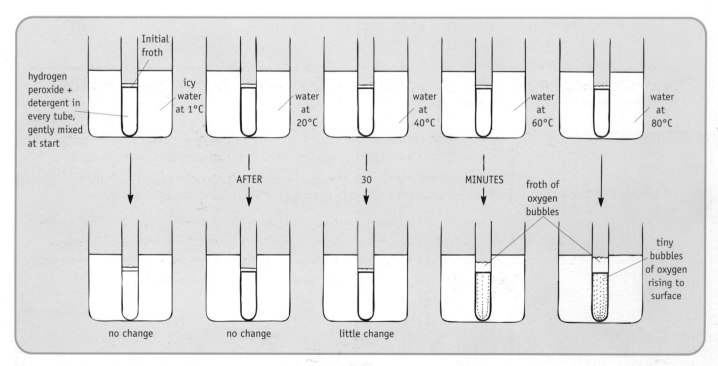

Figure 3.1 *Investigating the effect of heat on the breakdown of hydrogen peroxide*

It is therefore concluded that breakdown of hydrogen peroxide is promoted by heat energy and that its rate of breakdown increases as the temperature increases.

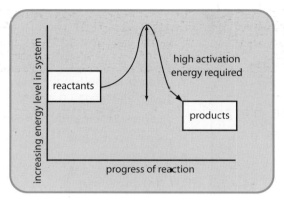

Figure 3.2 *Uncatalysed reaction*

Activation energy

The rate of a chemical reaction is indicated by the amount of chemical change that occurs per unit time. Such a change may involve the joining together of simple molecules into more complex ones or the splitting of complex molecules into simpler ones. In either case the energy needed to make the chemical reaction proceed is called its **activation energy**.

This energy input often takes the form of heat energy and the reaction only proceeds at a high rate if the chemicals are raised to a high temperature (see Figure 3.2).

Effect of manganese dioxide on the breakdown of hydrogen peroxide

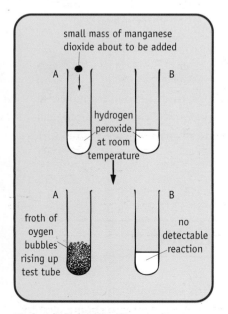

Figure 3.3 *Effect of a catalyst*

In the experiment shown in Figure 3.3, the bubbles forming the froth in tube A are found to relight a glowing splint. This shows that oxygen is being released during the breakdown of hydrogen peroxide. In tube B, the control, the breakdown process is so slow that no oxygen can be detected.

It is concluded therefore that manganese dioxide (which remains chemically unaltered at the end of the reaction) has increased the rate of this chemical reaction, which would otherwise have only proceeded very slowly. A substance that has this effect on a chemical reaction is called a **catalyst**.

Properties and functions of a catalyst

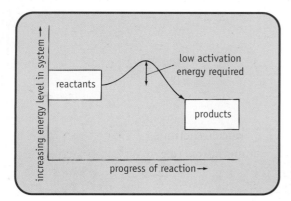

Figure 3.4 *Catalysed reaction*

A **catalyst** is a substance that:
- lowers the energy input (activation energy) required for a chemical reaction to proceed (Figure 3.4)
- speeds up the rate of a chemical reaction
- takes part in the reaction but remains unchanged at the end of it.

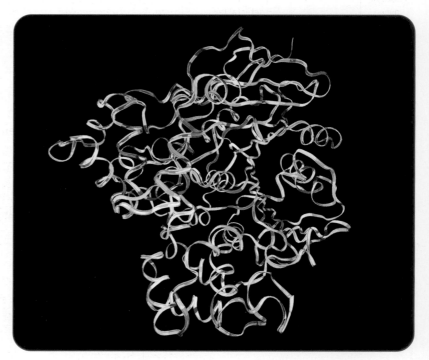

Figure 3.5 *Computer-generated model of an enzyme molecule*

Biological catalysts

Since living cells cannot tolerate the high temperatures needed to make chemical reactions proceed at a rapid rate, they employ catalysts. Biological catalysts are called **enzymes**. Enzymes are made of **protein** (see Figure 3.5) and occur naturally in all living cells.

Effect of catalase on the breakdown of hydrogen peroxide

Catalase is an enzyme made by living cells. It is especially abundant in fresh liver cells. In the experiment shown in Figure 3.6, the bubbles produced in tube C are found to relight a glowing splint. This shows that oxygen is being released during the breakdown of hydrogen peroxide. In

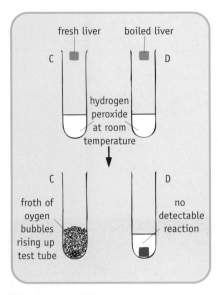

fresh liver   boiled liver

C ▯   ▯ D

hydrogen
peroxide
at room
temperature

C   D

froth of
oygen
bubbles
rising up
test tube

no
detectable
reaction

Figure 3.6 *Effect of catalase*

tube D, the control, the breakdown process is so slow that no oxygen can be detected.

It is concluded that the enzyme catalase has increased the rate of this chemical reaction, which would otherwise have proceeded only very slowly.

$$\text{hydrogen peroxide} \xrightarrow[\text{(enzyme)}]{\text{catalase}} \text{oxygen + water}$$
$$\text{(substrate)} \qquad\qquad\qquad \text{(end products)}$$

Importance of enzymes

Enzymes speed up the rate of all biochemical reactions, allowing them to proceed rapidly at the relatively low temperatures (e.g. 5–40°C) needed by living cells to function properly. In the absence of enzymes, biochemical pathways such as respiration and photosynthesis would proceed so slowly that life as we know it would cease to exist.

Testing Your Knowledge

1 a) What is meant by the term *rate of reaction*? (1)
  b) Suggest units that could be used to express the rate at which hydrogen peroxide breaks down. (1)
  c) Describe the rate at which this reaction would proceed at
    (i) 10°C;
    (ii) 100°C. (2)
  d) Explain the difference between these two rates in terms of energy input. (2)
  e) State a further way in which the rate of breakdown

of hydrogen peroxide can be increased without increasing its temperature. (1)

2 a) State THREE properties of a catalyst. (3)
  b) What general name is given to biological catalysts? (1)

3 a) Where are enzymes found in a living organism? (1)
  b) Of what type of organic substance are enzymes composed? (1)
  c) Briefly explain why enzymes are needed for the functioning of all living cells. (2)

Specificity

Active site

At some point on the surface of the protein molecule that makes up an enzyme there is an **active site** (see Figure 3.7). The shape of the active site is determined by its chemical structure which results from bonding between amino acids in the enzyme molecule.

Mechanism of action

An enzyme is able to act on only one type of substance (its **substrate**) since this is the only substance whose molecules exactly fit the enzyme's active site. The enzyme is said therefore to be **specific** to its substrate and the substrate's molecular shape is said to be **complementary** to the enzyme's active site.

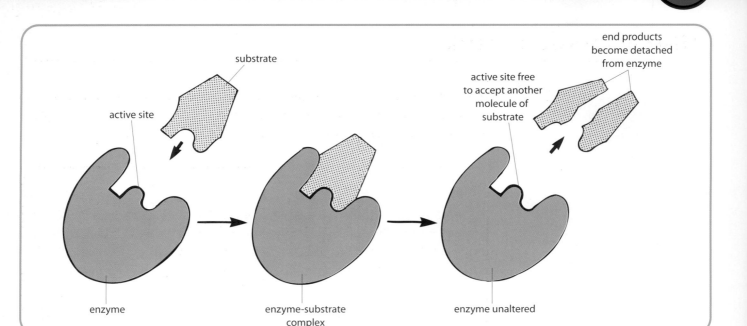

Figure 3.7 *Lock and key mechanism of enzyme action (degradation of complex substrate)*

Enzymes are thought to operate by a **lock-and-key** mechanism. This model of enzyme action proposes that the substrate combines with the enzyme at its active site in a precise way, just as a lock and key fit together. Figure 3.7 shows how the two combine briefly as the enzyme–substrate complex, allowing the reaction to occur. Following catalytic activity, the **end products** become detached from the active site, leaving the enzyme unaltered and free to combine with another molecule of substrate.

Some enzymes promote the **breakdown** (**degredation**) of complex molecules to simpler ones (Figure 3.7); others promote the **building up** (**synthesis**) of complex molecules from simpler ones (Figure 3.8).

Figure 3.8 *Lock and key mechanism of enzyme action (synthesis of complex product)*

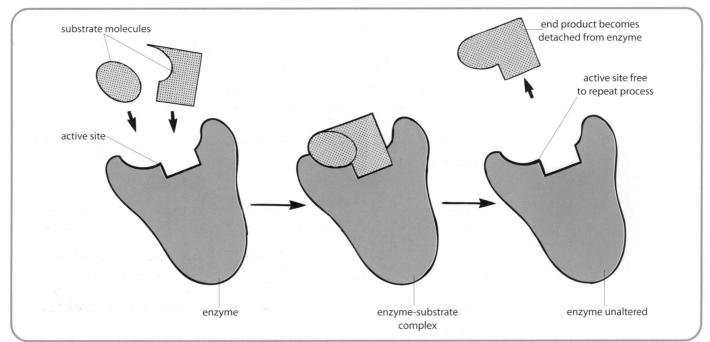

Effect of amylase on starch

In the experiment shown in Figure 3.9, the conditions in the two test tubes are identical at the start of the experiment except that tube A receives the enzyme amylase whereas tube B (the control) receives water.

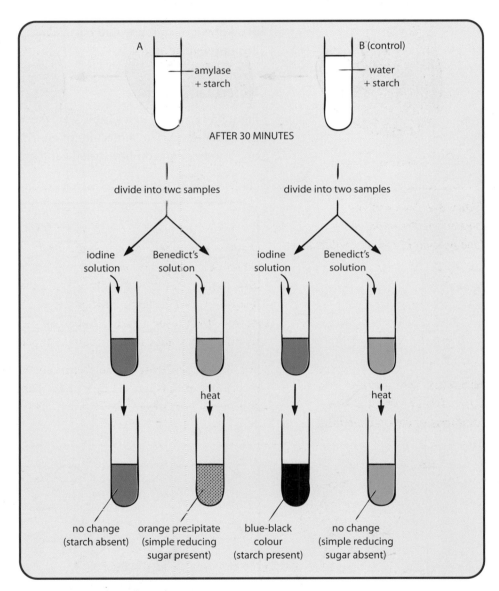

Figure 3.9 *Action of amylase*

From the results it is concluded that in tube A **amylase** has promoted the chemical **breakdown** (**degradation**) of starch to maltose as in the equation:

$$\text{starch} \quad \xrightarrow[\text{(enzyme)}]{\text{amylase}} \quad \text{maltose (simple sugar)}$$
$$\text{(substrate)} \qquad\qquad\qquad \text{(end product)}$$

In tube B, the control, which lacks the enzyme, no detectable reaction has occurred.

Importance of control

A **control** is a copy of the experiment in which all factors are kept exactly the same except the one being investigated in the original experiment. When the results are compared, any difference found between the two must be due to that one factor. For example, in the above experiment we can conclude that the enzyme has increased the rate of breakdown of starch.

If a control had not been set up, it would be valid to suggest that the breakdown of starch would have taken place rapidly whether the enzyme had been present or not.

Molecular model

Figure 3.10 shows, in a simple way, the action of amylase on starch at a molecular level. The molecules of enzyme and substrate are complementary to one another and combine briefly at the enzyme's active site.

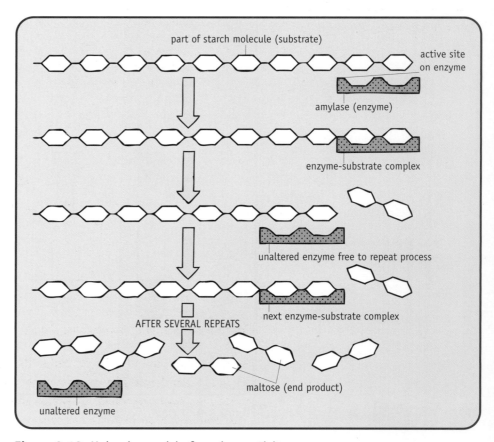

Figure 3.10 *Molecular model of amylase activity*

Effect of phosphorylase on glucose-1-phosphate

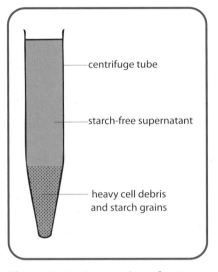

Figure 3.11 *Preparation of potato extract*

A sample of potato extract is prepared by liquidising a mixture of fresh potato tuber and water and then centrifuging the mixture until the supernatant (see Figure 3.11) is starch-free.

This starch-free potato extract is added to an active form of glucose (called glucose-l-phosphate) in each of the four dimples in row A of a tile, as shown in Figure 3.12. Rows B and C are controls. One dimple for each condition is tested at three-minute intervals with iodine solution. Starch is found to be formed in row A only.

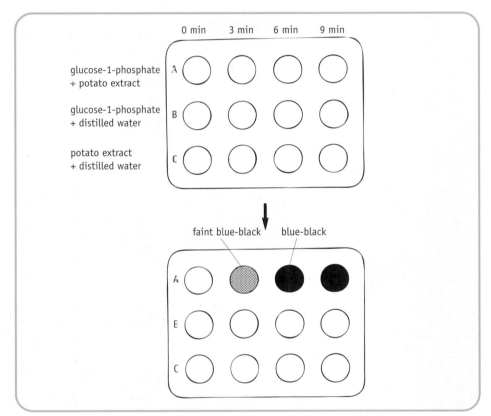

Figure 3.12 *Action of potato phosphorylase*

From this experiment it is concluded that a substance present in potato extract has promoted the **building up** (**synthesis**) of starch from glucose. The substance present in potato extract is an enzyme called **potato phosphorylase**. Its action can be summarised as follows:

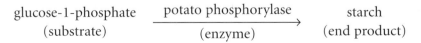

$$\underset{\text{(substrate)}}{\text{glucose-1-phosphate}} \xrightarrow[\text{(enzyme)}]{\text{potato phosphorylase}} \underset{\text{(end product)}}{\text{starch}}$$

Molecular model

Figure 3.13 shows, in a simple way, the action of phosphorylase on glucose-l-phosphate at a molecular level. The molecules of enzyme and substrate are complementary to one another and combine briefly at the enzyme's active site.

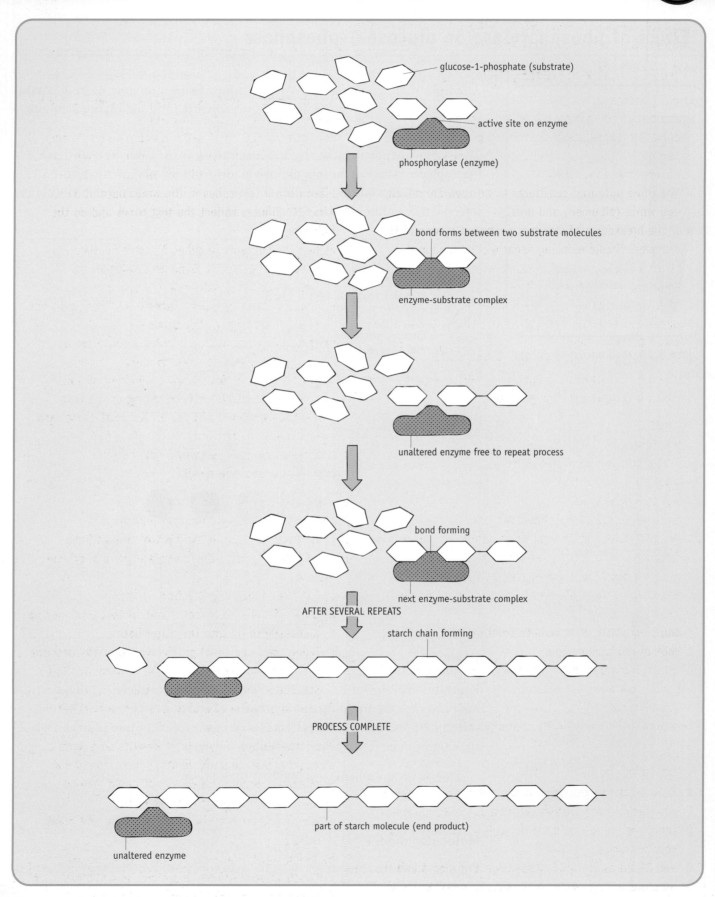

Figure 3.13 *Molecular model of phosphorylase activity*

Specificity of enzymes

INFORMATION

- You are going to carry out an experiment to demonstrate the specificity of two enzymes, amylase and pepsin (so this is really two experiments being done at the same time).
- The three potential substrates to be used are starch, egg white (albumen) and urea.
- If the breakdown of starch is promoted by one of the enzymes, simple reducing sugar will be produced, which will give a positive result with Benedict's solution.
- If breakdown of egg white (albumen) is promoted by one of the enzymes, the substrate's cloudy appearance will change to clear.
- If breakdown of urea is promoted by one of the enzymes, traces of ammonia gas will be released, which will make red litmus paper turn blue.

YOU NEED

6 test tubes

2 rubber stoppers to fit test tubes

6 labels or 1 marker pen

1 test tube stand

1 dropping bottle of 1% starch suspension

1 dropping bottle of 5% egg white (albumen) suspension

1 dropping bottle of 1% urea solution

1 dropping bottle of 1% amylase solution

1 dropping bottle of 1% pepsin solution (acidified with HCl)

1 dropping bottle of Benedict's solution

1 book of red litmus paper

1 250 ml glass beaker (to use as a simple water bath)

1 Bunsen burner

1 tripod stand

access to a thermostatically controlled water bath at 37°C

WHAT TO DO

1 Read all the instructions in this section and prepare your results table before carrying out the experiment.

2 Label the six test tubes A, B, C, D, E and F and add your initials.

3 Add 50 drops of starch suspension to tubes A and D, 50 drops of egg white suspension to tubes B and E and 50 drops of urea solution to tubes C and F.

4 Add 20 drops of amylase to tubes A, B and C and 20 drops of pepsin to tubes D, E and F.

5 For both tubes C and F use a rubber stopper to trap a piece of moist red litmus paper inside the tube so that it dangles above the liquid contents.

6 Place the six test tubes in the water bath at 37°C.

7 After 30 minutes collect the test tubes and do the following:

   a) test the contents of tubes A and D for simple reducing sugar using Benedict's solution and a simple hot water bath

   b) check the appearance of tubes B and E for evidence of a change from cloudy to clear

   c) check the litmus paper in tubes C and F for a change in colour from red to blue.

8 With your six test tubes lined up in front of you, complete your results table, referring to the test tubes, their contents and the results of the tests and checks that you carried out.

9 If other students have carried out the same experiment, pool the results.

REPORTING

Write up your report by doing the following:

1 Copy the title given at the start of this activity.

2 Put the subheading 'Aim' and state the aim of your experiment.

3 a) Put the subheading 'Method'.

   b) Draw a labelled diagram of the six test tubes set up and ready to go into the water bath.

   c) Using the impersonal passive voice, briefly describe the experimental procedure that you followed and state how you obtained your results.

4 Put the subheading 'Results' and draw a final version of your table of results.

5 Put a subheading 'Analysis of Results' and write a short paragraph to state what you have found out from a study of your results. This should include answers, in sentences, to the following:

   a) Which substrate(s) did amylase act on and break down?

   b) Which substrate(s) did amylase fail to act on?

   c) Which substrate(s) did pepsin act on and break down?

   d) Which substrate(s) did pepsin fail to act on?

6 Put a subheading '**Conclusion**'. Based only on the results of this experiment, draw a conclusion about the specificity of enzymes.

7 Put a final subheading '**Evaluation of Experimental Procedure**' and then answer the following:
   a) By what means was each part of the experiment controlled so that within it there was only one factor being varied at a time?
   b) Identify a source of error present in the procedure.

c) State how you could eliminate this source of error.
d) State how you could check the reliability of the results.
e) Feel free to comment on either of the following if you have an additional point that you wish to make:
   (i)   further improvements that you would include in a repeat of the experiment;
   (ii) limitations of the equipment.

Testing Your Knowledge

1 a) What determines the shape of an enzyme's active site? (1)
   b) What name is given to the type of substance on which an enzyme acts? (1)
   c) Why is an enzyme said to be specific in its relationship with its substrate? (1)
   d) According to the lock-and-key hypothesis, how is an enzyme thought to act? (2)

2 a) Give a named example of an enzyme that promotes
   (i)   the degradation of a substance;
   (ii) the synthesis of a substance. (2)

   b) For each of these enzymes, summarise the reaction that it promotes in a word equation. (2)

3 a) When preparing potato extract for the phosphorylase experiment, the mixture is centrifuged until the supernatant is starch-free. Explain the reason for this procedure. (1)
   b) Explain why controls B and C are included in the experiment shown in Figure 3.12. (2)

4 Figure 3.14 shows a model of enzyme molecular activity. Which substance (enzyme or substrate) is equivalent to the lock and which to the key? (1)

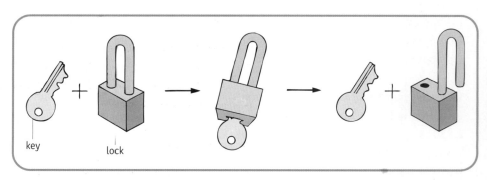

key   lock

Figure 3.14

Factors affecting enzyme activity

To function efficiently an enzyme requires a suitable temperature, a suitable pH and an adequate supply of substrate.

Effect of temperature

The graph in Figure 3.15 summarises the general effect of **temperature** on enzyme activity. At very low temperatures, enzyme molecules are **inactive** but undamaged. At low temperatures, enzyme and substrate molecules move around **slowly** in their surrounding medium. They meet only rarely and the rate of enzyme activity remains low.

As the temperature increases, the two types of molecule move about at a **faster rate** and more molecular collisions occur. A greater number of enzyme–substrate complexes are formed and the rate of reaction increases.

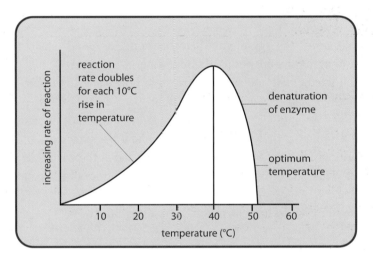

Figure 3.15 *Effect of temperature on enzyme activity*

Optimum

Since 40°C (approximately) is the temperature at which most molecular collisions occur yet molecular damage has barely begun, this is the temperature at which the reaction works best. It is called the enzyme's **optimum** temperature.

Denaturation

An increase in temperature makes an enzyme's atoms vibrate. At temperatures above 40°C, its atoms vibrate so much that some of the chemical bonds holding its amino acids together break and the molecule begins to come apart. Soon the shape of its active site (which is determined by a specific arrangement of amino acids) is changed and the enzyme is unable to 'lock' onto its substrate. An enzyme in this damaged state is said to be **denatured**. It is permanently inactive.

Beyond 40°C, as more and more molecules of enzyme become denatured and inactive, the rate of the reaction decreases rapidly. At temperatures of about 55–60°C, enzyme activity is found to have come to a complete halt. This is because all the enzyme molecules have become denatured.

Planning and designing an investigation into the effect of temperature on the activity of lipase

INFORMATION

- Lipase is an enzyme that promotes the digestion (breakdown) of fats as in the equation:

$$fats \xrightarrow{\text{lipase}} fatty\ acids + monoglycerides$$

- Creamy milk is a rich source of fat.
- Lipase works better if bile salts are present.
- Universal indicator is orange in acidic conditions, yellow in weakly acidic conditions and green in neutral conditions.

WHAT TO DO

1 Write a heading that states the aim of your investigation.
2 Make up your plan of action by answering the following in sentences:
   a) Name the variable factor that you would alter; state how many conditions of the variable factor you would include in your design and how you would do this.
   b) Identify three other variable factors that you would keep constant in your investigation.
   c) Draw a labelled diagram to show your design set up and ready to start.
   d) State the observations that you would make at the start and at the end of the experiment to obtain your set of results.
   e) Explain how you would draw your conclusions about the effect of temperature on the activity of lipase from your results.
   f) Compare your design with Specimen Answer 1 on pages 56–7.

Effect of pH

The symbol **pH** refers to the concentration of **hydrogen ions** present in a solution. The greater the number of hydrogen ions present, the lower the pH and the more acidic the solution.

The shape of an enzyme molecule is partly due to the presence of hydrogen bonds which hold its chain(s) of amino acids together. Exposure of an enzyme to an extreme of pH (such as the very high concentration of hydrogen ions present in a concentrated acid) normally results in bonds being broken and the enzyme becoming denatured.

Optimum

Each enzyme works best at a particular pH (its **optimum** pH) as shown in Figure 3.16. Most enzymes function within a working pH range of about 5–9 with an optimum around pH 7 (neutral). However, there are exceptions. Pepsin, which is secreted by the stomach's gastric glands, works best in strongly acidic conditions of pH 2.5; alkaline phosphatase, which plays a role in bone formation, works best at pH 10.

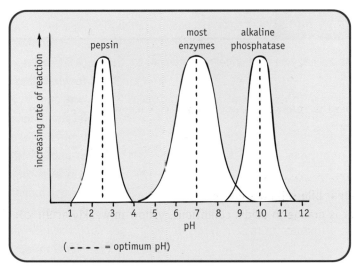

Figure 3.16 *Effect of pH on enzymes*

Planning and designing an investigation into the effect of pH on the activity of catalase

INFORMATION

● Catalase is an enzyme that promotes the breakdown of hydrogen peroxide to water and oxygen as in the equation:

$$\text{hydrogen peroxide} \xrightarrow{\text{catalase}} \text{water} + \text{oxygen}$$

● The oxygen is released as bubbles which can be gathered and sustained as a froth in a measuring cylinder by adding one drop of detergent to the hydrogen peroxide solution.
● Fresh liver is a rich source of catalase.
● The pH of a liquid can be altered by adding drops of acid or alkali as required.

WHAT TO DO

1 Write a heading that states the aim of your investigation.
2 Make up your plan of action by answering the following in sentences:
  a) Name the variable factor that you would alter; state how many conditions of the variable factor you would include in your design and how you would do this.
  b) Identify three other variable factors that you would keep constant in your investigation.
  c) Draw a labelled diagram to show your design set up and ready to start.
  d) State the measurements that you would make at the start and at the end of the experiment to obtain your set of results.
  e) Explain how you would draw your conclusions about the effect of pH on the activity of catalase from your results.
  f) Compare your design with Specimen Answer 2 on pages 57–8.

Testing Your Knowledge

1 Salivary amylase is an enzyme that digests starch in the human mouth.
   a) (i) Compare the rate of reaction of amylase at 5°C and 20°C.
      (ii) Explain the difference in rate in terms of behaviour of enzyme and substrate molecules. (3)
   b) (i) Compare the rate of the reaction at 40°C and 60°C.
      (ii) Explain the difference in rate in terms of the molecules. (3)
2 a) Which temperature given in question 1 is closest to the optimum for most enzymes? (1)
   b) (i) Suggest a temperature at which an enzyme is inactive but capable of activity if the temperature changes.

   (ii) Explain your answer. (2)
3 a) To what does the symbol pH refer? (1)
   b) Why does exposure to concentrated acid normally result in denaturation of an enzyme? (1)
4 a) Explain the meaning of the term *optimum condition* as applied to the activity of an enzyme. (1)
   b) Which of the following is the optimum range of pH for the human enzyme pepsin?
      A 2–3 **B** 4–5 **C** 6–7 **D** 7–8 (1)
5 Copy the following sentences choosing the one correct answer from each bracketed choice of three. Enzyme molecules are made of (fat/protein/carbohydrate) and are produced by (all/most/a few) living cells. Most enzymes work best at around (20°C/40°C/60°C) and are described as biological (substrates/products/catalysts). (4)

Specimen Answer 1

1 To investigate the effect of temperature on the activity of lipase.
2 a) The variable factor to be altered is temperature; four temperatures (2°C, 20°C, 37°C and 70°C) are included in the design by using a beaker of crushed ice, a beaker of water at 20°C and two thermostatically controlled water baths at 37°C and 70°C.
   b) The three factors kept constant from tube to tube are: volume of each chemical used; concentration of each chemical used; time that test tubes are left in containers at different temperatures.
   c) The design of the experiment is shown in Figure 3.17.
   d) The colour of the contents of each tube is noted at the start and at the end of the experiment. The results are recorded in a table.
   e) If a change from green (neutral) to orange (acidic) is found to have occurred, it can be concluded that the fat has been digested to fatty acids and that the temperature used has been especially favourable for lipase activity.

   If a change from green to yellow (weakly acidic) is found to have occurred, it can be concluded that little fat has been digested and that only limited activity of lipase has occurred at that temperature.

   If no colour change is found to have occurred, it can be concluded that at that temperature the enzyme has been inactive or denatured.

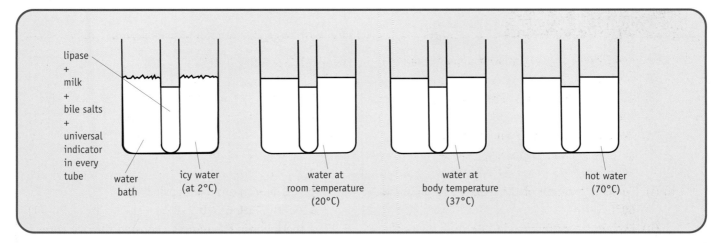

Figure 3.17 *Specimen Answer 1*

1 To investigate the effect of pH on the activity of catalase.
2 a) The variable factor to be altered is pH; three pH values (4, 7 and 9) are
   included in the design by adding drops of acid or alkali as required and
   checking each pH using pH paper.
  b) The three variable factors kept constant are: mass of liver used each
   time; volume of liquid added to each cylinder; concentration of hydrogen
   peroxide used.
  c) The design of the experiment is shown in Figure 3.18.
  d) The volume of froth of oxygen bubbles produced in each cylinder is
   measured to obtain a set of results.
  e) If a large volume of froth is found to have been produced, it can be
   concluded that the pH condition has been favourable for the activity of
   catalase.
   If little or no froth has been produced, it can be concluded that there
   has been little or no activity by catalase because the pH condition has
   been unfavourable.

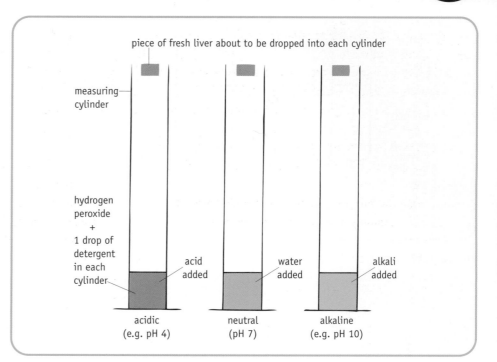

piece of fresh liver about to be dropped into each cylinder

measuring cylinder

hydrogen peroxide + 1 drop of detergent in each cylinder

acid added

water added

alkali added

| acidic (e.g. pH 4) | neutral (pH 7) | alkaline (e.g. pH 10) |

Figure 3.18 *Specimen Answer 2*

Applying Your Knowledge

1 Match the terms in list X with their descriptions in list Y.

List X	List Y
1) active site	**a)** term used to describe the condition of a factor at which an enzyme works best
2) catalyst	**b)** building up of large complex molecules from simpler ones by an enzyme-controlled reaction
3) denatured	**c)** type of organic chemical of which enzymes are composed
4) digestion	**d)** protein made by living cells that acts as a biological catalyst
5) enzyme	**e)** substance that increases the rate of a chemical reaction and remains unaltered
6) optimum	**f)** region of an enzyme where the complementary surface of the substrate becomes attached
7) product	**g)** complementary relationship of molecular structure allowing an enzyme to combine with one type of substrate only
8) protein	**h)** substance on which an enzyme acts, resulting in the formation of an end product
9) specificity	**i)** enzyme-controlled breakdown of large complex molecules to simpler ones
10) substrate	**j)** term used to describe the state of an enzyme that has been permanently destroyed
11) synthesis	**k)** substance formed as the result of an enzyme acting on its substrate

2 Hydrogen peroxide is a chemical that breaks down very slowly into oxygen and water. The experiment shown in Figure 3.19 (on the opposite page) was set up to investigate the effect of catalase (an enzyme present in living cells) on the rate of this chemical reaction.

a) Identify the substrate in this experiment.   (1)

b) (i) Which TWO tubes received cells that provided a supply of active catalase?

(ii) Describe the effect that this enzyme has on the rate of breakdown of hydrogen peroxide.

(iii) Name ONE of the end products formed and describe how it was identified.   (4)

c) Which TWO tubes received denatured enzyme?   (1)

d) Explain at molecular level why a denatured enzyme is ineffective. (2)

3 Figure 3.20 shows three separate stages that occur during an enzyme-controlled reaction.

a) What name is given to the complex indicated by the letter Z? (1)

b) Using the three letters given, indicate the correct sequence in which the three stages would occur if:

(i) the enzyme promotes the breakdown of a complex molecule to simpler ones

(ii) the enzyme promotes the building up of a complex molecule from simpler ones. (2)

4 Molecules of simple sugar can be built up into starch molecules. The rate of this chemical reaction is speeded up by the enzyme phosphorylase. Describe the mechanism by which this enzyme is thought to work, using all of the following terms: *active site on enzyme, enzyme, enzyme–substrate complex, lock and key, product, simple sugar, starch, substrate*. (4)

5 Figure 3.21 (on the next page) shows two boiling tubes set up to investigate the effect of the enzyme lipase on the fat in milk.

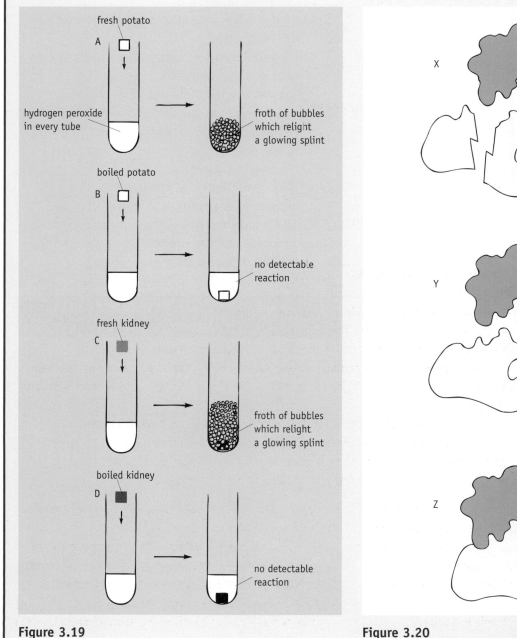

Figure 3.19

Figure 3.20

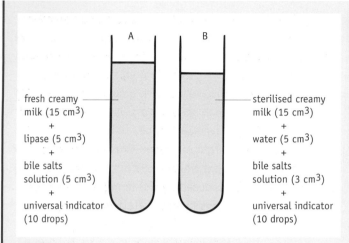

Figure 3.21

a) State TWO ways in which tube B must be altered to make it a valid control. (2)

b) Explain why the use of boiled lipase in the control would be preferable to water. (1)

6 In an investigation into the action of the enzyme urease, three versions of the apparatus shown in Figure 3.22 were set up. Urease activity is indicated by the release of ammonia gas which turns red litmus paper blue. The results of the experiment are summarised in Table 3.1.

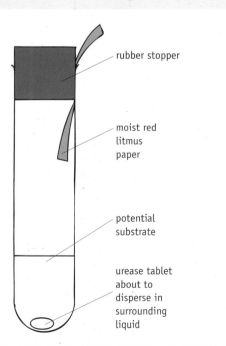

Figure 3.22

test tube	A	B	C
potential substrate	cloudy egg white suspension	urea solution	starch suspension
colour of litmus at start	red	red	red
colour of litmus after 30 minutes	red	blue	red

Table 3.1

a) What was the one variable factor investigated in this experiment? (1)

b) Name THREE factors that must be kept constant when setting up tubes A, B and C. (3)

c) Suggest why the three test tubes should be shaken gently at the start of the experiment. (1)

d) (i) Which test tube showed evidence of urease activity?

(ii) How could you tell? (1)

e) What characteristic of enzymes does this experiment demonstrate to be true of urease? (1)

7 Table 3.2 gives the results from an experiment set up to investigate the effect of temperature on the action of a digestive enzyme.

temperature (°C)	mass of substrate broken down (mg/h)
0	0
5	1
10	4
15	8
20	14
25	22
30	28
35	31
40	32
45	29
50	18
55	0

Table 3.2

a) Present the data as a line graph. (3)

b) (i) Explain what is meant by the term *optimum temperature*.

(ii) State the optimum temperature for the action of this enzyme. (2)

c) (i) By how many times was the rate of enzyme activity greater at 30°C than at 20°C?

(ii) Explain the difference in terms of rate of molecular movement and frequency of collision between enzyme and substrate molecules at these two temperatures. (3)

d) Which rise in temperature of 5°C brought about the biggest increase in rate of enzyme activity? (1)

e) At the temperature range 50°–55°C, the molecules are still gaining energy so why does the reaction come to a halt at 55°C? (1)

f) Predict the mass of substrate that would be broken down at 75°C. (1)

8 The graph in Figure 3.23 shows the effect of pH on the activity of three enzymes X, Y and Z.

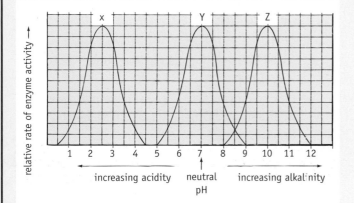

Figure 3.23

a) State the working range of pH for each of the enzymes. (3)

b) What generalisation can be drawn about:

(i) the breadth of working range of pH of each of the enzymes?

(ii) the extent to which they all share the same actual pH working range? (2)

c) State the optimum pH for each of the enzymes. (3)

d) Suggest which of the enzymes would show optimum activity in

(i) the human mouth;

(ii) the human stomach. (2)

e) Describe ONE adverse effect that a sudden change in pH can have on an enzyme's molecular structure. (1)

9 Explain each of the following in terms of enzymes:

a) Fevers that raise the body temperature to over 42°C are normally fatal to human beings. (2)

b) Vinegar (an acid) is used to preserve food against attack by micro-organisms. (2)

c) Cheese kept in a warm room turns mouldy much more quickly than cheese kept in a refrigerator. (2)

10 Gastric juice is produced by the lining of the human stomach. It contains hydrochloric acid and pepsin, a protein-digesting enzyme. Egg white is rich in protein. Four experiments were set up as shown in Figure 3.24 in an attempt to investigate the influence of certain abiotic factors on the activity of pepsin.

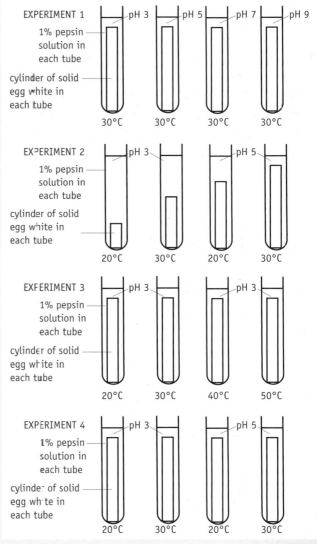

Figure 3.24

a) In what way would a cylinder of solid egg white become altered if pepsin successfully promoted the digestion of protein? (1)

b) Which of the four experiments is a valid investigation into the influence of temperature on pepsin activity? (1)

c) Which of the four experiments is a valid investigation into the effect of pH on pepsin activity? (1)

d) (i) Which experiment really consists of two variable factors being investigated at the same time?

(ii) Predict which combination of these two factors would be most effective at promoting the activity of pepsin.

(iii) Explain your answer to (ii). (4)

e) (i) Which experiment attempts to investigate the effect of mass of substrate on enzyme activity?

(ii) Why is the set-up invalid as it stands?

(iii) Draw a diagram showing the four tubes correctly set up ready to investigate the effect of mass of substrate on pepsin activity. (4)

11 One gram of chopped raw liver was added to hydrogen peroxide solution at different pH values and the time taken to collect 1 cm³ of oxygen was noted in each case. The results are given in Table 3.3.

pH of hydrogen peroxide solution	time to collect 10 cm³ of oxygen (s)	rate of breakdown of hydrogen peroxide (cm³/s)
6	50	0.2
7	40	
8	20	
9		1.0
10	25	
11		0.2
12	100	

Table 3.3

a) Copy and complete Table 3.3. (3)

b) Present the results as a line graph by plotting rate of breakdown of hydrogen peroxide against pH. (3)

c) From your graph state the pH at which the enzyme was
(i) most active;
(ii) least active. (2)

d) Of the pH values used in this experiment, which is the optimum for the enzyme present in the liver cells? (1)

e) How could you obtain an even more accurate measurement of the optimum pH at which this enzyme works? (1)

Word bank

breakdown	membrane
burst	osmosis
catalysts	permeable
cell	pH
complex	plasmolysed
concentration	products
denatured	protein
diffusion	selectively
energy	shrinks
enzyme	simpler
gain	site
gradient	specific
hypertonic	speeds
hypotonic	turgid
isotonic	unchanged
lock	walls
lose	waste
low	water

Table 3.4 *Word bank for Chapters 2 and 3*

What You Should Know

(Chapters 2 and 3) (See Table 3.4 for word bank)

1 Diffusion is the movement of molecules of a substance from a high to a _____ concentration down a _____ gradient. _____ is the means by which useful substances enter and _____ materials leave the cells of a living organism.

2 The cell membrane is freely _____ to small molecules such as oxygen, CO_2 and water. Some larger molecules diffuse through the membrane slowly; others are too large to pass through by diffusion. The membrane is said, therefore, to be _____ permeable.

3 When two solutions are of equal water concentration, they are said to be isotonic. When two solutions differ in water concentration, the one with the higher water concentration is said to be _____ and the one with the lower water concentration is said to be _____.

4 Osmosis is the movement of _____ molecules across a selectively permeable _____ from a region of high water concentration to a region of lower water concentration, i.e. down a concentration _____. _____ is a special case of diffusion.

5 When placed in a hypotonic solution, cells _____ water by osmosis. Animal cells swell up and may _____; plant cells swell up and become _____ but are prevented from bursting by the presence of cell _____.

6 When placed in a solution _____ to their contents, cells remain unchanged.

7 When placed in a hypertonic solution, cells _____ water by osmosis. An animal cell _____; the contents of a plant cell pull away from the cell wall resulting in the cell becoming _____.

8 A catalyst is a substance that lowers the _____ input required for a chemical reaction to occur. By doing this it _____ up the chemical reaction yet remains _____ at the end of the reaction.

9 Enzymes are biological _____. They are made of _____ and are needed for the normal functioning of every living _____.

10 The active _____ on an enzyme molecule is _____ to the molecular structure of its substrate, allowing the two to combine like a _____ and key. Following catalytic activity at the active site, the end _____ become detached leaving the _____ unaltered.

11 Some enzymes promote the chemical _____ (degradation) of complex molecules to _____ ones; others promote the building up (synthesis) of _____ molecules from simpler ones.

12 To function efficiently an enzyme requires an appropriate _____ and a suitable temperature. At temperatures above 55°C most enzyme molecules are inactive since they have become _____.

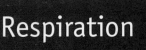 **Respiration**

Energy content of food

Heat energy is measured in units called **joules** (J) and **kilojoules** (kJ). 1000 joules = 1 kilojoule. The chemical energy contained in a food can be changed into heat energy and measured. 4.2 kJ is the quantity of heat energy needed to raise the temperature of 1000 g of water by 1°C.

In the experiment shown in Figure 4.1, 50 cm³ of water is poured into a long boiling tube and its initial temperature recorded (e.g. 20°C). 1 g of food (e.g. peanut) is ignited and held under the tube until the food stops burning.

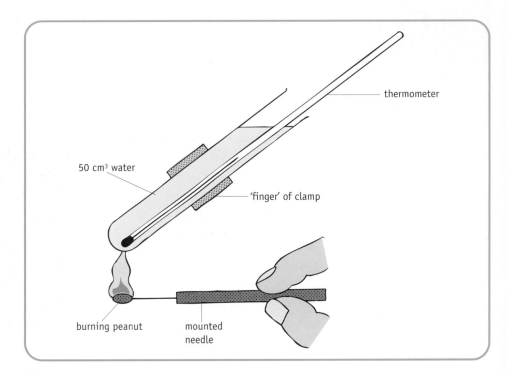

Figure 4.1 *Measuring the energy content of a food*

The final temperature of the water is recorded (e.g. 46°C). The energy released by the food is calculated using the following formula:

$$\text{energy released (kJ)} = \frac{4.2\ MT}{1000}$$

where M = mass of water (g) and T = rise in temperature (°C). Thus the energy released by 1 g of peanut in the above example is

$$\frac{4.2 \times 50 \times (46 - 20)}{1000} = 5.46 \text{ kJ}$$

Food calorimeter

The energy content of a food can be measured more accurately using a **food calorimeter** (Figure 4.2). Since this apparatus has been designed so that in total it is equivalent to 1000 g water, the energy released by 1 g food in kilojoules is:

$$\frac{4.2\ MT}{1000}$$

$$= \frac{4.2 \times 1000 \times T}{1000}$$

$$= 4.2 \times T$$

where T = rise in temperature in °C.
Table 4.1 gives the energy values of a variety of foodstuffs.

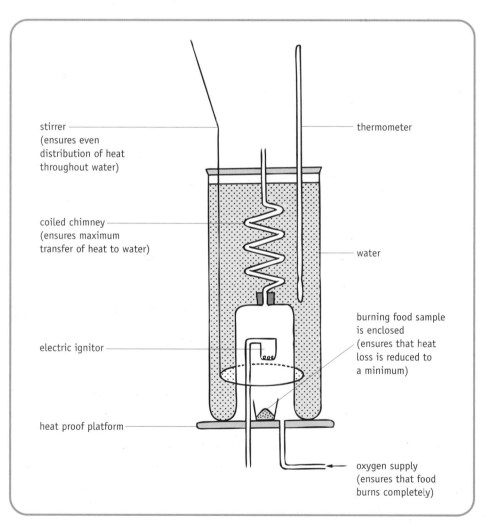

Figure 4.2 *Food calorimeter*

Activity

Selecting and presenting data on the energy content of foods

WHAT TO DO

Study Tables 4.1 and 4.2 and then attempt the following questions.

food	energy content (kJ/g)	protein (%)	fat (%)	carbohydrate (%)
chocolate	24	9	37	54
corn oil	38	0	100	0
egg white	19	100	0	0
ice cream	8	4	11	20
lard	37	0	100	0
olive oil	39	0	100	0
peas (boiled)	2.1	5	0	7.8
potatoes (boiled)	3.3	1.4	0	19.5
potatoes (chipped)	9.9	4	9	37
sucrose	19	0	0	100

Table 4.1 *Energy content of foods*

person	approximate daily energy requirement (kJ)
infant	3 000
2-year-old	5 000
6-year-old	7 500
10-year-old	8 000
15-year-old girl	9 500
15-year-old boy	12 000
woman (light work)	9 500
woman (heavy work)	12 000
man (light work)	11 500
man (heavy work)	13 000
man (very heavy work)	16 000
adult patient in hospital	7 500
old person (retired)	8 500

Table 4.2 *Daily energy requirements*

1 a) Identify the THREE foods in Table 4.1 that are composed exclusively of fat and calculate the energy content of fat based on these three foods.

b) Identify the food composed exclusively of (i) protein and (ii) carbohydrate and for each of these classes of food state its energy content.

c) Express the energy content of protein:fat:carbohydrate as a simple whole number ratio.

2 By how many times is the energy content of chipped potatoes greater than that of boiled potatoes?

3 Table 4.2 shows the approximate energy requirements of different persons and age groups.

a) What mass (in g) of ice cream as his only foodstuff would a man doing very heavy work have to eat to meet his daily energy needs?

b) What mass (in kg) of chocolate as her only foodstuff would a woman doing heavy work need to eat to meet her daily energy requirements?

c) Read the following life story and then draw a line graph to show the changes in the woman's energy requirements over the course of her life.

Annie was born in 1920. She left school at age 14 and began work as a shop assistant. At age 18 she became an auxiliary nurse in a geriatric ward of a hospital. This work often involved helping to lift and carry patients. She studied at night school and in 1950 became a midwife. She was off work for a year during 1955 suffering and recovering from tuberculosis. She then resumed her career as a midwife and retired at age 60. She died in 1995.

Energy release

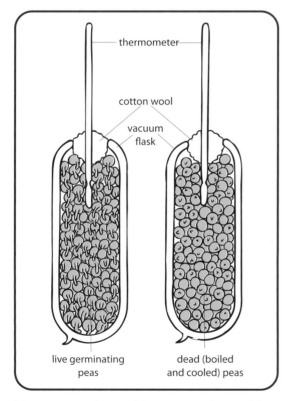

Figure 4.3 *Release of heat energy by respiring plants*

Glucose

Glucose is an energy-rich food (energy value per gram = 17 kJ). It is the most common end product resulting from the digestion of complex carbohydrates (e.g. starch) in the human body and other living organisms. Glucose is a main source of energy in a living cell.

Respiration

When a food such as glucose is burned in the laboratory in air or oxygen, its chemical energy is rapidly released as heat and light. However, in a living cell, the chemical energy stored in glucose is not released as rapidly. Instead release takes place relatively slowly in a controlled way. The process of energy release in a living cell is called **respiration**. It consists of a lengthy series of biochemical reactions each controlled by an intracellular enzyme.

Release of heat energy

Living cells need the energy released during respiration for a variety of reasons (see page 69). Some of the chemical energy in food is converted to **heat energy** during respiration and released from the cells. This is demonstrated by the experiment shown in Figure 4.3, in which the live peas bring about an increase in temperature after a few days.

In humans, generation of heat energy by respiring cells (e.g. active muscle cells) is necessary to maintain the body at 37°C, the optimum temperature for enzyme action.

Adenosine triphosphate (ATP)

Effect of ATP on muscle fibre

In the experiment shown in Figure 4.4 only **ATP** is found to bring about contraction of the muscle fibres. It is therefore concluded that ATP is able to provide immediately the **energy** required for muscle contraction whereas glucose, despite being an energy-rich compound, is unable to do this.

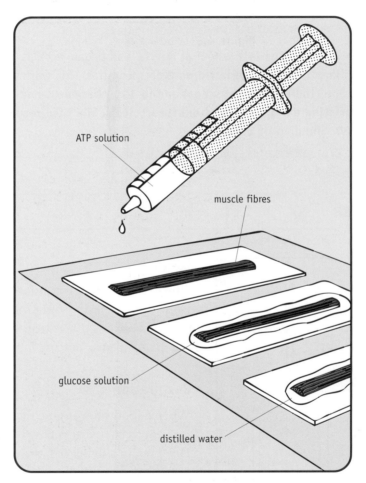

ATP solution

muscle fibres

glucose solution

distilled water

Figure 4.4 *Investigating the effect of ATP on muscle*

Structure of ATP

A molecule of **ATP** is composed of **adenosine** and three **inorganic phosphate** (**Pi**) groups, as shown in Figure 4.5

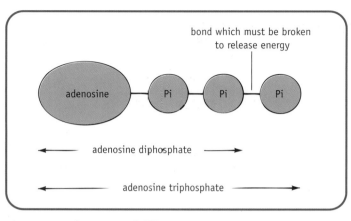

Figure 4.5 *Structure of ATP*

Energy stored in an ATP molecule is released when the bond attached to the terminal phosphate is broken down by enzyme action. This results in the formation of adenosine diphosphate (ADP) and Pi.

On the other hand, energy is required to regenerate ATP from ADP and Pi by an enzyme-controlled process called **phosphorylation**. This reversible reaction is summarised by the equation:

$$
\begin{array}{ccc}
\text{ATP} & \xrightarrow{\text{breakdown releasing energy}} & \text{ADP} + \text{Pi} \\
\text{(high energy state)} & \xleftarrow{\text{build-up requiring energy}} & \text{(low energy state)}
\end{array}
$$

Role of ATP

When an energy-rich substance such as glucose is gradually broken down during respiration in a living cell, it releases energy which is used to produce ATP. As a result many molecules of ATP are present in every living cell.

ATP can rapidly revert to ADP and Pi. This makes energy immediately available for energy-requiring processes such as

- muscular contraction
- cell division
- synthesis of proteins
- transmission of nerve impulses.

ATP is therefore important because it acts as the **link** between energy-releasing reactions and energy-consuming reactions. It provides the means by which chemical energy is transferred from one type of reaction to the other in a living cell, as shown in Figure 4.6

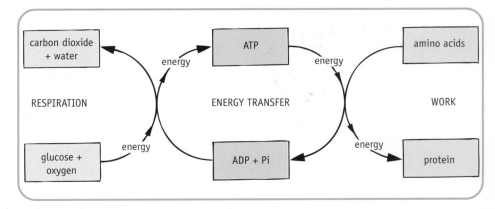

Figure 4.6 *Transfer of chemical energy by ATP*

Testing Your Knowledge

1 Name the units used to measure the energy contained in a foodstuff. (1)

2 a) Which sugar is the most common source of energy in living cells? (1)

b) What name is given to the series of enzyme-controlled reactions by which energy stored in sugar is released? (1)

c) Why is production of heat energy by respiring cells in the human body important? (1)

3 a) Give the full name of the chemical represented by the letters ATP. (1)

b) Draw a simple labelled diagram of an ATP molecule. (2)

4 a) In the two-way equation that follows, which numbered arrow represents (i) energy release? (ii) energy uptake?

$$ADP + Pi \; \underset{2}{\overset{1}{\rightleftharpoons}} \; ATP$$

(2)

b) Identify the source of energy needed to drive this process of energy uptake. (1)

c) The energy released during the process is put to many uses by living cells. Name TWO of these uses. (2)

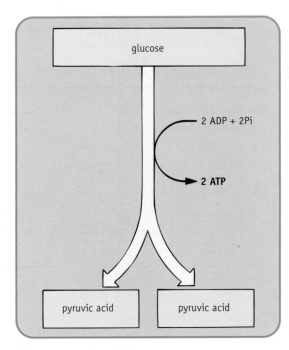

Figure 4.7 *Glycolysis*

Chemistry of respiration

Respiration is the process by which chemical **energy** is released during the breakdown of a foodstuff such as glucose. It occurs in every living cell and involves the regeneration of the high energy compound **ATP** by a complex series of biochemical reactions.

Glycolysis

Within a cell, the process of respiration begins with each molecule of **glucose** being broken down by a series of enzyme-controlled steps to form two molecules of **pyruvic acid** (see Figure 4.7). This respiratory pathway, involving 'glucose-splitting', is called **glycolysis**. It results in the production of two ATP per molecule of glucose.

Aerobic respiration

When oxygen is present, **aerobic** respiration occurs and each molecule of pyruvic acid is further broken down by many more enzyme-controlled steps. This respiratory pathway is illustrated in a simplified form in Figure 4.8 (on the following page). It results in the formation of carbon dioxide, water and 18 molecules of ATP per molecule of pyruvic acid. Therefore the complete breakdown of one molecule of glucose during aerobic respiration generates 38 (2 + 18 + 18) molecules of ATP. These high energy ATP molecules make energy available for use by the cell.

Aerobic respiration of one molecule of glucose can be summarised as follows:

$$\text{glucose} + \text{oxygen} + 38\text{ADP} + 38\text{Pi} \rightarrow CO_2 + \text{water} + 38\text{ATP}$$

Testing Your Knowledge

1 Give the meaning of the term *respiration* and state where this process occurs in living organisms. (2)
2 The respiratory pathway begins with the process of 'glucose-splitting'. Give the correct name of this enzyme-controlled process and identify the TWO different products that result from it. (3)
3 What chemical substance must be present (in addition to glucose) for aerobic respiration to proceed? (1)

4 a) How many molecules of ATP are produced from the complete breakdown of one molecule of pyruvic acid? (1)
   b) State the TWO additional products of this aerobic pathway. (2)
5 Write a word equation to summarise the complete breakdown of one molecule of glucose by aerobic respiration. (1)

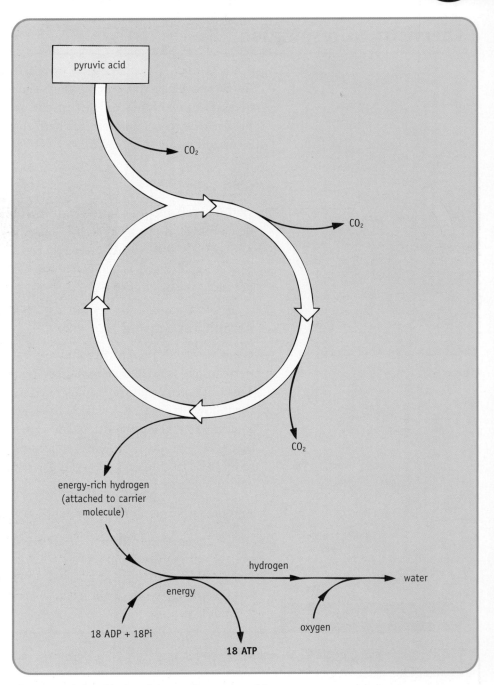

Figure 4.8 *Aerobic breakdown of pyruvic acid*

Anaerobic respiration

This is the process by which a little energy is derived from the **partial breakdown** of glucose in the **absence** of oxygen. Glycolysis (Figure 4.7) occurs as normal in the cell and two ATP molecules are formed. However, the pathway involving the complete breakdown of pyruvic acid to carbon dioxide and water (Figure 4.8) cannot proceed in the absence of oxygen. Instead the pyruvic acid undergoes one of the following pathways depending on the type of living organism involved.

Anaerobic respiration in animals

Figure 4.9 represents this process in animal cells such as skeletal muscle tissue in the human body. During vigorous muscular activity when oxygen supply cannot meet demand, cells undergo anaerobic respiration and **lactic acid** is formed.

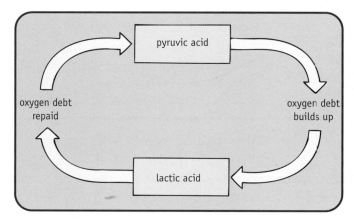

Figure 4.9 *Anaerobic respiration in animal cells*

Figure 4.10 *Repayment of oxygen debt*

As the concentration of lactic acid builds up in the muscle tissue (and bloodstream), it causes discomfort and pain. This reduces the efficiency of the muscles causing them to suffer **muscle fatigue**.

Oxygen debt

Since oxygen is required to remove the lactic acid, the body is said to build up an **oxygen debt** during anaerobic respiration. The debt is repaid when oxygen becomes available during a rest period (see Figure 4.10). The lactic acid is converted to pyruvic acid (which can then enter the aerobic pathway). Thus the anaerobic conversion of pyruvic acid to lactic acid in animal cells is a **reversible** process.

Anaerobic respiration in plants

Root cells of plants in waterlogged soil and yeast cells deprived of oxygen during wine-making are unable to respire aerobically. Such plant cells are dependent on anaerobic respiration (Figure 4.11).

Since a carbon atom is lost as CO_2 gas each time a molecule of pyruvic acid is broken down to **ethanol**, anaerobic respiration in plant cells is an **irreversible** process.

Comparison of aerobic and anaerobic respiration

Table 4.3 compares the two forms of respiration with reference to one molecule of glucose.

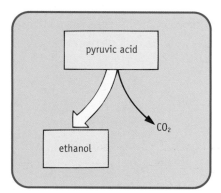

Figure 4.11 *Anaerobic respiration in plant cells*

	aerobic respiration	anaerobic respiration
need for oxygen	oxygen always required	oxygen never required
energy yield	efficient method of respiration, releasing 38 ATP per molecule of glucose	inefficient method of respiration, releasing 2 ATP per molecule of glucose
degree of breakdown of glucose	glucose completely broken down	glucose partially broken down
end products	carbon dioxide and water	lactic acid in animal cells; ethanol and carbon dioxide in plant cells

Table 4.3 *Comparison of aerobic and anaerobic respiration*

Testing Your Knowledge

1 Define the term *anaerobic respiration*. (1)

2 a) Construct a word equation to summarise anaerobic respiration in skeletal muscle tissue. (2)

 b) The following four statements refer to anaerobic respiration in human muscle tissue. Rewrite them in the correct order.
 Lactic acid builds up.
 Oxygen supply becomes limited
 Muscles suffer fatigue.
 Lactic acid is produced. (1)

 c) Identify TWO conditions necessary for an oxygen debt in the body to be repaid. (2)

3 Is anaerobic respiration in animal cells reversible? Explain your answer. (2)

4 a) Construct a word equation to summarise anaerobic respiration in yeast cells. (2)

 b) Is this process reversible? Explain your answer. (2)

5 Give THREE differences between the breakdown of a glucose molecule in a root cell respiring (i) aerobically in well aerated soil and (ii) anaerobically in flooded soil. (3)

Practical Activity

Measuring an earthworm's rate of respiration

Information

- A **respirometer** (Figure 4.12) is a piece of apparatus designed to measure rate of respiration (e.g. as volume of oxygen consumed per hour).
- Sodium hydroxide is a chemical that absorbs carbon dioxide (in this case the carbon dioxide given out by the animal respiring during the experiment).
- Oxygen taken in by the animal causes a decrease in volume of the enclosed gas, so the coloured liquid rises up the tube.

You need

2 respirometers
2 clamp stands
2 large beakers of water at room temperature
1 large healthy earthworm
1 marking pen
1 electronic balance

What to do

1 Set up two respirometers with an equal mass of sodium hydroxide pellets in each and the syringe plungers inserted to 0 on the scale.

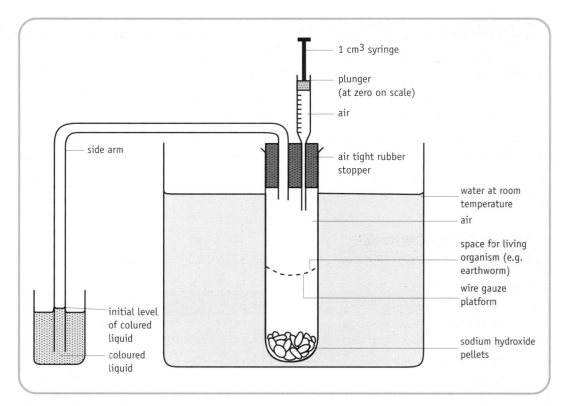

1 cm³ syringe

plunger (at zero on scale)

air

side arm

air tight rubber stopper

water at room temperature

air

space for living organism (e.g. earthworm)

wire gauze platform

initial level of colured liquid

coloured liquid

sodium hydroxide pellets

Figure 4.12 *Respirometer*

2 Clamp each tube into position in a large beaker of water at room temperature, with its side arm out of the beaker and its open end dipped into the coloured liquid.

3 Insert an earthworm into the first respirometer and leave the second respirometer empty since it is the control.

4 Allow two minutes for the respirometers to equilibrate and then mark the initial level of coloured liquid on each side arm.

5 After 30 minutes use the syringe to inject air and force the level of coloured liquid in each respirometer's side arm back to the start.

6 Record the volume of air needed in each case by referring to the scale on the syringe.

7 Convert the results to volume of oxygen consumed by the earthworm per hour.

8 Discuss the following two points of experimental design with the members of your group.

a) Why is it better scientific practice to carry out the experiment with the respirometer tubes immersed in water at room temperature rather than in air at room temperature?

b) Strictly speaking, it could be argued that the observed difference between the respirometers was not caused by the earthworm respiring but was simply due to the fact that some space was occupied in the first respirometer only. How could the control be improved so that this argument could be dismissed?

Activity

Planning and designing an investigation into anaerobic respiration in yeast

Information

- Yeast is a unicellular fungus that can respire both aerobically and anaerobically.
- Yeast is able to use 5% glucose solution as its energy source.
- Bicarbonate indicator is a chemical that varies in colour depending on the concentration of carbon dioxide that it contains (see Table 4.4).

What to do

1 Write a heading that states the aim of the investigation.

2 Make up your plan by answering the following in sentences.

   a) Describe how dissolved oxygen would be removed from the glucose solution.

relative CO_2 concentration	colour of bicarbonate indicator solution
high (above atmospheric)	yellow
medium (atmospheric)	red
low (below atmospheric)	purple

Table 4.4 *Bicarbonate indicator range*

   b) Name the most suitable type of container for use in this experiment, keeping in mind that if heat energy is released it will be lost from a normal glass test tube or flask.
   c) Describe how your design would allow any change in temperature to be recorded.
   d) Describe how your design would allow any release of carbon dioxide to be observed.
   e) What chemical would be added to the surface of the glucose solution to prevent oxygen from the air reaching the yeast cells and allowing them to respire aerobically?
   f) Draw a labelled diagram to show your design set up and ready to start.
   g) Compare your design with Figure 1.6 on page 5.
   h) State the one way in which the control would differ from the design that you have drawn.
   i) State the observations and measurements that would be made at the start and at the end of the experiment to obtain a set of results.
   j) Describe how the presence of alcohol in the glucose solution at the end of the experiment could be verified (see pages 5–6 for help).
   k) Name THREE changes that would have to occur during the experiment to allow you to conclude that yeast cells had respired anaerobically.
   l) Give the word equation of anaerobic respiration in yeast cells.

Applying Your Knowledge

1 Match the terms in list X with their descriptions in List Y.

List X
1) adenosine triphosphate
2) aerobic respiration
3) anaerobic respiration
4) chemical energy
5) ethanol
6) fatigue
7) glucose
8) glycolysis
9) heat energy
10) kilojoule
11) lactic acid
12) oxygen
13) oxygen debt
14) phosphorylation
15) pyruvic acid

List Y
a) unit used to measure energy
b) carbohydrate that acts as a source of energy in living cells
c) process by which high energy ATP is made from low energy ADP and Pi
d) stage of respiratory pathway common to aerobic and anaerobic respiration
e) chemical formed from the splitting of glucose during glycolysis
f) type of respiration that occurs in the presence of oxygen and produces much energy
g) type of respiration that occurs in the absence of oxygen and produces little energy
h) chemical formed during anaerobic respiration in animal cells
i) state that develops in the body during anaerobic respiration and is repaid during a rest period
j) state of muscle resulting from lack of oxygen and build-up of lactic acid
k) chemical formed during anaerobic respiration in plant cells
l) form of energy released from cells during respiration
m) chemical element needed for aerobic respiration
n) form of energy present in food before respiration takes place
o) high energy compound synthesised using energy from respiratory breakdown of glucose

2 Figure 4.13 (on the next page) shows six versions of the apparatus set up during an investigation to compare the energy content of two different foods.
a) Explain fully why a valid comparison cannot be made between set-ups:
(i) A and B; (ii) A and C; (iii) A and E. (3)
b) Describe THREE changes that would need to be made to set-up C so that it could be fairly compared with D. (3)
c) The experiment was carried out using set-ups B and F and the results in Table 4.5 were obtained.

food	water temperature (°C)		temperature rise (°C)	energy content (kJ)
	at start	at finish		
peanut	25	47		
glucose	26	38		

Table 4.5

(i) Calculate the energy content of the two foodstuffs in kilojoules per gram using the formula given on page 64.

(ii) A data book was found to give the energy content per gram for peanut as 25 kJ and for glucose as 17 kJ, measured using a food calorimeter. The fact that the results in Table 4.5 are much lower than these is accounted for by the fact that food burned in air under a beaker loses heat to the surroundings and does not burn completely to ashes. Study the diagram of the food calorimeter (Figure 4.2) and explain how use of it largely overcomes these two shortcomings. (3)

3 The data in Table 4.6 show the average energy expended in kJ per minute by people of differing body mass engaged in different activities.
a) On average how many kJ would be expended by a 63 kg person rowing vigorously for 15 minutes? (1)
b) Cornflakes contain 15 kJ per gram. How many grams would be needed by a 57 kg person to fuel half an hour of aerobic dancing? (1)
c) During which of the following activities would a person weighing 85 kg expend most energy? (Show your working in each case.)

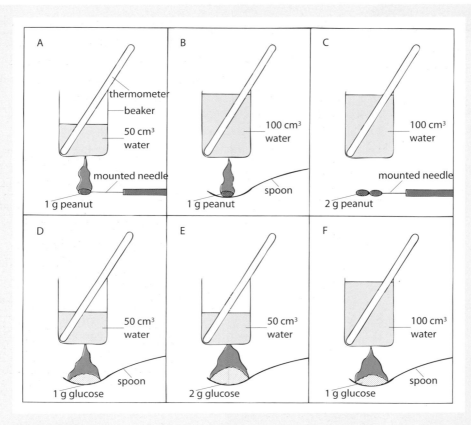

Figure 4.13

	average energy expended in kJ per minute for range of body mass (kg)			
activity	50–59	60–69	70–79	80–89
walking (3 mph)	16.4	18.9	22.3	24.4
walking (4 mph)	18.9	23.1	25.6	28.6
rowing gently	16.4	18.9	22.0	24.5
rowing vigorously	36.0	40.8	48.3	53.2
square dancing	23.1	26.3	31.1	33.6
aerobic dancing	24.0	27.6	32.9	36.0
jogging (5.5 mph)	36.0	41.0	48.3	53.2
jogging (8 mph)	43.7	50.0	59.2	65.1

Table 4.6

   (i)   Jogging at 8 mph for 20 minutes.

   (ii) Jogging at 5.5 mph for 25 minutes.

   (iii) Walking at 4 mph for 50 minutes.   (3)

4 Figure 4.14 gives a summary of the chemistry of respiration in a yeast cell.

a) Name process P.   (1)

b) Identify:

   (i)   the number of molecules that should have been inserted in boxes Q and T

   (ii) the names of the products that should have been inserted in boxes R and S.   (4)

c) Predict the fate of substance S.   (1)

d) State the effect of a high concentration of substance R on a yeast cell.   (1)

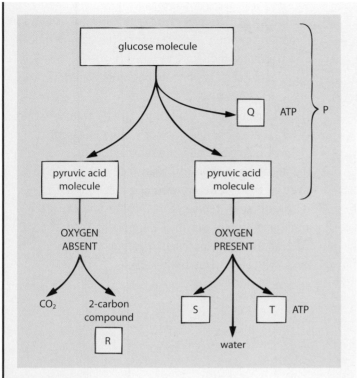

Figure 4.14

e) State the total number of molecules of ATP gained from
   (i)   the partial breakdown of one glucose molecule during anaerobic respiration
   (ii) the complete breakdown of one glucose molecule during aerobic respiration. (2)

5 The data in Table 4.7 refer to average values for athletes in training for different events.
   a) Copy and complete the table. (8)
   b) Look at Figure 11.9 (on page 233) of a stacked bar chart and then present the data from the two right-hand columns of your table in this format. (4)

6 The graph in Figure 4.15 shows the volume of carbon dioxide released by yeast cells during the fermentation of glucose.

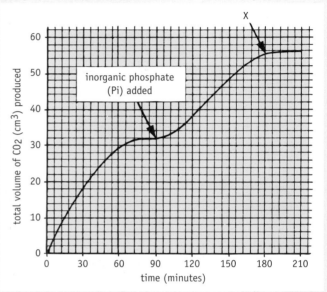

Figure 4.15

a) Is this process an example of aerobic or anaerobic respiration? (1)
b) For which specific biochemical process does a yeast cell require a supply of inorganic phosphate (Pi)? (1)
c) Name the substance to which the inorganic phosphate added at 90 minutes became chemically combined. (1)

athletic event	volume of oxygen needed for event (l)	volume of oxygen consumed during event (l)	oxygen debt (l)	percentage of energy obtained for event from aerobic respiration	percentage of energy obtained for event from anaerobic respiration
100 metres	10	0.5	9.5	5	95
800 metres	25	8	17	32	
1500 metres	36	18	18		
10000 metres	150	135			
marathon (42186 metres)	700			98	2

Table 4.7

d) (i) State what happened to the rate of CO_2 production following the addition of inorganic phosphate at 90 minutes.
   (ii) Suggest a reason for this change. (2)
e) Give TWO possible explanations for the decline in CO_2 production at point X on the graph. (2)

7 Figure 4.16 shows the equipment needed to investigate aerobic respiration in a stick insect using a simple respirometer.
   a) Draw a diagram of the respirometer set up ready to begin with the insect in place. (4)
   b) State the function of the sodium hydroxide pellets in this experiment. (1)

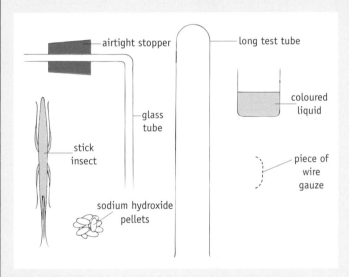

Figure 4.16

   c) What change in the apparatus indicates that the insect is using up oxygen? (1)
   d) In what way would the control differ from the experimental set-up that you have drawn? (1)

8 Figure 4.17 shows the process of energy transfer in skeletal muscle tissue. Copy the diagram and complete it by adding four arrow heads and supplying the words missing at positions 1 to 6. (8)

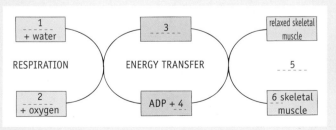

Figure 4.17

9 The following list gives five stages that occur during aerobic respiration. Put them into the correct order.
   1) Water produced.
   2) Molecule of 3-carbon compound broken down by enzyme action.
   3) Molecule of glucose split into two by enzyme action.
   4) CO_2 and a large quantity of energy released.
   5) Small quantity of energy released. (1)

10 The apparatus shown in Figure 4.18 was set up to measure the respiratory rate of dandelion leaf tissue at two different temperatures. Metal foil was placed round both tubes and they were maintained at 15°C for 2 hours. The experiment was repeated at 25°C. Table 4.8 summarises the results.

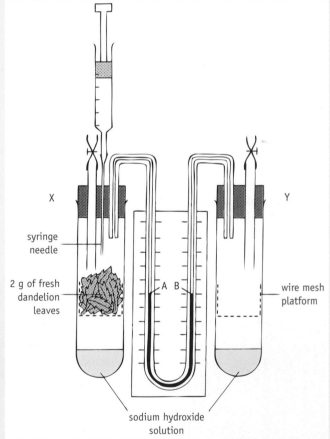

Figure 4.18

temperature (°C)	volume of air needed to return level A to starting point (ml)
15	2.8
25	4.8

Table 4.8

a) Describe how tubes X and Y could be kept at the required temperature during each part of the experiment. (1)

b) Why were tubes X and Y covered with metal foil? (1)

c) Why does level A rise during the experiment? (1)

d) By what means is level A returned to its starting point after two hours in each experiment? (1)

e) Calculate the respiratory rate (in millilitres of oxygen consumed per gram of fresh tissue per hour) of dandelion leaves at the two temperatures. (2)

f) What conclusion can be drawn from the data about the effect of temperature on the respiratory rate of dandelion leaf tissue? (1)

g) How could the reliability of the results at each temperature be improved without making any changes to the apparatus? (1)

h) In what way could control tube Y be improved? (1)

(5) Photosynthesis

Energy fixation

Photosynthesis is a series of enzyme controlled reactions which allows green plants to make their own food. This process involves the capture of light energy from the sun. Light energy is trapped ('fixed') by the green pigment **chlorophyll**. Chlorophyll is found in discus-shaped structures called **chloroplasts** present in green leaf cells (see Figure 5.1).

The light energy is converted to chemical energy in the form of ATP. This ATP is then used to produce carbohydrate food such as glucose.

Raw materials

A **carbohydrate** is a compound containing the chemical elements carbon (C), hydrogen (H) and oxygen (O) combined together using energy. The production of a carbohydrate food such as glucose requires a supply of the raw materials water and carbon dioxide to be available to supply the necessary chemicals.

During photosynthesis, molecules of carbon dioxide (CO_2) combine with hydrogen from water (H_2O) in the presence of chlorophyll and light energy to form carbohydrates such as glucose ($C_6H_{12}O_6$) and oxygen, a by-product of the process.

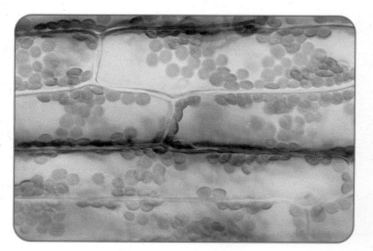

Figure 5.1 *Chloroplasts in leaf cells*

Diffusion

The process of diffusion is essential to photosynthesis for the movement of molecules of
- carbon dioxide from the external environment into the green leaf cells
- oxygen out of the green leaf cells into the external environment.

Summary

The process of photosynthesis is summarised in Figure 5.2 and by the following equation:

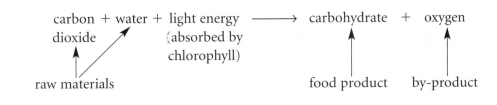

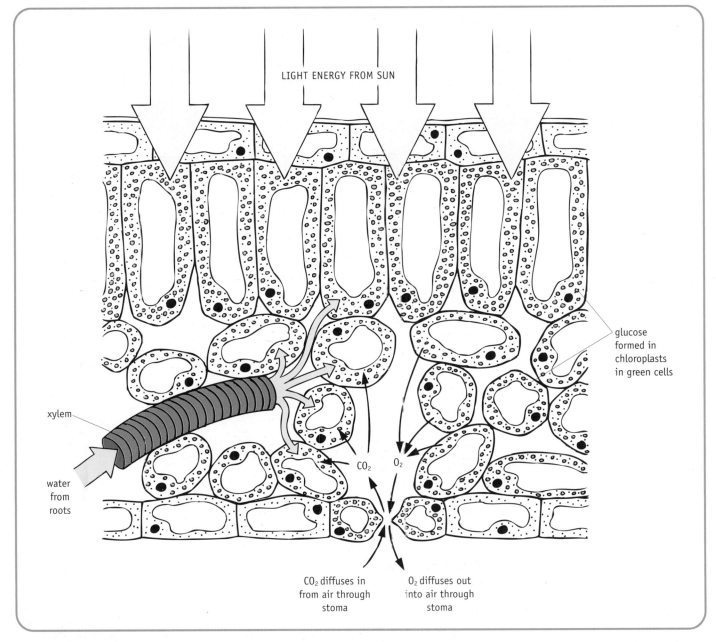

Figure 5.2 *Photosynthesis in a leaf of a land plant*

Practical Activity and Report

Comparing photosynthesis in light and dark conditions

INFORMATION

- Many plants convert excess glucose made during photosynthesis into starch. The presence of starch in a leaf is regarded as evidence that photosynthesis has occurred.
- Iodine solution turns starch blue–black.
- Green chlorophyll is insoluble in water but soluble in alcohol (ethanol).

YOU NEED

1 small sheet of metal foil

2 paper clips

1 pair of scissors

1 healthy pot plant (e.g. Geranium) that has been in darkness for 2 days

1 cork borer

1 Bunsen burner

1 tripod stand

1 250 ml glass beaker (for use as a simple hot water bath)

2 test tubes

1 dropping bottle of iodine solution

1 dropping bottle of alcohol (ethanol)

1 cavity tile

WHAT TO DO

1 Read all the instructions in this section and prepare your results table before carrying out the experiment.

2 Set up the plant as shown in Figure 5.3 with a leaf completely deprived of light energy.

3 Leave the plant in bright light for 2 days.

4 Cut out a disc from a leaf that has been in light using a cork borer.

5 Test the disc for starch by following the procedure shown in Figure 5.4 and using 50 drops of alcohol in the test tube.

6 Repeat the procedure using a disc from the leaf that has been in darkness.

7 Keep a note of your results in your table which should refer to the colour of each leaf disc before and after the addition of iodine solution.

8 If other students have carried out the same experiment, pool the results.

REPORTING

Write up your report by doing the following:

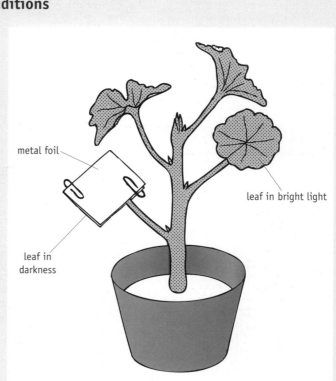

Figure 5.3 *Investigating the need for light*

1 Copy the title given at the start of this activity.

2 Put the subheading '**Aim**' and state the aim of your experiment.

3 a) Put the subheading '**Method**'.

  b) Draw a labelled diagram of your apparatus set up to extract chlorophyll from a leaf disc.

  c) Using the impersonal passive voice, briefly describe the experimental procedure that you followed and state how you obtained your results.

4 Put the subheading '**Results**' and draw a final version of your table of results.

5 Put a subheading '**Analysis of Results**' and write a sentence to state what you have found out from your results.

6 Put a subheading '**Conclusion**' and draw a conclusion about the ability of a plant to photosynthesise in a) the presence and b) the absence of light.

7 Put a final subheading '**Evaluation of Experimental Procedure**' and then answer the following questions in sentences:

a) By what means was the experiment controlled so

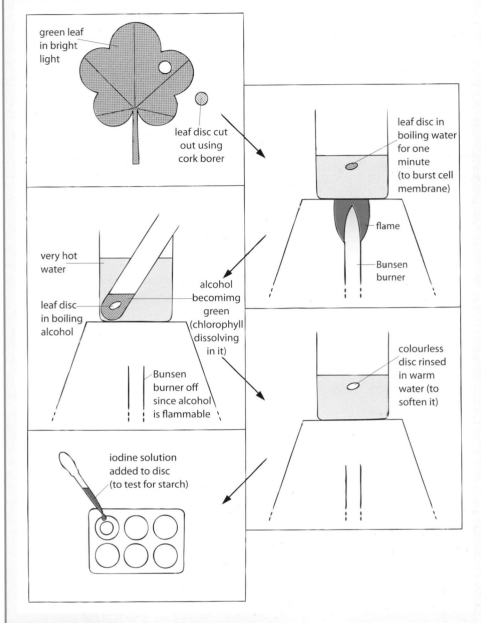

Figure 5.4 *Testing a leaf disc for starch*

that there was only one factor being varied at a time?

b) It could be argued that photosynthesis in a leaf is prevented by contact with metal foil and not by lack of light energy. How could the experimental design be improved to overcome this criticism?

c) How could the reliability of the results be checked?

d) Feel free to comment on any of the following if you have an additional point you wish to make:
   (i)   possible sources of error
   (ii) further improvements that you would include in a repeat of the experiment
   (iii) limitations of the equipment.

Practical Activity and Report

Comparing photosynthesis in the presence and absence of carbon dioxide

INFORMATION

- Many plants convert excess glucose made during photosynthesis into starch. The presence of starch in a leaf is regarded as evidence that photosynthesis has occurred.
- Iodine solution turns starch blue–black.
- Green chlorophyll is insoluble in water but soluble in alcohol (ethanol).
- Concentrated sodium hydroxide solution absorbs carbon dioxide from air.

YOU NEED

2 healthy pot plants (e.g. Geranium) that have been in darkness for 2 days
1 cork borer
1 Bunsen burner
1 tripod stand
1 250 ml glass beaker (for use as a simple hot water bath)
2 test tubes

1 dropping bottle of iodine solution
1 dropping bottle of alcohol (ethanol)
1 cavity tile
1 small beaker containing 50 ml concentrated sodium hydroxide
1 small beaker containing 50 ml cold tap water
2 plastic bags
2 elastic bands

WHAT TO DO

1 Read all the instructions in this section and prepare your results table before carrying out the experiment.
2 Set up the experiment as shown in Figure 5.5 and leave the plants in bright light for 2 days.
3 Cut out a disc from a leaf of plant A using a cork borer.
4 Test the disc for starch by following the procedure shown in Figure 5.4.
5 Repeat the procedure using a disc from plant B.

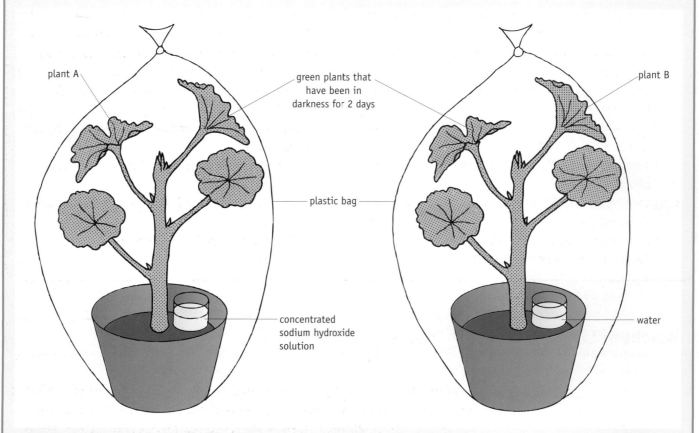

plant A

green plants that have been in darkness for 2 days

plant B

plastic bag

concentrated sodium hydroxide solution

water

Figure 5.5 *Investigating the need for CO$_2$*

6 Keep a note of your results in your table which should refer to the colour of each leaf disc before and after the addition of iodine solution.

7 If other students have carried out the same experiment, pool the results.

REPORTING

Write up your report by doing the following:

1 Copy the title given at the start of this activity.

2 Put the subheading 'Aim' and state the aim of your experiment.

3 a) Put the subheading 'Method'.
   b) Draw a labelled diagram of your apparatus set up to extract chlorophyll from a leaf disc.
   c) Using the impersonal passive voice, briefly describe the experimental procedure that you followed and state how you obtained your results.

4 Put the subheading 'Results' and draw a final version of your table of results.

5 Put a subheading 'Analysis of Results' and write a sentence to state what you have found out from your results.

6 Put a subheading 'Conclusion' and draw a conclusion about the ability of a plant to photosynthesise in a) the presence and b) the absence of carbon dioxide.

7 Put a final subheading 'Evaluation of Experimental Procedure' and then answer the following questions in sentences:
   a) By what means was the experiment controlled so that there was only one factor being varied at a time?
   b) How could the reliability of the results be checked?
   c) Feel free to comment on any of the following if you have an additional point you wish to make:
      (i) possible sources of error
      (ii) further improvements that you would include in a repeat of the experiment
      (iii) limitations of the equipment.

Testing Your Knowledge

1 a) Identify the type of energy needed by green plants to make their own food. (1)
   b) What name is given to this food-manufacturing process? (1)

2 a) Name the green pigment in a plant that captures light energy. (1)
   b) Exactly where in a plant cell is this green pigment found? (1)
   c) Into which type of energy is light converted during photosynthesis? (1)

3 a) Identify the THREE chemical elements present in a carbohydrate. (3)
   b) From which TWO raw materials does the plant obtain a supply of these elements? (2)

   c) Name the by-product of photosynthesis. (1)
   d) Give the word equation for photosynthesis. (2)

4 a) Name the complex carbohydrate most commonly stored in leaf cells following photosynthesis. (1)
   b) Which chemical reagent is used to test for this carbohydrate? (1)
   c) When testing a leaf for this carbohydrate, why is it necessary to:
      (i) immerse the disc in boiling water for one minute?
      (ii) immerse the disc in boiling alcohol for several minutes?
      (iii) rinse the disc in warm water for a few seconds? (3)

Biochemistry of photosynthesis

Photosynthesis is a complex biochemical pathway which consists of a series of enzyme-controlled reactions. The process of photosynthesis falls into two separate parts: a light-dependent stage called **photolysis** and a temperature-dependent stage called **carbon fixation**.

Photolysis

Solar (light) energy is trapped by chlorophyll in chloroplasts and converted into chemical energy. This process involves several important events which are summarised in Figure 5.6.

Light energy is used to split molecules of water into **hydrogen** and **oxygen**. This process is called **photolysis** of water. The oxygen is released as a by-product. The hydrogen combines with a hydrogen acceptor. When loaded with hydrogen, the hydrogen acceptor is said to be **reduced**.

Chlorophyll also makes energy available for the regeneration of **ATP** from ADP and Pi. This process is called **photophosphorylation**.

The hydrogen held by the reduced hydrogen acceptor and the energy held by the ATP at the end of the light-dependent stage are essential for use in carbon fixation, the second stage of photosynthesis.

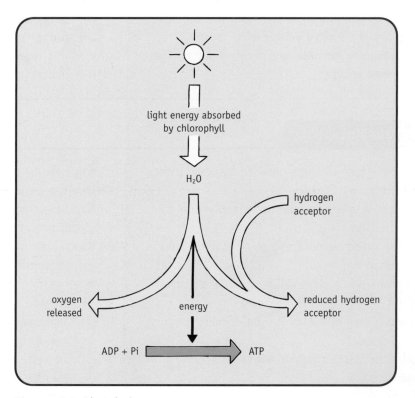

Figure 5.6 *Photolysis*

Carbon fixation

Carbon fixation, the second stage of photosynthesis, also occurs in the chloroplasts. It consists of several enzyme-controlled reactions which take the form of a cycle. Figure 5.7 shows a simplified version of this biochemical pathway. It provides the means by which **carbon** (in carbon dioxide) becomes 'fixed' into carbohydrate by combining with the hydrogen produced during the photolysis of water. Carbon fixation results in the formation of **glucose**. The energy needed to drive this process is supplied by the ATP produced during photosynthesis.

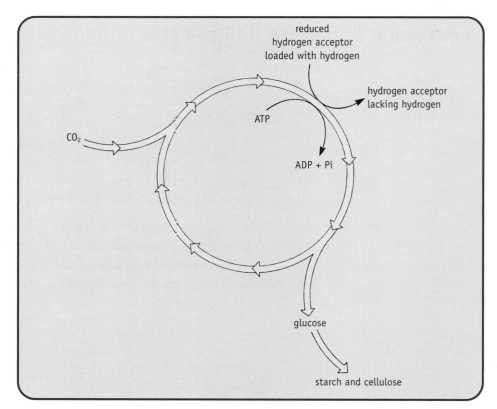

Figure 5.7 *Carbon fixation*

Figure 5.8 gives a summary of the biochemistry of photosynthesis in a chloroplast.

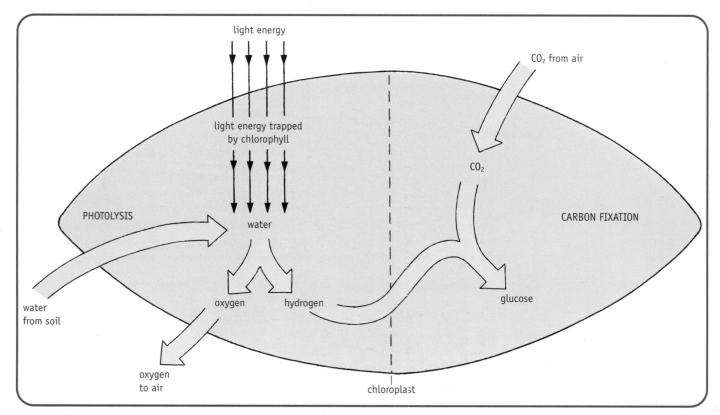

Figure 5.8 *Summary of the biochemistry of photosynthesis*

Conversion of glucose into complex carbohydrate

As a plant grows, it continues to make **glucose** by photosynthesis. Some of this sugar is broken down again when required to provide the plant with energy for growth and reproduction. Some of the remaining glucose molecules become linked into long chains and packed together into spherical **starch** grains found in the cell's cytoplasm (see Figures 5.9, 5.10 and 5.11). This starch is the plant's store of food and can be converted back to sugar for energy when needed. Starch is therefore called a **storage** carbohydrate.

Other glucose molecules are built into long chains of **cellulose**. These are gathered together to form ribbon-like fibres used to build cell walls. Cellulose is therefore called a **structural** carbohydrate. The conversion of glucose to starch and cellulose involves chemical reactions controlled by enzymes.

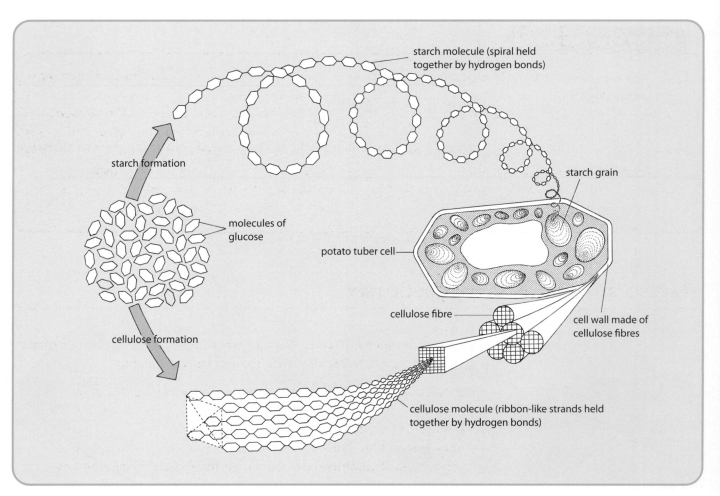

Figure 5.9 *Formation of storage and structural carbohydrate*

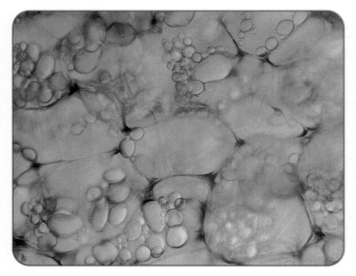

Figure 5.10 *Potato tuber cells in water*

Figure 5.11 *Potato tuber cells in iodine solution*

Testing Your Knowledge

1 a) What name is given to the light-dependent stage of photosynthesis? (1)
   b) What effect does light energy have on molecules of water during this stage? (1)

2 a) Describe the fate of the hydrogen and oxygen released as a result of photolysis. (2)
   b) What is meant by the term *photophosphorylation*? (2)

3 a) What name is given to the second stage of photosynthesis which takes the form of an enzyme-controlled cycle? (1)
   b) Which TWO substances from the light-dependent stage must be present for this cycle to turn? (2)
   c) Identify the end product of this cycle. (1)

4 Name (i) the structural and (ii) the storage carbohydrate formed in plant cells using glucose molecules made by photosynthesis. (2)

Factors affecting photosynthetic rate

Several environmental factors affect the rate of photosynthesis. These include light intensity, carbon dioxide concentration and temperature.

The **rate** of photosynthesis can be estimated by measuring one of the following:

- **evolution of oxygen** per unit time
- **uptake of carbon dioxide** per unit time
- **production of carbohydrate** (as increase in dry weight) per unit time.

Investigating the effect of varying light intensity

Elodea **bubbler experiment**

The number of **oxygen bubbles** released per minute by the cut end of an *Elodea* stem indicates the rate at which photosynthesis is proceeding. At first the lamp (see Figure 5.12) is placed exactly 100 cm from the plant and the number of oxygen bubbles released per minute counted. The lamp is then moved to a new position (say 60 cm from the plant) and the rate of bubbling noted (once the plant has had a short time to become acclimatised to this new higher light intensity).

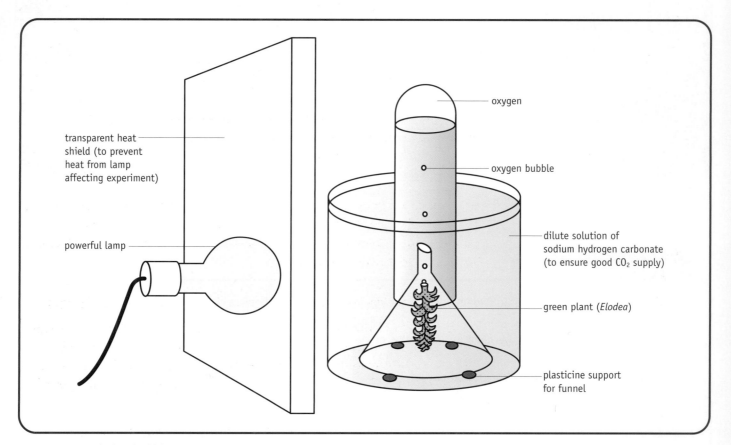

Figure 5.12 *Elodea bubbler experiment*

The process is repeated for lamp positions even nearer the plant, as shown in Table 5.1. When this typical set of results is displayed as a graph (Figure 5.13), it can be concluded that as light intensity increases, photosynthetic rate also increases until it reaches a maximum of 25 bubbles per minute at around 64 units of light.

distance from plant (cm)	units of light (calculated using mathematical formula)	number of oxygen bubbles released per minute
100	4	4
60	11	10
40	25	19
30	45	24
25	64	25
20	100	25

Table 5.1 *Elodea bubbler results*

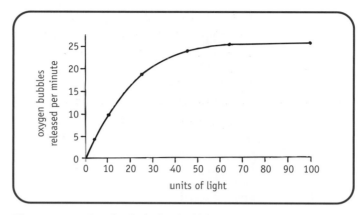

Figure 5.13 *Graph of Elodec bubbler experiment results*

Limiting factors

Further increase in light intensity does not increase the photosynthetic rate in the above experiment. This is because shortage of some other factor (e.g. concentration of carbon dioxide) is now holding up the process. This factor in short supply is called a **limiting factor**.

Investigating the effect of varying carbon dioxide concentration

In this experiment the concentration of carbon dioxide made available to the *Elodea* plant is gradually increased by adding appropriate masses of sodium hydrogen carbonate to the water. The number of oxygen bubbles released is counted as before. The lamp is kept in one position to give uniform light of medium intensity.

The graph of a set of results (Figure 5.14) shows that when the plant is supplied with a carbon dioxide concentration of only 1 unit, photosynthesis is **limited** by this low concentration of carbon dioxide to 3 oxygen bubbles per minute. When carbon dioxide concentration is increased to 2 units, photosynthesis increases to 6 bubbles per minute but no further since carbon dioxide concentration becomes limiting again.

A further increase in carbon dioxide concentration to 3 units brings about a further increase in photosynthetic rate. However, beyond this point the graph

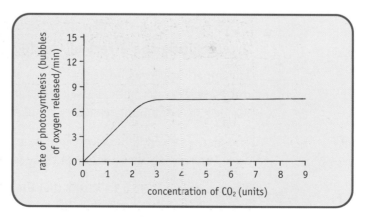

Figure 5.14 *Variation in CO_2 concentration at one light intensity*

levels out and any further increase in carbon dioxide concentration does not affect photosynthetic rate. This is because some other factor (e.g. light intensity) is now limiting the process.

Effect of varying carbon dioxide concentration at different light intensities
In Figure 5.15, graph ABC represents a repeat of Figure 5.14 and graph ADE represents the results from a further *Elodea* bubbler experiment using the same plant in conditions of constant high light intensity.

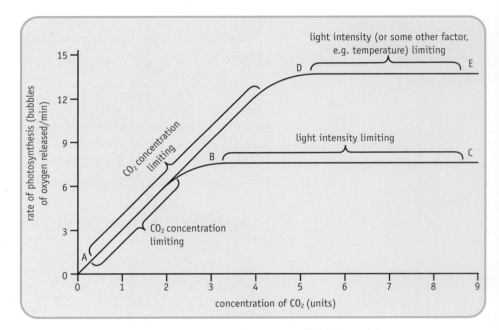

Figure 5.15 *Variation in CO₂ concentration at two light intensities*

In this second experiment an increase in carbon dioxide concentration to 4 and then 5 units brings about a corresponding increase in photosynthetic rate in each case. This is because carbon dioxide concentration is still the **limiting factor** up to 5 units of carbon dioxide when light intensity is higher.

However, beyond 5 units of carbon dioxide, the graph levels off again since some other factor (such as light intensity or temperature) has become limiting.

Investigating the effect of varying temperature

The apparatus shown in Figure 5.12 is adapted for use in this experiment by using a large water bath whose temperature is under thermostatic control. Plastic bags containing ice cubes are used to create low temperatures.

The plant is given light of constant high intensity and a rich supply of carbon dioxide to ensure that neither of these factors limits the process.

The graph in Figure 5.16 shows that the photosynthetic rate rises to an **optimum** at around 35°C (though this varies from species to species and is often lower than 35°C). Beyond the optimum, the photosynthetic rate drops rapidly. This is because photosynthesis consists of many reactions controlled by enzymes that are denatured at high temperatures.

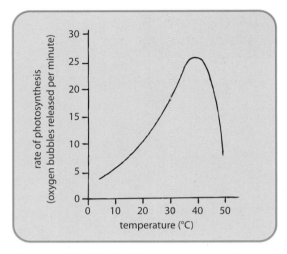

Figure 5.16 *Effect of temperature on photosynthetic rate*

Early crops in horticulture

If one of the above three factors is limiting the rate of photosynthesis by a crop of plants such as wheat in a field, there is no economically viable way for the farmer to solve the problem. However, if a factor is limiting the rate of photosynthesis by a crop of plants such as tomatoes in a greenhouse, steps can be taken to improve the situation.

If **temperature** is the only limiting factor then on very cold, bright, winter days, for example, the temperature of the greenhouse can be increased using a heating system. When **light intensity** is the only limiting factor then on dull, mild, winter days, for example, the horticulturist can employ electric lighting to increase light intensity. The daily period of illumination during the winter and/or early spring can also be extended to produce an early crop.

When **carbon dioxide concentration** is the limiting factor, extra supplies can be added to the air in a greenhouse. This creates a **carbon dioxide enriched environment** which increases the yield of photosynthetic product in most species. By using a combination of these techniques, the horticulturist can produce early crops and maximise photosynthetic yield.

Testing Your Knowledge

1 State THREE ways in which rate of photosynthesis can be measured. (3)
2 a) State TWO environmental factors that can be demonstrated (by the bubbler experiment) to affect the rate of photosynthesis by the water weed *Elodea*. (2)

b) As the distance between the powerful lamp and the *Elodea* plant is gradually decreased, what effect does this have on:
(i) the intensity of the light reaching the plant?
(ii) the number of bubbles of oxygen released by the plant per minute? (2)

3 Consider the information in Table 5.2.

ingredients needed to make one loaf	Baker A's stock	Baker B's stock
500g flour	5 kg flour	5 kg flour
30g fat	60g fat	300g fat
10g yeast	50g yeast	50g yeast
5g sugar	40g sugar	15g sugar

Table 5.2 *Factors limiting bread-making*

a) Which ingredient limits Baker A's bread production to two loaves? (1)

b) Which ingredient limits Baker B's bread production to three loaves? (1)

4 a) What is meant by the term *limiting factor*? (1)

b) Identify TWO factors that can limit the process of photosynthesis. (2)

5 When no other factors are limiting the process of photosynthesis, what effect on photosynthetic rate is brought about by raising the temperature of the plant from

(i) 15°C to 35°C;

(ii) 35°C to 55°C? (2)

6 a) Explain why it is economically viable to supply tomato plants but not wheat plants with a carbon dioxide enriched environment. (2)

b) What benefit would a horticulturist gain by providing a crop of plants in a greenhouse with supplementary heating and lighting during winter? (1)

Applying Your Knowledge

1 Match the terms in list X with their descriptions in list Y.

List X
1) ATP
2) carbon dioxide
3) carbon fixation
4) chemical
5) cellulose
6) chlorophyll
7) chloroplast
8) glucose
9) hydrogen
10) light
11) limiting factor
12) oxygen

13) photophosphorylation
14) photolysis
15) starch
16) water

List Y
a) green pigment that traps light energy
b) production of ATP using energy trapped during a light-dependent reaction
c) complex carbohydrate stored in plant cells
d) high energy compound that provides energy to drive carbon fixation
e) raw material that supplies carbon atoms to be fixed into carbohydrate
f) light-dependent stage of photosynthesis
g) stage of photosynthesis that results in formation of carbohydrate
h) by-product of photolysis of water needed for aerobic respiration
i) raw material that becomes split into oxygen and hydrogen during photosynthesis
j) discus-shaped structure in a cell of a green leaf
k) 6-carbon sugar molecule formed by photosynthesis
l) general term for a factor whose restricted supply prevents an increase in the rate of a process
m) product of photolysis of water needed during carbon fixation
n) complex carbohydrate used to construct plant cell walls
o) form of energy trapped by chlorophyll for photosynthesis
p) form of energy contained in carbohydrate formed by photosynthesis

2 The graph in Figure 5.17 shows the variation in carbon dioxide concentration of the air among the leaves of a potato crop during three days in summer.

a) On which day and between which times (to the nearest hour) did the concentration of CO_2 drop at the fastest rate? (1)

b) Which physiological process was responsible for each decrease in CO_2 concentration in the graph? (1)

c) At approximately what time did (i) day break? (ii) night fall? (2)

d) Suggest which day was probably the least sunny. Explain your answer. (2)

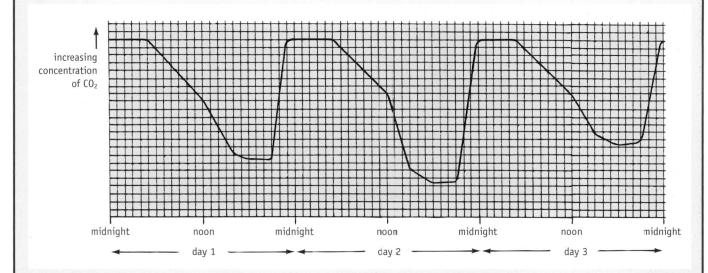

Figure 5.17

e) Redraw the axes and then make a simple sketch to give an idea of how the CO_2 concentration would be likely to vary during three dull, cold days in winter. (2)

3 Table 5.3 refers to the uptake of carbon dioxide by two green plants.

plant	total volume of CO_2 entering plant daily (mm³)	number of leaves	average area of one leaf (mm²)
X	24 000	12	500
Y	40 000	5	1000

Table 5.3

a) Calculate the daily rate of diffusion of CO_2 into plant X (in mm³ CO_2 per mm² of leaf). (1)

b) By how many times is the daily rate of diffusion of CO_2 into plant Y greater than that into plant X? (1)

4 The graph in Figure 5.18 (on the next page) shows the changes in oxygen concentration in the water of two marine rock pools during a sunny summer's afternoon and early evening. Pool X was shaded in the afternoon from 14.00 hours onwards. Pool Y was shaded until 14.00 hours.

a) Identify the dominant physiological process occurring between 12.00 and 14.00 hours in (i) pool X; (ii) pool Y. (2)

b) Identify the dominant physiological process occurring between 14.00 and 17.00 hours in (i) pool X; (ii) pool Y. (2)

c) Identify the dominant physiological process occurring between 18.00 and 19.00 hours in (i) pool X; (ii) pool Y. (2)

d) At which time of day shown on the graph would the water from pool Y be richest in carbon dioxide? Explain your answer. (2)

e) In what way would the graph for pool X have differed if the day had been cloudy between 12.00 and 14.00 hours? (1)

5 The plant in Figure 5.19 (on the next page) was destarched before being set up as shown. This demonstration really consists of three separate experiments being done on the plant at the same time.

a) After the plant has been in bright light for three days, which two discs should be tested for starch in order to show that:
 (i) light is essential for photosynthesis?
 (ii) chlorophyll is essential for photosynthesis?
 (iii) carbon dioxide is essential for photosynthesis? (3)

b) Explain your answer in each case. (3)

6 A horticulturist grew specimens of a species of tomato plant in seven greenhouses containing different concentrations of carbon dioxide in the air surrounding the plants. After a standard length of time, he measured the mass of sugar produced at each concentration of carbon dioxide, as shown in Table 5.4.

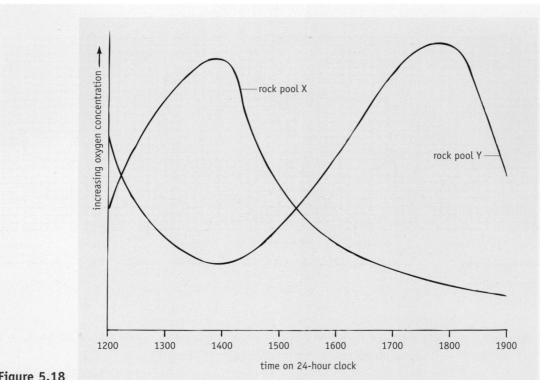

Figure 5.18

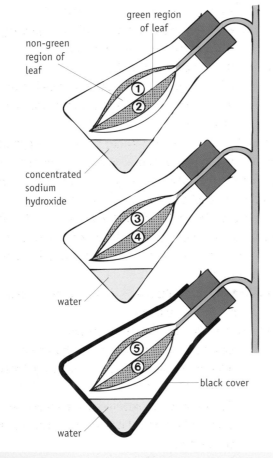

Figure 5.19

a) (i)   Which factor is used to measure photosynthesis in this experiment?
   (ii)   What additional information would you need in order to express this factor as the rate of photosynthesis?
   (iii) Name another TWO factors that could be used to measure the rate of photosynthesis. (4)

b) Express the mass of sugar produced per kilogram of dry plant at a carbon dioxide concentration of 70 units as a percentage of the dry mass of plant material. (1)

c) Using graph paper, construct a line graph of the results. (2)

d) From your graph, state what the sugar production would be at 15 units of carbon dioxide. (1)

e) (i)   Explain how the graph shows that carbon dioxide was the factor limiting photosynthesis between concentrations 0 and 50 units.
   (ii)   Explain how the graph shows that carbon dioxide was **not** the limiting factor from a carbon dioxide concentration of 50 units onwards.
   (iii) Name the factor that could have been

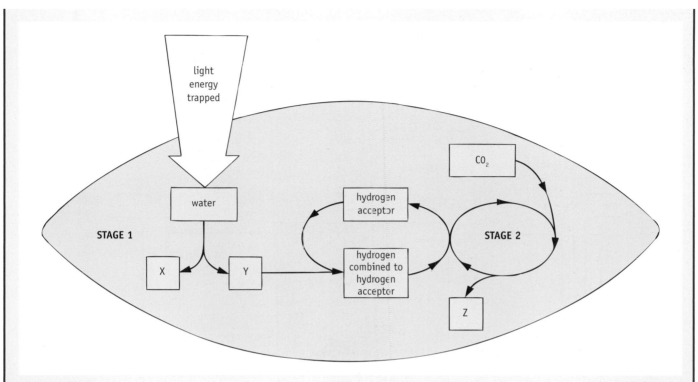

light
energy
trapped

water

STAGE 1

X

Y

hydrogen
acceptor

hydrogen
combined to
hydrogen
acceptor

CO₂

STAGE 2

Z

Figure 5.20

carbon dioxide concentration (units)	sugar production (g/kg of dry plant)
0	0.0
10	1.7
20	3.3
30	4.8
40	5.7
50	6.0
60	6.0
70	6.0

Table 5.4

limiting the process of photosynthesis from 50 units of carbon dioxide onwards. (3)

7 Figure 5.20 gives an illustrated summary of the biochemistry of photosynthesis.
   a) (i) Name the discus-shaped organelle within which photosynthesis takes place.
   (ii) Give the colour and name of the chemical substance responsible for trapping light energy. (3)
   b) (i) Identify substances X, Y and Z.

(ii) Which of these is a carbohydrate?
(iii) Which of these is a by-product of photolysis that diffuses out of the plant? (5)
c) (i) Name stages 1 and 2.
   (ii) Which high energy substance, needed to drive stage 2, has been omitted from the diagram? (3)

8 Figure 5.21 (on the next page) shows the apparatus set up to investigate the effect of temperature on the rate of photosynthesis by the water weed *Elodea*. The initial level on the measuring cylinder was recorded at 14.00 hours on Monday April 7; the final level was recorded at 18.00 hours on Friday 11 April. The apparatus was illuminated continuously using a powerful lamp.

a) Name the gas released by *Elodea* that collects in the measuring cylinder. (1)
b) Calculate the rate of photosynthesis. (1)
c) (i) Describe how the apparatus should be adapted to measure rate of photosynthesis at 25°C, 40°C and 55°C.
   (ii) Explain why an interval of time must be allowed between the readings taken at each temperature. (2)
d) (i) Predict at which of the temperatures given

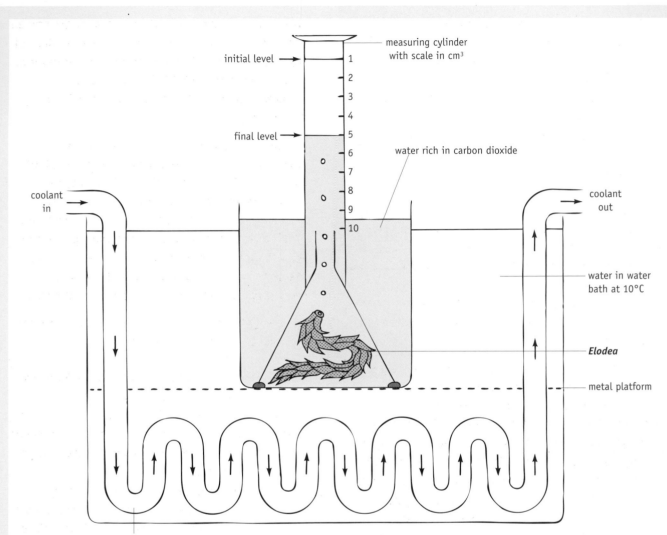

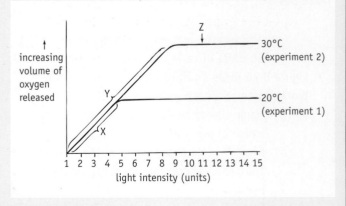

Figure 5.21

in part c) (i) the photosynthetic rate would be poorest.

    (ii) Explain your choice. (2)

e) (i) What is the reason for using water rich in carbon dioxide?

    (ii) What chemical solution could have been used to enrich the water with carbon dioxide? (2)

f) Describe how the apparatus could be employed to measure the effect of light intensity on rate of photosynthesis. (2)

9 The graph in Figure 5.22 shows the results from two experiments set up to investigate the rate of photosynthesis of a water plant exposed to different environmental conditions. The CO_2 concentration of

pond water was maintained at a high optimum level throughout the two experiments.

Figure 5.22

a) In experiment 1, the temperature was kept constant at 20°C.
   (i) Name the environmental factor that was varied.
   (ii) Name a second environmental factor that was kept constant. (2)

b) In the second experiment, the temperature was kept constant at 30°C. Suggest how this was done. (1)

c) What factor was limiting the rate of photosynthesis at region X on the graph? (1)

d) State the light intensity at which a temperature of 20°C first began to limit photosynthetic rate. (1)

e) State the factor which limits photosynthetic rate
   (i) in region Y;
   (ii) at point Z on the graph for experiment 2. (2)

10 Read the passage and answer the questions that follow it.

Successful plant growth depends on a variety of processes such as uptake of water and mineral salts, photosynthesis, cell division and differentiation, all of which respond to changes in environmental factors. As long ago as 1843 a scientist named Liebig stated in his 'law of the minimum' that 'the yield of a crop is determined by the factor which is relatively the minimum.' This law is found to be just as valid today. Agricultural scientists still attempt to regulate the factors that affect growth. One way to do this is to determine the effect caused by altering one factor while other conditions are kept constant.

The temperature of a greenhouse can be regulated by using thermostatically controlled heaters when the outside temperature is low and by using air-conditioning when the weather is hot. Warm temperatures applied continuously do not,

however, necessarily result in optimum growth. In tomatoes, growth, as measured by the height and dry mass of plant, is found to be much greater at a day temperature of 20°C and a night temperature of 10°C than at 20°C continuously day and night.

In addition, caution must be exerted when drawing conclusions from experiments on plants since it is often impossible to alter only one variable factor at a time. An alteration in light intensity, for example, affects stomatal opening, which in turn affects carbon dioxide uptake and water balance. Furthermore, each plant species seems to be a law unto itself with respect to its precise responses to environmental conditions.

a) Name TWO 'environmental factors' (1st sentence) to which the processes affecting plant growth could respond. (2)

b) (i) What name did Liebig give to the 'principle of limiting factors'?
   (ii) Explain what this law means. (3)

c) With reference to a normal day and night cycle of events in the life of a tomato plant, state when
   (i) photosynthesis;
   (ii) aerobic respiration would be occurring.
   (iii) What effect does an increase in temperature from 10°C to 20°C have on a plant's rate of respiration?
   (iv) Suggest why tomato plants kept at a night temperature of 10°C rather than 20°C gain more mass. (4)

d) Explain why an increase in light intensity does not necessarily constitute a simple increase in the one factor that had been limiting photosynthesis. (2)

Word bank

ADP	lactic
ATP	light
aerobic	limited
anaerobic	oxygen
cellulose	phosphate
chlorophyll	photolysis
CO2	photosynthesis
energy	pyruvic
enzyme	respiration
ethanol	starch
fatigue	temperature
fixation	thirty-eight
glucose	transfer
glycolysis	two
heat	wall
hydrogen	water
kilojoules	

Table 5.5 *Word bank for Chapters 4 and 5*

What You Should Know

(Chapters 4 and 5) (See Table 5.5 for word bank)

1 Energy is measured in units called _____.

2 Glucose is the main source of _____ in a living cell. The chemical energy stored in glucose is released by a series of reactions controlled by enzymes. This biochemical process is called _____ and some of the energy is released as _____.

3 ATP is a high-energy compound able to release and _____ energy when it is required for cellular processes.

4 ATP is regenerated from _____ and inorganic phosphate using energy released from the breakdown of _____ during respiration.

5 _____ is a respiratory pathway common to aerobic and anaerobic respiration. It involves the breakdown of glucose to _____ acid.

6 In the presence of oxygen, _____ respiration occurs. In plant and animal cells, water and CO_2 are formed and _____ molecules of ATP are produced per molecule of glucose.

7 In the absence of oxygen, _____ respiration occurs. In animal cells, pyruvic acid is reversibly converted to _____ acid which causes muscle _____. In plant cells, pyruvic acid is irreversibly converted to _____ and CO_2. In each case only _____ molecules of ATP are produced per molecule of glucose.

8 Green plants use sunlight as their source of energy to make food. This process is called _____. It is the means by which the raw materials _____ and CO_2 are converted to the end products glucose and _____ by a series of _____-controlled reactions.

9 During the first stage of photosynthesis called _____, light energy is captured by green _____ in chloroplasts and used to split molecules of water into oxygen and _____. Some of the energy is used to regenerate _____ from ADP and inorganic _____.

10 During the second stage of photosynthesis, called carbon _____, hydrogen from photolysis is combined with _____ to form glucose. This process requires energy supplied by ATP.

11 Some glucose molecules formed during photosynthesis are converted to _____ and stored; others are converted to _____ and become part of the cell_____.

12 Photosynthesis is affected by environmental factors such as _____ intensity, CO_2 concentration and _____. Its rate is therefore _____ by whichever of these factors is in short supply.

Unit 2

Environmental Biology and Genetics

Factors affecting biodiversity of species and inherited variation amongst the members of species are of considerable social and economic importance to the future wellbeing of humankind

6 Energy flow

Components of an ecosystem

An **ecosystem** is made up of a **community** of living things and their **habitats**. The community consists of several **populations** of plants, animals and micro-organisms, which interact with one another and their non-living environment forming a balanced biological unit. Figure 6.1 gives a summary of the components of an ecosystem.

Producers and consumers

All of the energy needed by living things in an ecosystem comes from the **sun**. Green plants (see Figure 6.2) are called **producers** because they are able to produce their own food by converting the sun's light energy into chemical energy (contained in food) by photosynthesis (see Chapter 5).

Animals (and non-green plants such as fungi) cannot produce their own food from sunlight. They are called **consumers** because they must consume plants or other animals in order to obtain the energy they need to stay alive and grow.

Among animals there are different types of consumer. A **herbivore** (e.g. sheep, Figure 6.3) eats plant material only. A **carnivore** (e.g tiger, Figure 6.4) eats animal material only. An **omnivore** (e.g. pig, Figure 6.5) eats a mixture of plant and animal material.

Predators and prey

An animal which hunts another animal for its food is called a **predator**. The hunted animal is called the **prey**.

Decomposers

Waste materials produced by an organism during its lifetime contain chemical elements. The organism's body is also composed of a variety of chemical substances. When the organism dies of old age (or part of its body is left uneaten by another organism or passes undigested through the body of a predator), the chemicals present in the dead parts are needed for use by other organisms in the ecosystem.

Decomposers are micro-organisms, such as bacteria and fungi (see Figure 6.6), which obtain their energy supply by breaking down waste materials and

component of ecosystem	description of component	appearance of component in pond ecosystem
habitat	the place where an organism lives (e.g. pondwater, garden soil, mountain top, bark of tree, small village)	 habitat 1 **pondwater** habitat 2 **mud**
population	a group of organisms of the same species (e.g. a pack of wolves, a shoal of herring, the human inhabitants of a town, the oak trees in a wood)	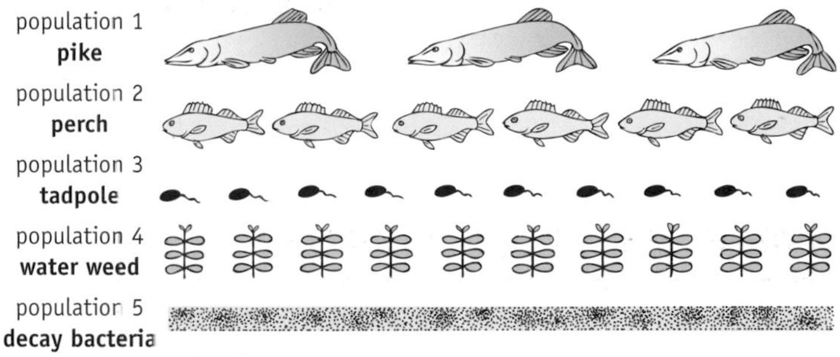 population 1 **pike** population 2 **perch** population 3 **tadpole** population 4 **water weed** population 5 **decay bacteria**
community	the sum total of all the populations of plants, animals and micro-organisms living together in an ecosystem	
ecosystem	the balanced interaction between the members of a community and their physical habitats	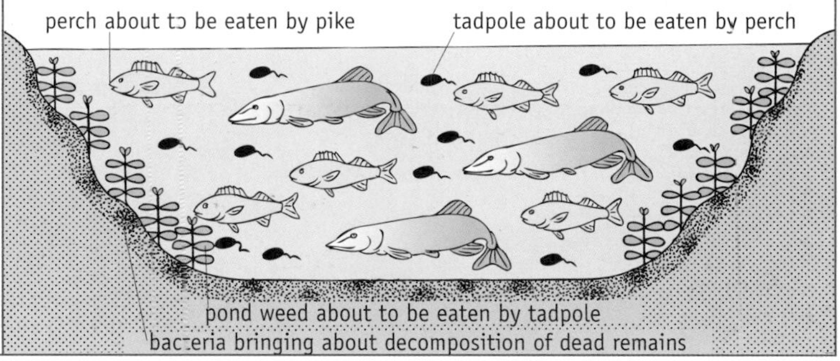 perch about to be eaten by pike tadpole about to be eaten by perch pond weed about to be eaten by tadpole bacteria bringing about decomposition of dead remains

Figure 6.1 *Components of an ecosystem*

dead bodies using enzymes. In doing so they also release chemical nutrients into the ecosystem that can be **recycled** and used by other organisms.

Figure 6.2 *Producers*

Figure 6.3 *Herbivore*

Figure 6.4 *Carnivore*

Figure 6.5 *Omnivore*

Figure 6.6 *Decomposers*

Figure 6.7 *Predator and prey*

Energy flow

When a plant is eaten by an animal, **energy** is transferred from the plant to the animal (the **primary consumer**). When the primary consumer is eaten by a second animal (the **secondary consumer**) energy is again transferred and so on through a series of organisms, as shown in Figure 6.8.

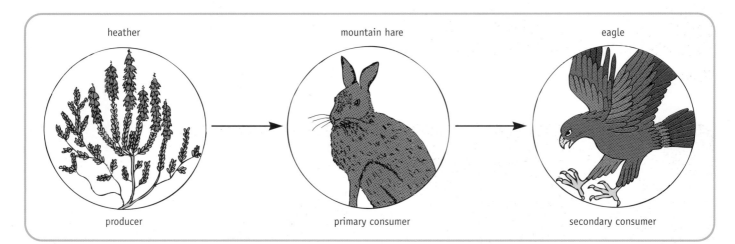

heather	mountain hare	eagle
producer	primary consumer	secondary consumer

Figure 6.8 *Moorland food chain*

Food chain

A relationship where one organism feeds on the previous one and in turn provides food for the next one in the series is called a **food chain**. A food chain always begins with a green plant. Each arrow in the food chain shown in Figure 6.8 indicates the direction of energy flow.

Food web

A food chain rarely occurs in isolation in nature because the producer is normally eaten by several animals which are in turn preyed upon by several different predators. For example, heather plants are eaten by grouse, moth larvae and many other animals. Hares are eaten by foxes, who also eat grouse. Grouse may also be eaten by eagles and falcons, and so on.

Under natural conditions an ecosystem really contains many inter-connecting food chains. This more complex relationship is called a **food web**. Two examples of food webs are shown in Figures 6.9 and 6.10, where each arrow again indicates the direction of energy flow.

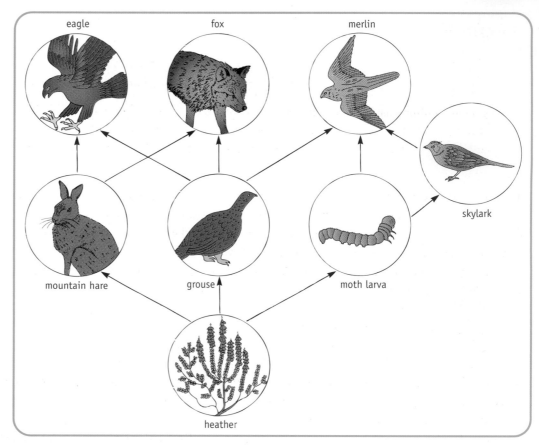

Figure 6.8 *Moorland food web (organisms not to scale)*

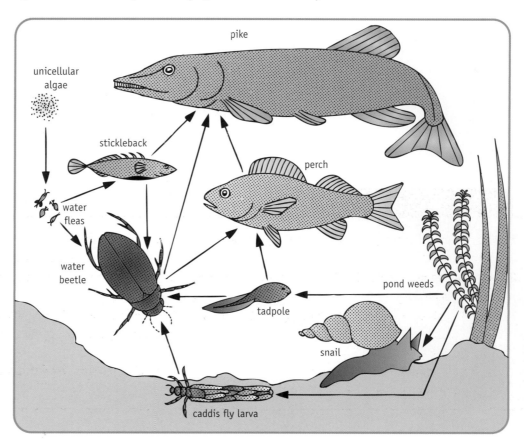

Figure 6.10 *Pond food web (organisms not to scale)*

Activity

Identifying an ecosystem's component parts and inter-relationships

INFORMATION

A rocky shore contains many pools and crevices among rocks and sandy inlets. These habitats are sometimes submerged by sea water and sometimes exposed to the air, depending on the state of the tide. An enormous variety of organisms live on rocky shores. Only a few representatives are shown in Figure 6.11 on page 110.

WHAT TO DO

1 a) Name the FOUR different populations found on the rocky shore shown in Figure 6.11.
   b) State which of these populations make up this ecosystem's community.
2 a) Identify the FOUR different habitats present in the rocky shore, shown in Figure 6.11.
   b) Name the TWO habitats occupied by brown seaweed.
   c) Suggest why brown seaweed is unlikely to be found growing in either of the other two habitats.
3 a) Give ONE way in which
      (i)   the crab depends on the limpet
      (ii) the limpet depends on the seaweed
      (iii) all of the organisms depend on the decay bacteria.
   b) Construct a food chain using THREE of the organisms shown in Figure 6.11.
4 Using tracing paper, combine TWO parts of Figure 6.11 to produce a diagram of a rock pool ecosystem.
5 Do some research (e.g. in the school library) and find out the names of at least TEN other plants and animals that could also be part of this ecosystem.
6 Construct a food web to include as many as possible of the organisms you named in question 5 above.

Testing Your Knowledge

1 What is the source of the energy entering a food chain or web? (1)
2 Describe the meaning of the terms *producer*, *consumer* and *decomposer*. (3)
3 Present each of the following groups of organisms as a food chain:
   a) lion, grass, zebra
   b) weasel, fieldmouse, owl, wheat
   c) greenfly, oak leaf, thrush, ladybird
   d) herring, animal plankton, human, plant plankton
   e) frog, hawk, grass, snake, grasshopper. (5)
4 Construct a food chain in which you are the final consumer. (2)
5 Copy and complete the oak tree food web shown in Figure 6.12 on page 111 using FIVE additional arrows. (5)
6 What does each arrow in a food chain or web mean? (1)

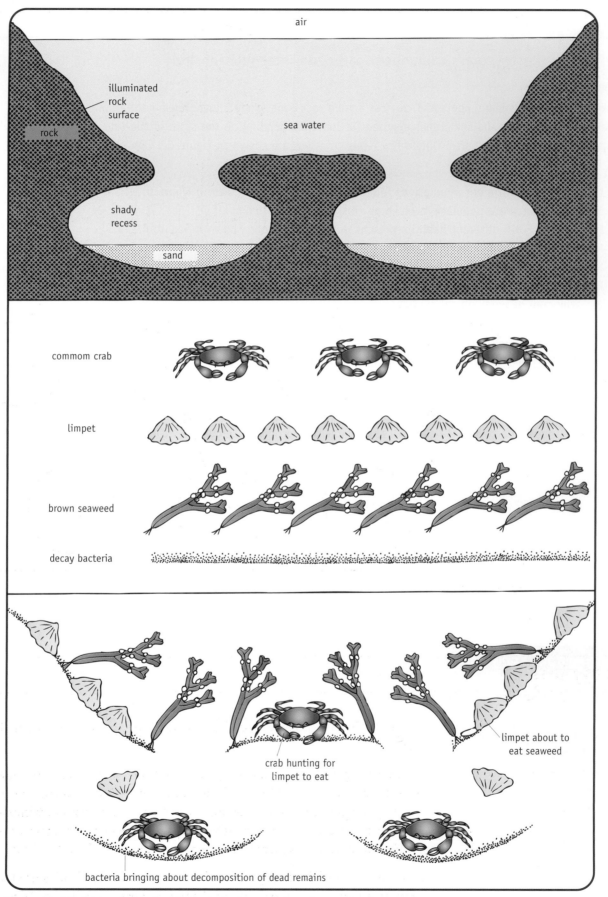

air

illuminated
rock
surface

rock

sea water

shady
recess

sand

commom crab

limpet

brown seaweed

decay bacteria

limpet about to
eat seaweed

crab hunting for
limpet to eat

bacteria bringing about decomposition of dead remains

Figure 6.11 *Parts of a rocky shore ecosystem*

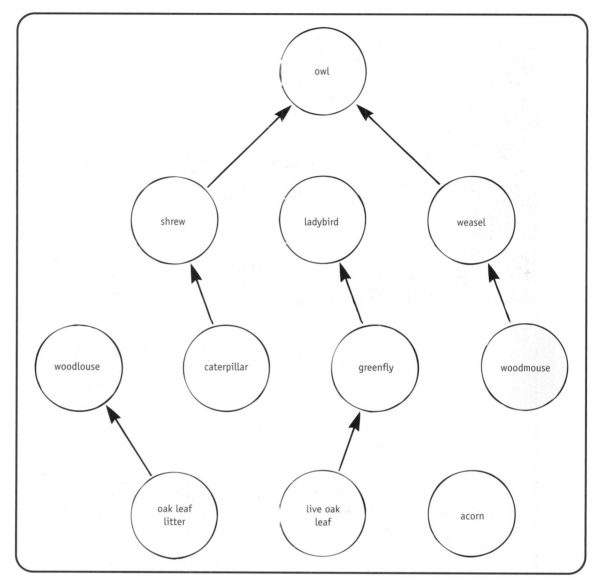

Figure 6.12

Energy loss

As energy flows through a food chain or web, a progressive loss occurs for two reasons.

Firstly, an organism uses up energy to build its body. However, this may include parts such as cellulose cell walls or bone or skin or horns which, when eaten by the next consumer, may turn out to have little or no nutritional value. These parts tend therefore to be left **uneaten** by the next consumer or to be expelled **undigested** as faeces. As a result, energy is lost from the food chain.

Secondly, most of the energy gained by a consumer in its food is used for **moving** about and, in warm-blooded animals, for keeping warm. Much energy is therefore lost as **heat** and only about 10% on average of the energy taken in by an organism is incorporated into its body tissues. Figure 6.13 shows the fate of energy as it is transferred along a marine food chain. Figure 6.14 shows an alternative way of presenting this information involving the flow and loss of energy in an ecosystem.

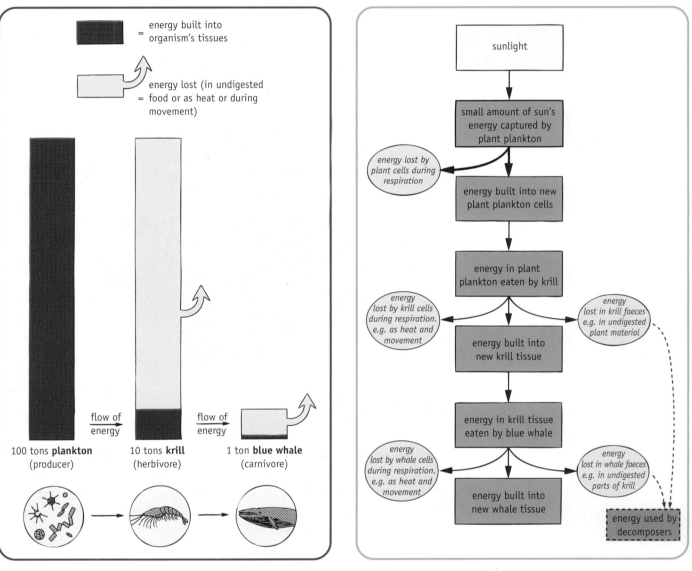

Figure 6.13 *Energy loss in a food chain*

Figure 6.14 *Energy transfer and loss*

Length of a food chain

More efficient use is made of food plants by humans consuming them directly rather than first converting them into animal products since this cuts out at least one of the energy-losing stages in the food chain.

Pyramid of numbers

Consider the following food chain:

alga ⟶ water flea ⟶ stickleback ⟶ pike

In terms of numbers, the producers (the algae) are found to be the most numerous followed by the primary consumer and so on along the chain with the final consumer being the least numerous. This numerical relationship is called a **pyramid of numbers** and is often illustrated in the form shown in Figure 6.15.

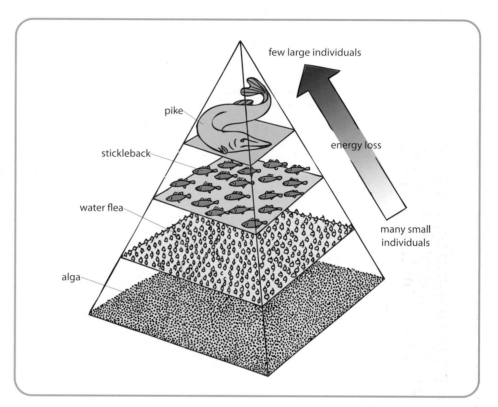

Figure 6.15 *Pyramid of numbers*

The relationship takes the form of a pyramid because (a) the energy loss at each link in the food chain limits the quantity of living matter that can be supported at the next level and (b) the final consumer tends to be larger in body size than the one below it and so on. A simpler way of representing a pyramid of numbers is shown in Figure 6.16.

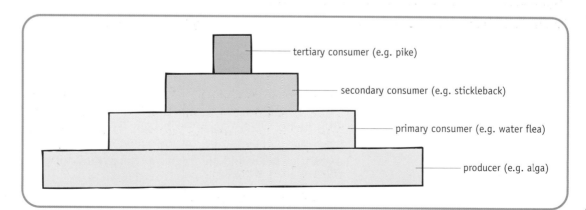

Figure 6.16 *Pyramid of numbers (alternative style)*

In some food chains the producer is a **single large plant** and the pyramid therefore takes a different form. Figure 6.17, for example, shows the 'pyramid' of numbers for the food chain:

oak tree $\longrightarrow$ caterpillar $\longrightarrow$ shrew $\longrightarrow$ owl

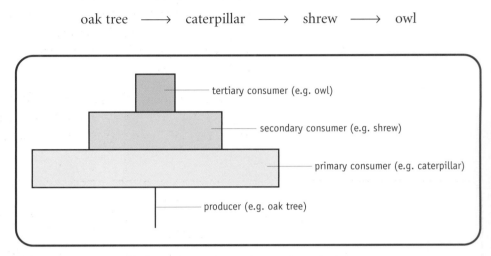

Figure 6.17 *Pyramid of numbers – exception to the rule*

Pyramid of biomass

The **biomass** of a population is its total mass of living matter. In a food chain the biomass of the producer is greater than that of the primary consumer, which in turn is greater than that of the secondary consumer and so on along the chain. Since biomass decreases at each level, it can also be represented as a pyramid.

Figure 6.18 shows a **pyramid of biomass** for a rocky shore ecosystem where the average mass of dry material per square metre (i.e. g/m^2) has been calculated.

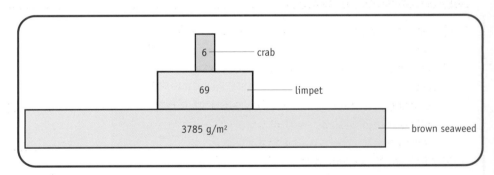

Figure 6.18 *Pyramid of biomass*

In some ecosystems where the plants have a **high turnover rate**, the 'standing crop' at any one given moment (e.g. the biomass of grass present in a grazed field) may happen to be low and under-represent the normal biomass of plant material. This would fail to give a true pyramid if represented diagrammatically.

In addition, two organisms with the same biomass may not have the same energy content since they may differ in **chemical composition**. Fat contains more than twice the energy present in protein and carbohydrate per unit mass.

Pyramid of energy

A more reliable comparison between the organisms found at the different levels in a food chain can be made based on **productivity**. This is measured as grams of dry mass per square metre per year and then converted into its energy equivalent in kilojoules per square metre per year. The results are used to construct a **pyramid of energy**, which illustrates the transfer of energy from one level to another.

A pyramid of energy always takes the form of a true pyramid since only a proportion of energy is transferred from one level to the next. Figure 6.19 shows a pyramid of energy for a river ecosystem.

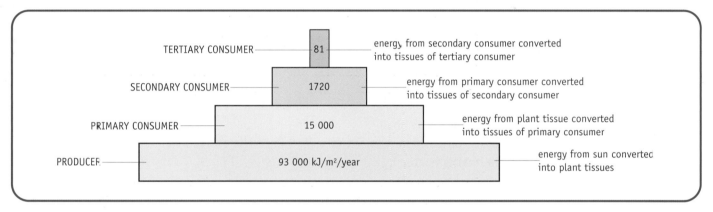

Figure 6.19 *Pyramid of energy*

Testing Your Knowledge

1 a) State two ways in which energy may be lost from a food chain or web. (2)

b) Explain how the energy may be lost in each case. (2)

2 a) Copy the pyramid of numbers shown in Figure 6.20 and complete it using the following organisms: water flea, pike, alga, stickleback. (2)

b) Which of these organisms is the secondary consumer? (1)

c) Which population of organisms in this pyramid contains most energy? (1)

d) Compared to the other organisms, what rule applies to the individual body size of the organisms occupying the top position in a food pyramid? (1)

e) Why do the numbers decrease towards the top of a food pyramid? (1)

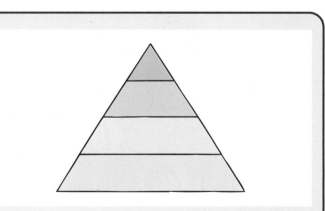

Figure 6.20

3 a) Give the meanings of the terms *pyramid of biomass* and *pyramid of energy*. (2)

b) State which of these is the more reliable method of representing the organisms in a food chain. Explain why. (2)

Applying Your Knowledge

1 Match the terms in list X with their descriptions in list Y.

List X
1) community
2) consumer
3) decomposer
4) ecosystem
5) food chain
6) food web
7) habitat
8) population
9) predator
10) prey
11) producer
12) pyramid

List Y
a) place where an organism lives
b) natural biological unit made of living and non-living parts
c) group of living organisms of one type
d) all of the populations of plants, animals and micro-organisms that live together in an ecosystem
e) micro-organism that obtains its energy by breaking down dead organic material
f) animal that hunts other animals for its food supply
g) relationship starting with a green plant followed by a series of animals, each of which feeds on the previous one
h) green plant that makes food by photosynthesis
i) general name for an organism unable to photosynthesise and dependent on a ready-made food supply
j) diagram that indicates the relative numbers, biomass or energy content of the organisms in a food chain
k) complex relationship composed of several inter-related food chains
l) animal that is hunted by other animals

2 Write a paragraph to describe the ecosystem shown in Figure 6.21. Your answer should include the correct use of the following terms: *community*, *habitat*, *population* and *ecosystem*. Give named examples where possible. (6)

3 Using the names given on the following page, match each predator shown in Figure 6.22 to its prey and present the information as a table.

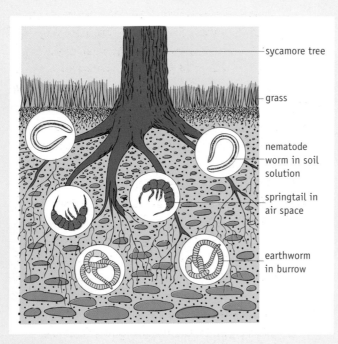

sycamore tree

grass

nematode worm in soil solution

springtail in air space

earthworm in burrow

Figure 6.21

Figure 6.22

cat, diving beetle, fox, frog tadpole, greenfly, hawk, ladybird, locust hopper, mouse, perch, pike, rabbit, scorpion, songbird, sperm-whale, squid. (5)

4 Figure 6.23 shows a simple food web.

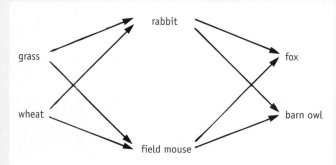

Figure 6.23

Decide whether each of the following statements, which refer to the food web, is true or false and then use T or F to indicate your choice. Where a statement is false, give the word(s) that should have been used in place of the word(s) in **bold print**.

a) Each arrow in the food web represents flow of **energy** from one organism to the next.

b) The primary consumers are **larger** in body size than the secondary consumers.

c) Foxes and barn owls are the **rarest** organisms in the ecosystem.

d) Both rabbits and **foxes** are examples of primary consumers.

e) The combined biomass of grass and wheat **exceeds** that of all the consumers in the food web.

f) The total quantity of energy that passes from plants to rabbits is **less** than the total quantity of energy that passes from rabbits to foxes.

g) The producer is always eaten by a **secondary** consumer. (7)

5 Figure 6.24 shows the fate of the energy present in the grass in a field when consumed by a cow.

a) Which lettered arrow represents the flow of energy from producer to consumer? (1)

b) Calculate the percentage of energy successfully converted from grass to the cow's body tissues. (1)

c) Name TWO ways in which energy could be lost by the cow at arrow X. (2)

d) Explain why the energy lost at arrow Y is not lost to the ecosystem as a whole. (1)

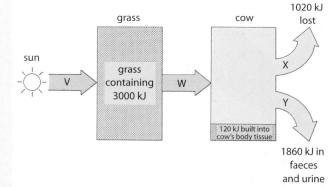

Figure 6.24

e) If the energy conversion efficiency of cow to human is 8%, how many of the cow's 120 kJ could have become built into human tissues? (1)

6 Imagine a sailor shipwrecked on a barren rocky island lacking top soil. From the ship's cargo he has managed to salvage one live hen and a bag of wheat grains.

a) To make these, his only resources, last for as long as possible, which of the following courses of acton should he take?

1) Eat the hen and then eat the wheat.

2) Feed all of the wheat to the hen, eat its eggs and then eat the hen when the wheat runs out.

3) Share the wheat with the hen and then eat the hen when the wheat is finished. (1)

b) Justify your choice of answer and explain why the other courses of action would be less successful. (2)

7 Figure 6.25 on page 118 shows the members of a food chain and, in brackets, the concentration in parts per million (ppm) of a non-biodegradable pesticide residue in their tissues. The water in their ecosystem contains 0.00005 ppm of pesticide.

a) (i) Referring to the types of living organism only by the names given in the diagram, draw a simple (two-dimensional) diagram to represent a pyramid of numbers for this food chain.

(ii) Which level in the pyramid would contain least energy?

(iii) Compared to the other organisms in the pyramid, which population would possess the greatest biomass? (4)

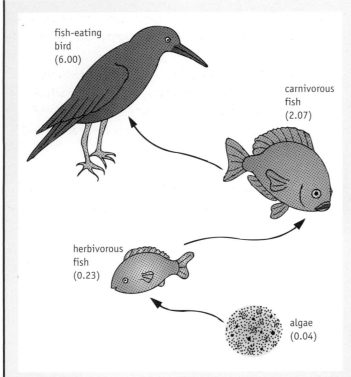

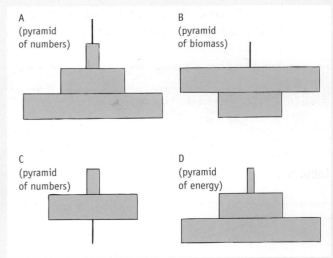

Figure 6.26

Figure 6.25

b) What is the original source of energy in this food chain? (1)

c) (i) By how many times did the concentration of pesticide increase between water and algae?

(ii) Between which two organisms did the pesticide concentration increase by nine times? (2)

8 a) Arrange each of the following groups of organisms into a food chain:

(i) owl, oak tree, woodmouse

(ii) sheep, human, grass

(iii) animal plankton, tuna, algae, anchovy

(iv) heather, eagle, mountain hare. (4)

b) Match each of these food chains with one of the following ecosystems: moorland, ocean, heavily grazed grassland, natural woodland. (2)

c) Match each food chain with one of the lettered pyramids in Figure 6.26 and justify your choice in each case. (4)

9 Read the passage and answer the questions that follow it.

Energy is lost at each link in a food chain. A certain mass of grain can support many people directly by providing them with cereal products (e.g. bread). On the other hand the same quantity of grain can support far fewer people indirectly by first feeding it to livestock and then providing humans with meat products.

The efficiency with which animals convert food into their own body tissues varies enormously. Carnivores are more efficient than herbivores since they consume an easily digested diet rich in protein. In addition, carnivores do not depend on the presence of energy-consuming micro-organisms in their gut to aid digestion of plant cell walls and their faeces contain less undigested material.

If an animal is endothermic ('warm-blooded'), it maintains its body temperature at around 37°C and therefore uses much of the energy in its food to stay warm in cold weather.

Ectothermic ('cold-blooded') animals do not maintain a body temperature above that of the environment and therefore more of the energy in their food is available for secondary productivity.

a) Allowing a 10% conversion rate each time, state the biomass of human tissue that could be produced from 1000 kg of cereal as (i) bread and (ii) pork chops from pigs fed on cereal. (2)

b) (i) Table 6.1 refers to two animals. Which is the salmon and which is the sheep? Give two reasons to support your choice.

(ii) What happens to the energy not converted into the animals' body tissues? (4)

animal	percentage energy absorbed from food	percentage energy built into tissues
X	36.6	5.2
Y	84.3	29.8

Table 6.1

c) Explain why intensive farming methods often include:
  (i) keeping animals indoors, especially during winter
  (ii) rearing animals in tightly confined spaces. (2)

d) Which of the following forms of Scottish agriculture would you expect to be more productive in terms of energy conversion into animal tissue per gram of feed?
  (i) trout fish farm
  (ii) ostrich farm.
  Explain your choice of answer. (2)

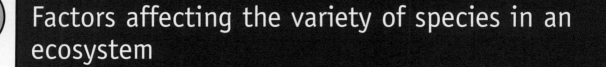

7 Factors affecting the variety of species in an ecosystem

Species

A **species** is a group of living organisms which are so similar to one another that they are able to interbreed and produce fertile offspring. This means that they in turn will be able to produce offspring on reaching sexual maturity.

Biodiversity

Biodiversity means the total variation that exists among all living things on Earth. Biodiversity includes the variation found **between** different species and the variation found **within** the same species. It has taken about 3–6 billion years for this vast assembly of varied life forms to evolve.

Number of different species

About 1.75 million different species have so far been studied by scientists but many more (estimates range from 10 to 100 million) await discovery and investigation. Scotland alone is thought to possess about 50 000 different land and freshwater species and about 39 000 species in the surrounding seas (see Figure 7.1).

Biodiversity within an ecosystem

An ecosystem is a natural biological unit made up of living and non-living parts. Within one ecosystem, biodiversity refers to the **range** of species present in that ecosystem's community.

Habitat and niche

An organism's **habitat** is the environment where it lives within an ecosystem. An organism's **ecological niche** is the role that it plays within a community. This refers to its whole way of life and includes the use that it makes of the resources in its environment. A few examples are given in Table 7.1.

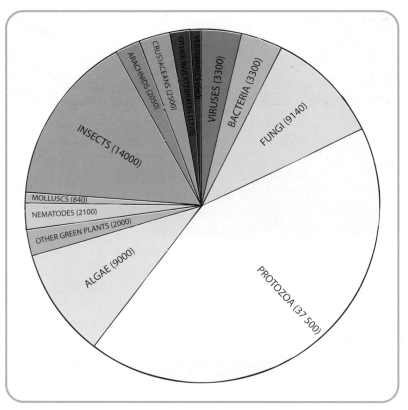

Figure 7.1 *Scottish biodiversity*

Adaptation

To succeed in a particular ecological niche, the members of a species must possess appropriate adaptations. An **adaptation** is an inherited characteristic that makes an organism well suited to survival in its environment.

organism	habitat	niche
bracken (see Figure 7.2)	open countryside	producer that rapidly spreads by growth of underground stems; dominates environment by choking out rivals; produces poison that deters grazing animals
alder plant (see Figure 7.3)	exposed river banks	producer whose roots resist underwater rot and survive in water-logged soil; able to exploit environment unavailable to other trees
red deer (see Figure 7.4)	woodland and moorland	herbivorous consumer showing population increase in absence of its extinct natural predator (wolf)
common seal (see Figure 7.5)	salt water and sand banks	fish-eating consumer with few serious rivals; temporary drop in numbers in recent years caused by viral disease
brown trout (see Figure 7.6)	fresh water	insect-eating consumer suffering intense competition from introduced rainbow trout
mole (see Figure 7.7)	underground burrow	nocturnal worm-eating consumer preyed on by owls

Table 7.1 *Niches of Scottish wildlife*

Figure 7.2 *Bracken*

Figure 7.3 *Alder plant*

Figure 7.4 *Red deer*

Figure 7.5 *Common seal*

Figure 7.6 *Brown trout*

Figure 7.7 *Mole*

Some adaptations involve the **structure** of the organism's body; others depend on the way in which the organism **behaves** in response to environmental stimuli.

Distribution

A species is only able to settle into an ecological niche and prosper if it is well adapted to that way of life. Often the very adaptations that suit it to a particular habitat or niche leave it ill-equipped to deal with other habitats or niches. It is for this reason that a species' adaptations play a major part in influencing its **distribution** in the ecosystem.

Desert plants

A **desert** is a habitat where the conditions are extremely dry and the soil lacks water. A normal land plant (e.g. an oak tree) would be unable to survive in the desert. Its leaves would present an enormous surface area of thin tissue from which water would be rapidly lost as water vapour by transpiration. At the same time its roots would be unable to find water in the ground to replace this loss.

A desert plant (e.g. cactus) is able to survive in such an extreme habitat because it possesses certain adaptations. Figure 7.8 shows the prickly pear plant. Its **leaf surface area is greatly decreased** by its leaves being reduced to protective spines. Its stem is divided into green lobes which carry out photosynthesis. Each of these portions of stem is fleshy and stores water in **succulent tissues**. Water loss from the stem is kept to a minimum by the **thick, waxy cuticle** which coats its outer surfaces.

Some cacti have very **long roots** for reaching supplies of subterranean water. Others possess extensive systems of **superficial roots** which grow parallel to the soil surface. These enable the plant to absorb maximum quantities of water on the rare occasions when rain does fall.

Figure 7.8 *Prickly pear plant*

Mosquitoes

Some organisms such as mosquitoes are adapted to life in warm climates and cannot survive at low temperatures. Worldwide spread of mosquitoes is prevented by their inability to tolerate environmental temperatures of below 10°C (see Figure 7.9). If global warming takes place in the future, mosquitoes may be able to extend their range into southern parts of Britain.

Darwin's finches

Charles Darwin visited the Galapagos Islands in 1835. He found them to be inhabited by many unique life forms, including 13 different species of finch.

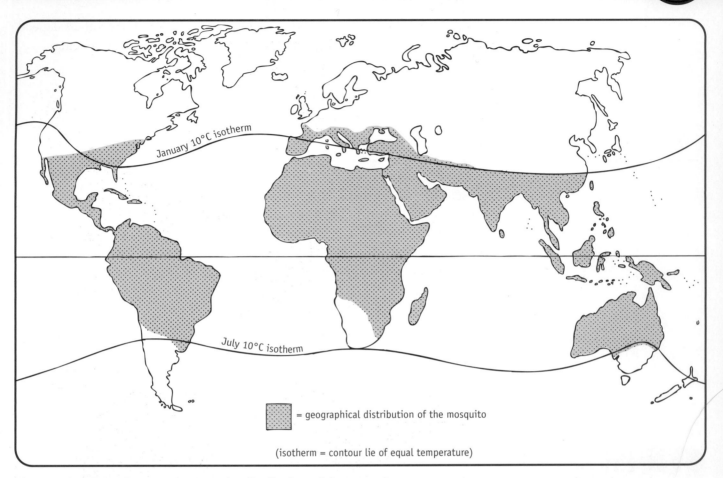

Figure 7.9 *Effect of temperature on the distribution of the mosquito*

Within the figure:

January 10°C isotherm

July 10°C isotherm

■ = geographical distribution of the mosquito

(isotherm = contour lie of equal temperature)

Figure 7.10 shows a few of these finches. They are found to vary greatly in **beak size** and **shape**. These differences make each species well adapted to its environment by enabling it to exploit a particular ecological niche. For example the insect-eating 'warbler' finch's small, sharp, pointed beak is ideally suited to picking insects out of narrow crevices in the bark of a tree. The seed-eating finch's large, blunt beak, on the other hand, is well suited to breaking open tough coats surrounding seeds and nuts.

A **stable ecosystem** (see Figure 7.11) contains a wide range of species of plants (**producers**), animals (**consumers**) and micro-organisms (**decomposers**) living together in one or more habitats. The members of the community are interdependent. The animals and micro-organisms depend on the plants directly or indirectly for food and oxygen. The plants depend on the micro-organisms to decompose wastes and release chemical nutrients. The animals depend on the plants for shelter and camouflage. The plants may need animals to bring about pollination and/or seed dispersal.

Figure 7.10 *Darwin's finches*

Testing Your Knowledge

1 a) Define the term *species*. (2)

 b) Explain what is meant by the expression *fertile offspring*. (1)

 c) When a lion is crossed with a tiger, the result is a sterile animal called a tiglon. Do a tiglon's parents belong to the same species? Explain your answer. (2)

2 Give the meanings of the terms *biodiversity*, *habitat*, *ecological niche* and *adaptation*. (4)

3 Copy and complete Table 7.2. (4)

organism	habitat	example of an adaptation that suits the organism to its habitat	reason why this adaptation is of survival value
prickly pear plant			
insect-eating 'warbler' finch			

Table 7.2

Biodiversity within a stable ecosystem

Thus a **stable ecosystem** is a delicately balanced biological unit which normally contains a **wide variety** of species. An example is shown in Figure 7.12 of a woodland ecosystem (where several seasons of the year are represented simultaneously).

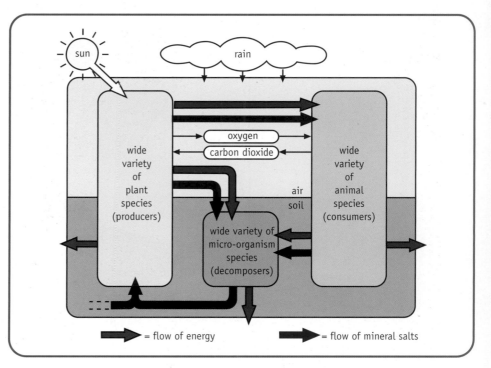

Figure 7.11 *Inter-relationships in a stable ecosystem*

Effect of grazing on the variety of species in an ecosystem

Rabbits

Natural grassland normally contains a rich variety of plant species. Some are especially sturdy and show vigorous growth; others are more delicate. Rabbits are relatively **unselective** grazers. By eating almost all types of plant growing on grassland, they maintain a high diversity of plant species since the vigorous grasses are kept in check.

If the disease myxomatosis or human activities remove rabbits from a piece of grassland, then the aggressive dominant grasses are no longer kept in check.

Figure 7.12 *Biodiversity in a woodland ecosystem*

They drive the other less vigorous species out, thereby reducing the variety present in that ecosystem.

When the rabbits return, they maintain a closely cropped area of grassland which promotes the re-emergence of **rich biodiversity** once more. This relationship is summarised in Figure 7.13. At very high intensities of grazing, the species diversity may be reduced slightly if the less vigorous species become overgrazed.

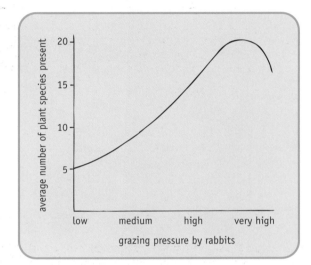

Figure 7.13 *Effect of grazing on species diversity*

Sheep

Although sheep are selective grazers, they also maintain species-rich communities since they select the competitive dominant grasses and hold them in check.

Effect of pollution on the variety of species in an ecosystem

Sulphur dioxide

When fossil fuels (such as coal and oil) are burned in power stations to generate electricity, **sulphur dioxide** (SO_2) gas is released into the air (see Figure 7.14). Car exhaust fumes also contain SO_2. Even in low concentrations this harmful gas aggravates human respiratory ailments and causes damage to many types of plant.

Lichens

A **lichen** is a simple plant composed of a fungus and an alga often found growing on the bark of trees and on rock surfaces (see Figure 7.15). Different species of lichen

Figure 7.14 *Air pollution*

vary in their sensitivity to SO_2. The variety of lichen species present in an ecosystem decreases as the concentration of SO_2 in the air increases (see Figure 7.16).

Acid rain

The combustion of coal by electricity-generating stations produces sulphur

Figure 7.15 *Lichens*

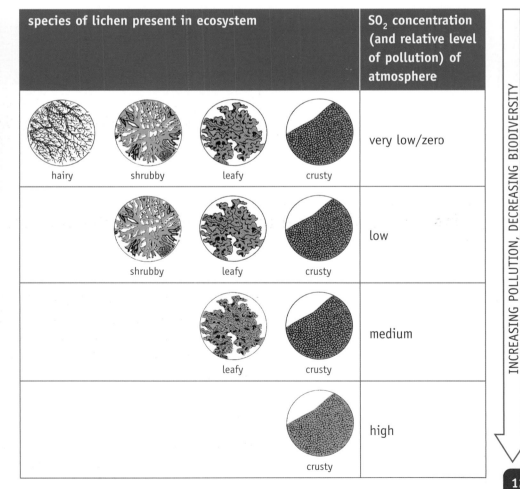

species of lichen present in ecosystem	SO_2 concentration (and relative level of pollution) of atmosphere
hairy shrubby leafy crusty	very low/zero
shrubby leafy crusty	low
leafy crusty	medium
crusty	high

INCREASING POLLUTION, DECREASING BIODIVERSITY

Figure 7.16 *Effect of air pollution on the variety of lichen species*

dioxide and nitrogen oxides which combine with water to form acids. When present in excess in the atmosphere, they lead to the formation of **acid rain** which has a devastating effect on several types of ecosystem and their communities.

Figure 7.17 shows the effect of decreasing the pH of loch water on the variety of fish species found in this aquatic ecosystem.

Acid rain also affects land environments. On Scottish moorlands, the roots of ling heather plants have been found to grow less well as acidification of the

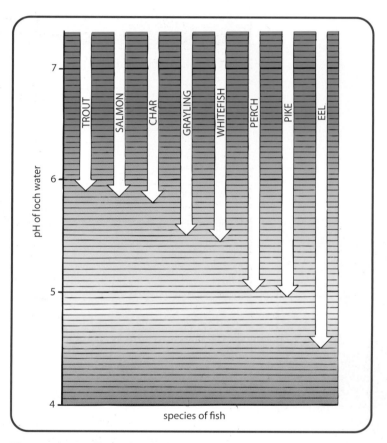

Figure 7.17 *Effect of pH on the variety of fish species*

ground increases. If this process continues, the species composition of the heather moorland (and the Scottish landscape) will be altered as ling heather declines.

Thermal pollution

Some types of electricity-generating power stations use local river water as a coolant. When the water is returned to the river it is considerably warmer and causes **thermal pollution**. The increase in temperature of river water is accompanied by a decrease in its oxygen content and a decrease in the variety of fish species in the river (Figure 7.18). Whereas all of the species shown in the diagram could be present in water containing 4 mg/l of oxygen, only tench would be found at 1 mg/l of oxygen.

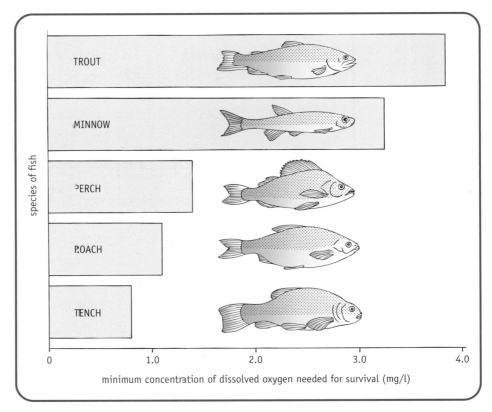

Figure 7.18 *Effect of oxygen concentration on the variety of fish species*

Sewage

In some densely populated areas, sewage works may become overloaded. As a result the effluent being discharged into the local river is rich in **untreated sewage**. This provides food for bacteria which rapidly multiply and use up the water's dissolved oxygen supply (Figure 7.19).

Effect of oxygen shortage on numbers of species

A clean water ecosystem normally contains a large number of different species of animals and plants. Although the clean water species, such as mayfly and stonefly nymphs, are the most numerous within this community, a few representatives of those species that can tolerate (or even thrive well in) polluted water are also present, but only in very small numbers owing to the intense competition for food and other resources.

This natural balance between the living organisms in the ecosystem is upset by a very high level of pollution since it completely eliminates all of the clean water species. This leads to a **decrease in species diversity** since the variety of species within the community is now reduced and only a few pollution

131

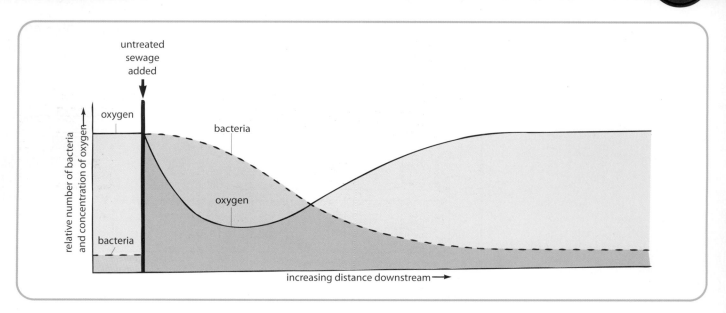

Figure 7.19 *Effect of sewage on oxygen concentration*

tolerant species (e.g. sludgeworms) survive. These animals increase dramatically in number owing to the lack of competition.

Further downstream where the water is less polluted, species such as waterlouse which can tolerate fairly high levels of pollution, are found to rise in number and become dominant in the ecosystem. This trend continues (see Figure 7.20) until eventually the river recovers its original state and clean water animals return and resume their dominant niche in the ecosystem. Table 7.3 summarises the changes in the numbers of different species (see also question 5 at the end of this chapter).

type of animal species	clean		state of water		
			very badly polluted	partly polluted	clean
clean water species	many large populations	organic effluent added	none	few small populations	many large populations
semi-tolerant species	few small populations		none	many large populations	few small populations
dirty water species	few small populations		many large populations	few small populations	few small populations
overall variety within ecosystem	large number of different species		very small number of different species	small number of different species	large number of different species

Table 7.3 *Effect of pollution on number and variety of species*

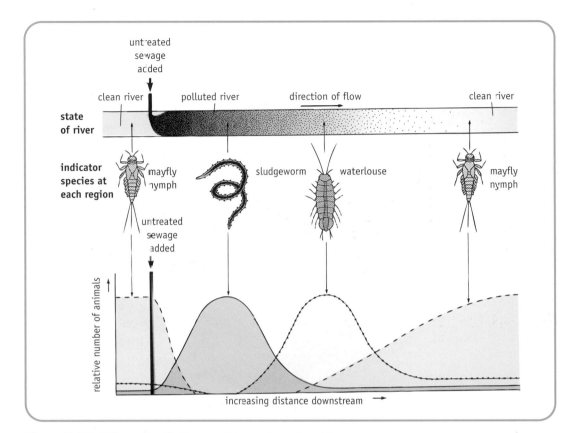

Figure 7.20 *Effect of pollution on the number of species*

Effects of human activity on the environment

Until about 10 000 years ago, human beings obtained food by hunting animals and gathering edible plants. The total human population was relatively small and made up a natural part of a stable ecosystem containing a wide variety of interdependent species. In these ancient times the environment remained healthy and unaffected by human activities.

Habitat destruction

Today there are so many human beings in the world that we no longer form a small natural part of a balanced ecosystem. As we strive to support the ever-increasing world population and its energy demands, we dominate the ecosystem. In doing so we pursue activities that cause **pollution** (see pages 128–132), **deforestation** and **desertification**, and eventually destroy natural habitats.

Deforestation

This term is used to refer to the complete clearing away of vast tracts of natural forest and the failure to plant new forest in their place.

Many disastrous consequences can result from intensive **deforestation**. Table 7.4 and Figure 7.21 give some examples of the possible far-reaching

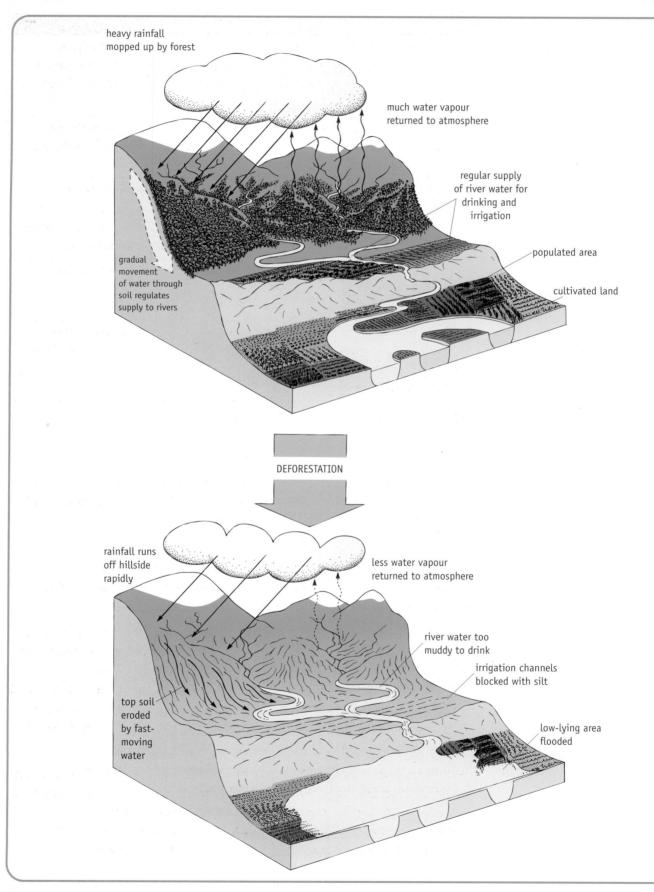

heavy rainfall
mopped up by forest

much water vapour
returned to atmosphere

regular supply
of river water for
drinking and
irrigation

populated area

cultivated land

gradual
movement
of water through
soil regulates
supply to rivers

DEFORESTATION

rainfall runs
off hillside
rapidly

less water vapour
returned to atmosphere

river water too
muddy to drink

irrigation channels
blocked with silt

top soil
eroded
by fast-
moving
water

low-lying area
flooded

Figure 7.21 *Deforestation*

effect of deforestation	consequence
loss of forest's retentive 'sponge' effect (especially during rainy season); flow of water to rivers no longer regulated	rivers fail to provide the regular supplies of water needed for human consumption and irrigation of crops
rapid rather than gradual run-off of rain water from hillsides	flooding of low-lying downstream areas which are often cultivated and inhabited
erosion of fertile top soil from hillsides by fast-moving water	soil fertility of hillsides reduced; rivers, lakes, irrigation channels and dams become blocked with silt; water becomes muddy and undrinkable
less water vapour returned to the atmosphere by evaporation and transpiration	reduction in rainfall, making local climate become drier
more CO_2 produced during burning of forest; less CO_2 absorbed from atmosphere in the absence of photosynthesising forest	extra CO_2 in the atmosphere may contribute to the 'greenhouse effect'

Table 7.4 Possible consequences of deforestation

consequences of this form of **environmental mismanagement**. Many of these have already occurred in various regions of the world and a large number of forest-dwelling species have already become extinct.

Desertification

In some dry parts of the world, the area of land covered by desert is increasing. When this process is largely the result of human activities it is called **desertification**.

The top part of Figure 7.22 shows a typical scene at the edge of a desert as it was in the 1960s. The rainfall is erratic but adequate to sustain the plants and maintain the water table. Marginal land (i.e. territory on the edge of a cultivated zone) and forests hold soil and form a windbreak. Some land is used for grazing. Traditional agricultural practices are employed on cropland. For example, crops are rotated allowing each piece of land in turn to lie fallow for a spell and recover its fertility.

The bottom part of Figure 7.22 shows the situation in the 1990s. The land has been lost to the desert and there is little chance of such land being returned to cultivation. The people can no longer feed themselves and their supply of drinking water has dried up. Out of desperation they may try to cultivate other areas of marginal land which in turn accelerates the process of desertification.

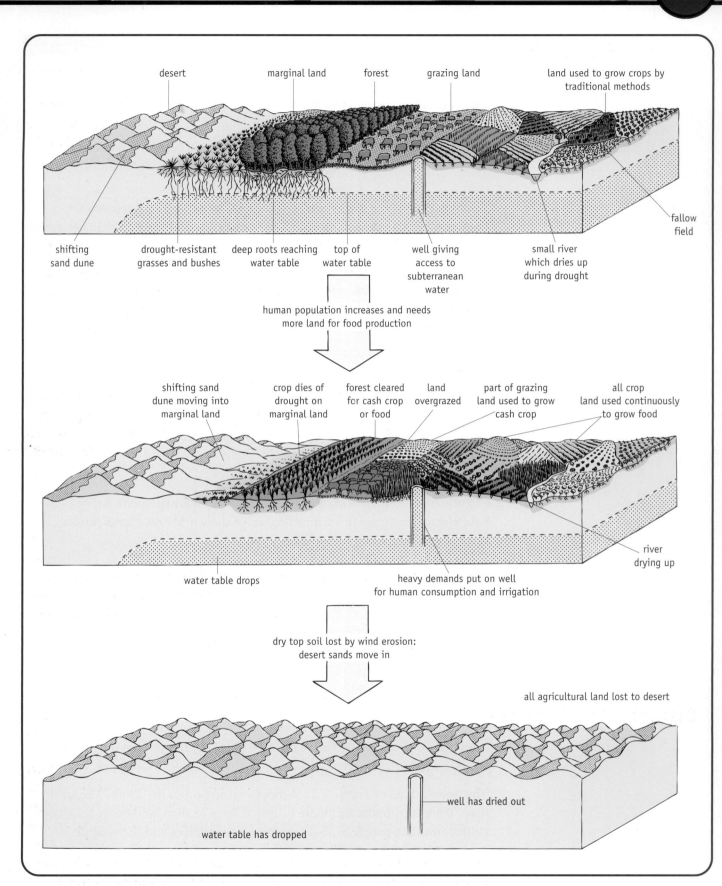

Figure 7.22 *Desertification*

Biodiversity in crisis

Ever since life evolved on Earth, new species better suited to the environment have appeared. Older, less successful, forms have died out. Fossil evidence shows that life on Earth has been punctuated by several waves of **mass extinction**. These are closely related to changes in global climate that have occurred.

When the environment becomes disrupted, some species which had been enjoying minimum success may find themselves well adapted to the new environment and evolve rapidly. Other species that were previously successful may suddenly find themselves at a major disadvantage and become extinct.

Human activities are causing the current wave of extinction to run at about **four hundred times** the natural rate. Since the year 1600, hundreds of birds and mammals have been wiped out by **over-hunting** and by **habitat destruction**. Figure 7.23 shows a few examples of endangered species.

Tiger

Figure 7.24 anticipates the effect of human activities on the distribution of the tiger. In addition to these 'high profile' animals, a much larger number of plants and invertebrate species (e.g. insects) are in critical danger as humans continue to chop down, plough up, pave over, drain, dam and pollute natural environments.

Medicinal leech

The loss of farm ponds which were used by farm animals, has led to the decline in recent years of the wild medicinal leech (see Figure 7.25). It is now confined to about 20 sites in Britain. However, major steps are being taken to prevent it from becoming extinct because it has been found to be useful as a tool in plastic surgery and as a source of anti-clotting agents. But what about the other 'low profile' species that have not yet proved useful to humankind? Are they to be allowed to vanish for ever?

Disruption of food webs

Under natural conditions, an ecosystem contains several interconnecting food chains that make up a **food web** (see Chapter 6). This complex relationship is in a finely balanced state in equilibrium with the environment. A certain mass of green plant material grows continuously and supports a fairly constant and large number of primary consumers which are consumed by a fairly constant but smaller number of secondary consumers, and so on. If a factor affects one type of organism in the food web, this is likely to have a **knock-on effect**, which disrupts the lives of several other species in the food web and brings about a change in population structure.

blue whale

tiger

Californian condor

brown bear

black rhinoceros

snow leopard

mountain gorilla

panda

Figure 7.23 *Endangered species*

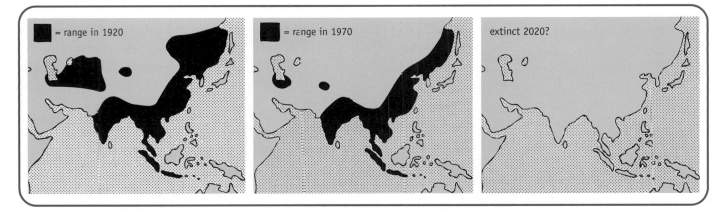

Figure 7.24 *Effect of human activities on the distribution of tiger (Panthera tigris)*

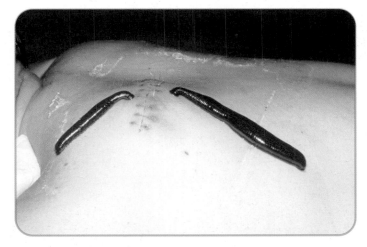

Figure 7.25 *Medicinal leeches*

Number of links

If a food web has only a few links then the effect on the remaining organisms of removing one species can be severe. Figure 7.26 shows a food web where rabbits are the main source of food for foxes and birds of prey. However, in 1954/5 a disease called myxomatosis wiped out almost the entire population of rabbits. As a result, more tree seedlings and grass grew but many more lambs than normal were attacked by foxes and birds of prey.

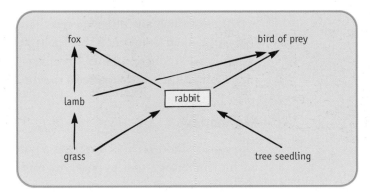

Figure 7.26 *Disturbing a food web with few links*

If a food web has many links, then removal of one species may not have such a drastic effect. For example, removal of cockles from the food web shown in Figure 7.27 would leave more plant plankton for other consumers but would probably not seriously alter the ecosystem. However, even this food web would be unable to resist disruption if it were subjected to continuous over-fishing by humans.

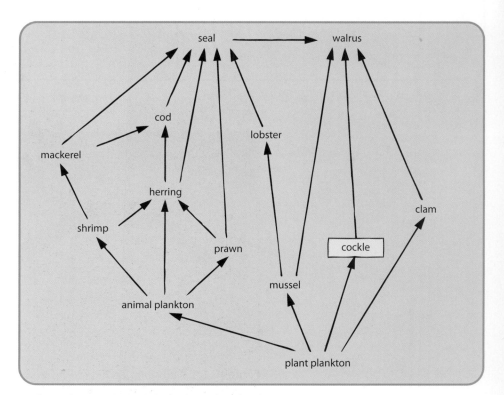

Figure 7.27 *Disturbing a food web with many links*

Importance of biodiversity

Aesthetics

The vast majority of people agree that our lives are enriched by observing the activities of the many other species that share our planet and by appreciating their individual beauty and their collective contribution to the landscape around us. A reduction in this biodiversity of plants and animals would inevitably result in a **decrease in the quality** of human life.

Economics

The world's biodiversity comprises a **natural storehouse** of **genetic variation**. Each member of a species possesses hundreds or even thousands of genes. Since two or more alleles (see Chapter 9) exist for most genes, the number of genetic permutations possible among the members of a species is enormous.

So far, humans have only scratched the surface of this vast resource. A few plant species have been developed on a large scale to feed the world. Several others have been discovered that produce medicines, drugs and useful chemicals. However, the vast majority of living species have not been fully tested for their potential uses, yet many of these are in danger of extinction. We run the risk of losing for ever the **potential riches** stored in their genes.

There is no way of knowing in advance which species are most likely to be of value to us. Unexpected ones can turn out to be of great use. Scientists have recently discovered, for example, that a fluorescent chemical made by a species of jellyfish (see Figure 7.28) can act as a marker for cancer cells.

Figure 7.28 *Fluorescent jellyfish*

Effect of selective breeding

Biodiversity within a plant species is reduced by **selectively breeding** those individuals with the highest yields (see Chapter 10). As the crop becomes more **uniform** genetically, it also becomes more **susceptible** to environmental change such as disease or global warming.

If many **wild varieties** of the species exist then there is a good chance that at least some of these will be **resistant** to the disease or be able to tolerate higher environmental temperatures. These wild varieties can be used to supply alternative alleles for breeding into existing domesticated varieties. The same principle applies to selective breeding of domesticated animals.

Biodiversity is essential to allow species to adapt to environmental change and to enable plant and animal breeders to develop improved strains of economically valuable species in the future.

Opportunists

If the process of habitat destruction continues, the future will be bleak. Species necessary for the survival of humankind will be lost and many ecological niches will become over-run by **opportunist** species such as cockroaches, rats and weed plants.

International concern

In recent years representatives of many of the world's governments have attended United Nations Conferences of Environment and Development. They have pledged that they will:

- conserve biodiversity
- use the components of biodiversity in a way that ensures that they will be available in the future
- share out the benefits of biodiversity between all nations.

However, nations have been unable to reach a binding global agreement on a policy to save the tropical rain forest and the future of biodiversity continues to look bleak.

Testing Your Knowledge

1 Does an unselective grazer tend to maintain or reduce the diversity of plant species on a piece of grassland? Explain your answer. (2)

2 Construct a table to show:
  (i) THREE types of pollution that reduce the variety of species in an ecosystem
  (ii) the ecosystems involved
  (iii) the species affected. (3)

3 a) Explain what is meant by the terms *deforestation* and *desertification*. (2)
  b) In general what effect does habitat destruction have on biodiversity? (1)

4 Predict the effect on the plants of removing the thrushes from the food web in Figure 7.29. Explain your answer. (2)

5 Figure 7.30 shows a food web from an ecosystem in Asia. The tigers were completely exterminated from an area inhabited by humans. Explain why this action turned out to be both advantageous and disadvantageous to the local people. (2)

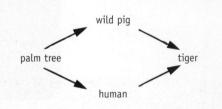

Figure 7.30

6 a) Give ONE aesthetic reason for conserving biodiversity. (1)
  b) Give TWO economic reasons for conserving biodiversity. (2)

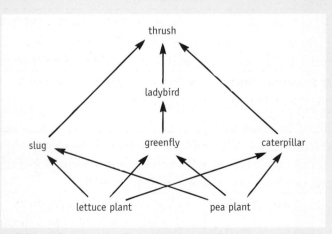

Figure 7.29

Behavioural adaptations in animals

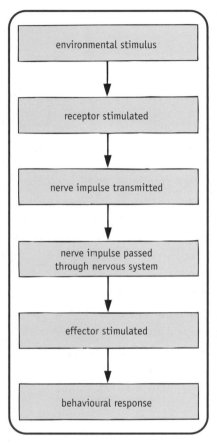

Figure 7.31 *Events leading to behavioural response*

In order to be able to behave in a certain way in response to an environmental **stimulus,** an animal must possess **receptors** (e.g. sense organs), a system of **internal communication** (e.g. the nervous system) and **effectors** (e.g. muscles). The flow chart in Figure 7.31 represents the sequence of events leading to a **behavioural response.**

Different species of animals respond to environmental stimuli in different ways. Given a range of light intensity in their environment, for example, some species congregate in the dimmest region while others move to the brightly lit area. Given a range of temperature, some species favour a cool environment, others a warm one. Given a range of air humidity, some gather in damp places, others in dry areas. Given a range of pH conditions, some move to acidic regions, some to neutral and others to alkaline conditions, and so on. But why?

Close examination of each species' behaviour shows that the behaviour is normally of **adaptive significance** (see Table 7.5). This means that the members of the species increase their chance of survival by behaving in a particular way. Their inherited behavioural adaptations make them, for example, speed up and move away from harmful environmental stimuli and slow down and congregate in favourable conditions.

animal	environmental stimulus	behavioural response	adaptive significance of behaviour
blowfly larva	range of light intensity	animals congregate in the most dimly lit region	animals obtain food, moisture and a hiding place in a normally dark environment
springtail	range of humidity	animals congregate in a damp region	animals' permeable skins prevented from losing water and causing animals to die of desiccation
flatworm	chemical molecules diffusing from food (e.g. liver)	animals move up a concentration gradient of chemical from low to high using chemoreceptors	animals increase their chance of locating food
Paramecium	range of pH	animals move to region of low pH	animals increase their chance of obtaining a food supply of decaying matter and bacteria
swallow	decreasing day lengths	animals migrate to warmer climate	animals avoid cold temperatures and shortage of food during winter

Table 7.5 *Adaptive significance of behaviour*

143

Practical Activity and Report

The response of woodlice to light and dark conditions

INFORMATION

- **A choice chamber** is a piece of apparatus that allows a scientist to study the effect of differences in an environmental factor on the behaviour of an animal.
- Woodlice are invertebrate animals that are preyed on by birds and other animals.

YOU NEED

1 choice chamber
1 sheet of nylon gauze
20 woodlice in a small lipped beaker
1 stopclock
1 sheet of light-proof material (e.g. black polythene)
1 metal spatula
1 lamp

WHAT TO DO

1 Read all the instructions in this section and prepare your results table before carrying out the experiment.
2 Set up the choice chamber as shown in Figure 7.32.
3 Quickly 'pour' the woodlice into the choice chamber through the central hole (using the spatula to direct any potential escapees into the hole).
4 Check that the light-proof cover is in place and covering exactly half of the choice chamber.
5 Position the lamp above the choice chamber and switch it on.
6 Start the clock and record the number of woodlice found in each side of the choice chamber at 30-second intervals for 5 minutes.
7 Keep a note of your results in your table.
8 If other students have carried out the same experiment, pool the results.

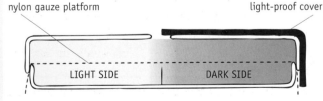
nylon gauze platform   light-proof cover
LIGHT SIDE   DARK SIDE

Figure 7.32 *Choice chamber for investigating the effect of light and dark*

REPORTING

Write up your report by doing the following:
1 Copy the title given at the start of this activity.
2 Put the subheading '**Aim**' and state the aim of your experiment.
3 a) Put the subheading '**Method**'.
   b) Draw a labelled diagram of the choice chamber set up and ready to receive the woodlice.
   c) Using the impersonal passive voice, briefly describe the experimental procedure that you followed and state how you obtained your results.
4 Put the subheading '**Results**' and draw a final version of your table of results.
5 Put a subheading '**Presentation of Results**' and present your results as a line graph.
6 Put a subheading '**Conclusions**' and write a short paragraph to state what you have found out from a study of your results. This should include answers to the following:
   a) What overall pattern or trend is shown by your results?
   b) In nature, woodlice are normally found under stones and rotting wood. Account for their presence in such habitats with reference to the results of your experiment and suggest why this behavioural adaptation is of survival value to woodlice.
7 Put a final subheading '**Evaluation of Experimental Procedure**' and then answer the following:
   a) Why were as many as 20 woodlice used per choice chamber?
   b) Why should the woodlice all belong to the same species?
   c) Identify a possible source of error in the experiment and explain how it could be put right.
   d) State how the reliability of the results could be checked.
   e) Feel free to comment on either of the following if you have an additional point that you wish to make:
      (i) further improvements that you would include in a repeat of the experiment
      (ii) limitations of the equipment.

Practical Activity and Report

Response of woodlice to relative humidity

INFORMATION

- A **choice chamber** is a piece of apparatus that allows a scientist to study the effect of differences in an environmental factor on the behaviour of an animal.
- Relative humidity refers to the quantity of water vapour present in air. Air in contact with water has a high relative humidity; air in contact with a drying agent has a lower relative humidity.
- A woodlouse's skin is permeable to water. When placed in an environment with a low relative humidity, the animal loses water as water vapour to the surrounding air.

YOU NEED

1 choice chamber
1 sheet of nylon gauze
20 woodlice in a small lipped beaker
1 stopclock
1 metal spatula
5 rubber stoppers (to fit the lid of the choice chamber)
drying agent (e.g. calcium chloride powder)
water

WHAT TO DO

1 Read all the instructions in this section and prepare your results table before carrying out the experiment.
2 Set up the choice chamber as shown in Figure 7.33 and leave it for 10 minutes to equilibrate (i.e. to allow a humidity gradient to develop between the two sides).
3 Remove the rubber stopper from the central hole and quickly 'pour' the woodlice into the choice chamber (using the spatula to direct any potential escapees into the hole).
4 Replace the rubber stopper in the central hole.
5 Start the clock and record the number of woodlice found in each side of the choice chamber at 30–second intervals for 5 minutes.
6 Keep a note of your results in your table.
7 If other students have carried out the same experiment, pool the results.

REPORTING

Write up your report by doing the following:
1 Copy the title given at the start of this activity.

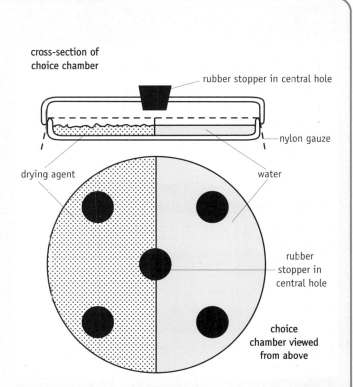

cross-section of choice chamber

rubber stopper in central hole

nylon gauze

drying agent

water

rubber stopper in central hole

choice chamber viewed from above

Figure 7.33 *Choice chamber for investigating the effect of humidity*

2 Put a subheading '**Aim**' and state the aim of your experiment.
3 a) Put the subheading '**Method**'.
   b) Draw a labelled diagram of the choice chamber set up and ready to receive the woodlice.
   c) Using the impersonal passive voice, briefly describe the experimental procedure that you followed and state how you obtained your results.
4 Put the subheading '**Results**' and draw a final version of your table of results.
5 Put a subheading '**Presentation of Results**' and present your results as a line graph.
6 Put a subheading '**Conclusions**' and write a short paragraph to state what you have found out from a study of your results. This should include answers to the following:
   a) What overall pattern or trend is shown by your results?
   b) In nature, woodlice are normally found under stones and rotting wood. Account for their presence in such habitats with reference to the results of your experiment and suggest why this behavioural adaptation is of survival value to woodlice.

7 Put a final subheading '**Evaluation of Experimental Procedure**' and then answer the following:
a) Why were as many as 20 woodlice used per choice chamber?
b) Why should the woodlice all belong to the same species?
c) Identify a possible source of error in the experiment and explain how it could be put right.
d) State how the reliability of the results could be checked.
e) Feel free to comment on either of the following if you have an additional point that you wish to make:
(i) further improvements that you would include in a repeat of the experiment
(ii) limitations of the equipment.

Testing Your Knowledge

1 What is meant by the term *behavioural adaptation*? (1)
2 a) Name THREE environmental stimuli to which animals respond. (3)
b) For each of these stimuli describe how a named animal responds and why this behavioural adaptation is of survival value. (6)
3 a) What name is given to the type of apparatus shown in Figure 7.34? (1)
b) Which environmental factor's effect on animal behaviour could be investigated using this apparatus? (1)

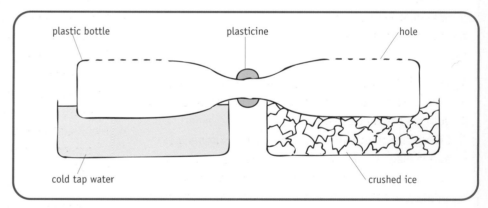

plastic bottle   plasticine   hole

cold tap water   crushed ice

Figure 7.34

Competition between plants

Plants growing in the same habitat compete with one another for factors such as *light*, *water* and *soil nutrients* if any one of these is in short supply.

Competition between members of the same species

Plants of the same species have exactly the same growth requirements. When grown together, they will be in direct competition with one another if any resource is limiting and this competition will be **intense**.

Commercial applications

Farmers use a drilling machine to space out crop seeds during planting. Horticulturalists mix tiny seeds with sand to sow them as thinly as possible and then thin out newly germinated seedlings to reduce competition.

Effect of competition on the growth of a population of cress seedlings

INFORMATION

100 cress seeds will germinate successfully on 2 g cotton wool soaked with 20 ml cold tap water in a yoghurt carton. There is little or no competition between the seedlings that develop. If they are left in a propagator at 24°C for 5 days, they grow tall and develop healthy green leaves.

YOU NEED

cotton wool
access to an electronic balance
5 150 ml yoghurt cartons
1 20 ml syringe
cress seeds
5 labels or 1 marker pen
access to a propagator at 24°C

WHAT TO DO

1 Read all the instructions in this section and prepare your table of results before carrying out the experiment.
2 Design your experiment by:
   a) establishing clearly the aim of your experiment
   b) deciding which variable factor you are going to alter
   c) deciding how many conditions of the variable factor you will include in your experiment and what these will be
   d) drawing a diagram of your experiment to show what it will look like set up and ready to go into the propagator.
3 Show the teacher your design.
4 Once the teacher has approved of your design, set up the experiment and leave the cartons in the propagator.
5 Construct a table of results which, for example, refers to:
   ● the number of seeds planted
   ● the number of healthy seedlings present after 5 days
   ● the percentage of healthy seedlings present after 5 days.
6 After 5 days examine your cartons and complete your table of results.

7 If other students have carried out the same experiment, pool the results.

REPORTING

Write up your report by doing the following:
1 Copy the title given at the start of this activity.
2 Put the subheading '**Aim**' and state the aim of your experiment.
3 a) Put the subheading '**Method**'.
   b) Include a labelled diagram of your approved design.
   c) Using the impersonal passive voice, briefly describe the experimental procedure that you followed and state how you obtained your results.
4 Put the subheading '**Results**' and draw a final version of your table of results.
5 Put a subheading '**Presentation of Results**' and present your results as a line graph by plotting, for example, the number of seeds planted against the percentage of healthy seedlings present after 5 days.
6 Put a subheading '**Conclusions**' and write a short paragraph to state what you have found out from a study of your results. This should include reference to:
   a) the overall pattern or trend in your results
   b) the effect of competition on the growth of a population of cress seedlings.
7 Put a final subheading '**Evaluation of Experimental Procedure**' and then answer the following:
   a) Identify one possible source of error and describe the steps that you would take to eliminate it in a repeat of the experiment.
   b) By what means were all the variable factors except the one under investigation controlled?
   c) State how you could check the reliability of the results.
   d) Feel free to comment on either of the following if you have an additional point you wish to make:
      (i) further improvements that you would include in a repeat of the experiment
      (ii) limitations of the equipment.

Competition between members of different species

Plants of different species that occupy the same habitat often differ from one another in growth form (e.g. rooting depth, leaf shape, etc.) and mineral requirements. Thus competition between members of different species is normally **less intense** than competition between members of the same species.

Nevertheless some species are able to become dominant at the expense of others. Conifer trees grown in closely packed plantations (see Figure 7.35) almost totally prevent growth of other plants on the forest floor by cutting out the light. In addition, the carpet of pine needles on the ground produces an acidic soil low in minerals that further inhibits the growth of potential competitors.

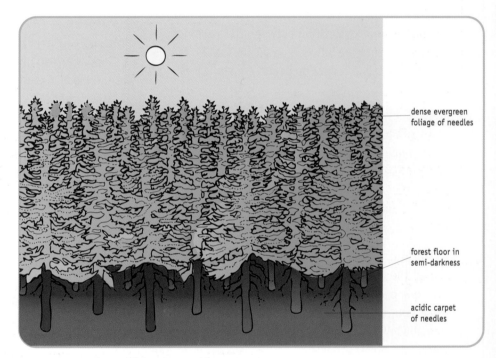

dense evergreen foliage of needles

forest floor in semi-darkness

acidic carpet of needles

Figure 7.35 *Competition by conifer trees*

Competition between animals

Animals compete with one another for resources such as **food**, **water** and **shelter** if these are in short supply.

Competition between members of the same species

Animals of the same species normally all need exactly the same resources. Therefore competition between them is **very intense** if a resource is limited. Such competition often results in the following behaviour.

Territoriality

Territoriality is the name given to behaviour that involves competition between members of the same species (especially birds) for territories.

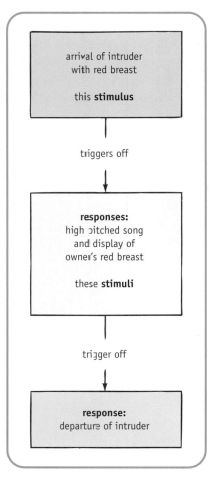

Figure 7.36 *Flow chart of territorial behaviour*

An animal's total range is the area that it covers during its lifetime. Within this range a male animal often establishes a smaller area called its **territory** which contains enough food for himself and eventually a mate and their young. He defends his territory fiercely using **social signals** (sign stimuli) as shown in Figures 7.36 and 7.37. He is most aggressive at the centre of his territory. The further he moves away from the centre the less likely he is to attack intruders. Eventually during such an attack there comes a point when he is equally likely to fight or turn and flee.

Such points of balance are established by 'pendulum' fighting (alternating attack and escape) and mark the boundary between the territories of two rivals.

Figure 7.37 *Territorial behaviour in robins*

Advantages of territorial behaviour

In addition to providing the animal with a safe place to breed, territorial behaviour spaces out the population in relation to the available food supply. This ensures that there will be enough food for the number of young produced.

Red grouse

Within a species, **territorial size** varies depending on the availability of food. The red grouse lives on moorland and feeds on the shoot tips and flowers of heather plants. Young heather plants are more nutritious than older ones. Rather than compete directly for food, the male red grouse claims a territory large enough to provide food for his dependants during the breeding season.

Each enclosed space in Figure 7.38 represents a red grouse's territory on the same piece of moorland over a period of four years. During years 1 and 4, food was plentiful and the birds only needed to defend small territories. During years 2 and 3, the heather was poor and a larger territory was required to supply a bird's needs. When times were lean, competition was more intense and weaker birds that failed to establish a territory did not breed.

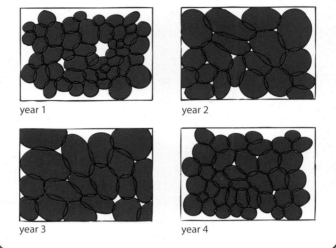

year 1   year 2

year 3   year 4

Figure 7.38 *Sizes of red grouse territories*

Competition between members of different species

When two different species occupy the same ecological niche, competition may become so fierce that one species tends to force the other out.

Squirrels

The introduction of the North American grey squirrel (see Figure 7.39) to Britain has resulted in the widespread decline (almost to extinction) of the red squirrel (see Figure 7.40). Both types of squirrel occupy a similar ecological niche in the woodland ecosystem. The grey squirrel is thought to have become so widely distributed and successful because it competes aggressively for food and is able to make use of a wider variety of foodstuffs.

Figure 7.39 *Grey squirrel*

Figure 7.40 *Red squirrel*

Figure 7.41 *Rainbow trout*

It is therefore continuing to populate areas at the expense of the more timid red squirrel.

Trout

A similar situation is developing in some river ecosystems. Rainbow trout that have been introduced from North America are invading the river habitat of the native brown trout. The American fish are more aggressive and greedy for food. The fierce competition that has resulted is forcing the brown trout into decline.

Reduction in competition

Competition between members of two different species occupying a similar niche is sometimes reduced by the rivals reaching a compromise. For example, they might eat different foods, seek their food at different times of the day, nest in slightly different habitats, and so on.

The common cormorant and the green cormorant are two species of sea bird which nest on cliffs and dive into the sea for fish (see Figure 7.42). They appear therefore to occupy exactly the same ecological niche. However, one species has not forced the other out because competition is minimised by the common cormorant feeding mostly on flat fish and crustaceans from the sea bed while the green cormorant feeds further out to sea on eels and other fish swimming in the upper water.

It is highly likely, of course, that either type of cormorant, in the absence of the other, would extend its range. Thus, under normal circumstances, each species is affecting the distribution of the other.

Predator-prey interactions

A delicate balance exists between populations of predators and their prey. An increase in number of prey (perhaps due to climatic conditions favouring growth of their plant food) leads to an increase in predation.

As the size of the prey population decreases, competition between predators for the remaining prey becomes more and more intense until eventually the number of predators drops. This in turn allows the prey population to build up again which leads to a corresponding increase in the predator population and so on. This series of events is summarised in Figure 7.43. The predator curve takes the same shape as that of the prey but lags behind it since time is required for each change to take effect.

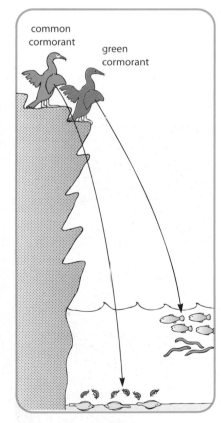

Figure 7.42 *Minimising competition*

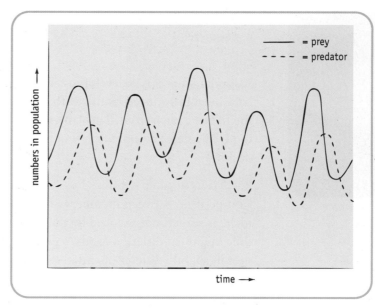

Figure 7.43 *Predator-prey interactions*

Testing Your Knowledge

1 a) Name TWO environmental factors for which
neighbouring plants could be competing. (2)

b) When a factor essential for plant growth is in short
supply, is competition between members of the
same species normally more or less intense than
competition between members of different species?
Explain why. (2)

2 a) Explain the difference between the terms *territory*
and *territoriality*. (2)

b) Briefly describe an example of territorial behaviour
with reference to a named animal. (3)

c) Does territoriality involve members of the same
species or different species? (1)

d) State TWO advantages of territoriality. (2)

3 With reference to cormorants, describe the means by
which competition between species is kept to a
minimum. Explain why this is of advantage to the
species involved. (3)

Applying Your Knowledge

1 Match the terms in list X with their descriptions in list Y.

List X
1) acid rain
2) adaptation
3) behavioural response
4) biodiversity
5) choice chamber
6) competition
7) deforestation
8) desertification

List Y
a) struggle for existence between members of a community caused by the limited supply
of an essential resource
b) loss of land to desert as a result of human activities
c) general term for contamination of the environment by harmful substances
d) group of interbreeding organisms whose offspring are fertile
e) poisonous gas that harms living things and reduces biodiversity
f) role played by a species within a community
g) inherited characteristic that makes an organism well suited to its environment
h) piece of apparatus used to study the effect of differences in an environmental factor
on an animal's behaviour

9) ecological niche
10) pollution
11) species
12) sulphur dioxide

i) harmful form of precipitation that affects plants and animals, reducing biodiversity
j) range of species in an ecosystem
k) complete clearing away of vast tracts of natural forest and the failure to renew them
l) reaction shown by an animal on being exposed to an environmental stimulus

2 Algae are simple green plants that normally live in aquatic ecosystems. *Pleurococcus* is an alga adapted to life on land provided that its habitat (e.g. the surface of a tree trunk) remains damp.
The graphs in Figure 7.44 show the results from an experiment in which the percentage *Pleurococcus* cover and light intensity were measured at 8 compass points round a tree.

a) Ten measurements of percentage *Pleurococcus* cover and light intensity were made at each compass point and averages calculated. Why is this better scientific practice than simply making one measurement each time? (1)

b) What percentage *Pleurococcus* cover was found to occur on the south-east (SE) side of the tree? (1)

c) At which compass point was the percentage *Pleurococcus* cover found to be (i) highest? (ii) lowest? (2)

d) What light intensity was recorded on the north-west (NW) side of the tree? (1)

e) At which compass point was light intensity found to be at **(i)** its highest level? **(ii)** its lowest level? (2)

f) (i) In general what relationship exists between the distribution of *Pleurococcus* cover and light intensity?

(ii) Give a possible explanation for this relationship. (3)

3 Two similar squares of turf (X and Y) were cut out of a piece of grassland and examined. Each was found to possess the same community of 20 different plant species. The squares were kept in a greenhouse under identical conditions except that X was regularly cropped whereas Y was left uncut. After three years one of the squares was found to possess 11 of the original species while the other still had all 20. Identify which square was which and explain why. (3)

4 This question refers to Figure 7.17 on page 130.

a) What is the lowest pH at which (i) grayling; (ii) char were found? (2)

b) Which one of the eight species of fish is least tolerant of acid conditions? (1)

c) Which one of the eight species of fish is most tolerant of acid conditions? (1)

d) How many of the eight species of fish were found in loch water of:
(i) pH 6.0   (ii) pH 5.6   (iii) pH 5.3
(iv) pH 4.8   (v) pH 4.3? (5)

e) Make a generalisation about the effect of pH on the variety of species of fish present in the loch. (1)

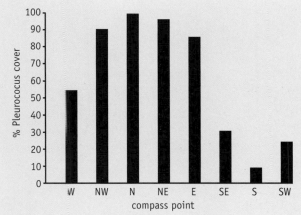

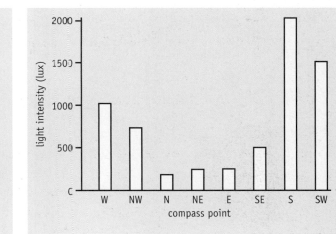

Figure 7.44

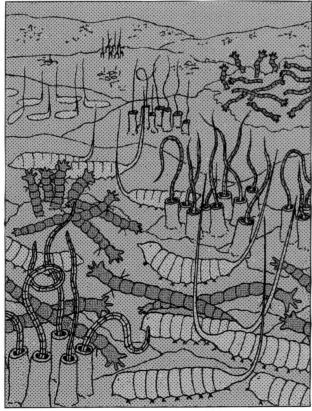

Figure 7.45

5 The upper part of Figure 7.45 shows a stable freshwater ecosystem; the lower part shows the same ecosystem after it has been polluted by sewage. Write a paragraph to describe the effect of pollution on this ecosystem. Your answer should include reference to:

(i) the variety of species present before and after pollution

(ii) the changes in population numbers brought about by the pollution. (5)

6 A coal-fired power station was known to release sulphur dioxide into the local atmosphere. A group of scientists carried out a survey by counting the number of species of lichen at five sample sites in a north-easterly line from the power station.

They repeated the procedure in a south-easterly line as shown in Figure 7.46. The diagram also gives details of an annual survey of wind direction for the area. Table 7.6 gives the results of the lichen survey.

a) How far from the power station was

(i) sample site A and (ii) sample site E? (2)

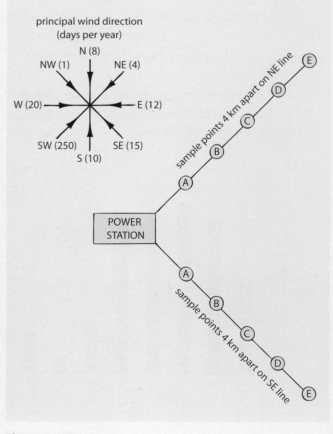

Figure 7.46

	average number of species of lichen at sampling site				
	A	**B**	**C**	**D**	**E**
NE line	2	2	3	3	4
SE line	3	5	6	8	11

Table 7.6

b) (i) Why were several counts taken at each sample site and an average calculated for the number of lichen species present?

(ii) What happens to the biodiversity of lichen species as the distance from the power station increases along the south-east (SE) sample line?

(iii) Explain why. (3)

c) (i) How many calm days lacking significant wind occurred during the year to which the wind survey refers?

(ii) In which direction did the wind blow least often?

(iii) In which direction did the wind blow most often? (3)

d) (i) Make a generalisation about the difference between the number of lichen species found on the north-east (NE) line compared with the south-east (SE) line.

(ii) Give a possible explanation for this difference based on the information given. (3)

7 The spider mite feeds on green leaves, making them turn yellow and fall off the plant. This tiny creature belongs to the food web shown in Figure 7.47 where * indicates organisms that are sensitive to a certain type of pesticide spray.

One year a farmer sprayed his crop with the pesticide and found that only a little of the crop was lost. The next year he found his crop to be plagued with a pest so he sprayed again. However, he lost most of the crop.

a) Under natural conditions which animals keep the numbers of plant-eating pests in check? (1)

b) Why was any of the crop lost following the first application of pesticide? (1)

c) Why was damage to the crop so severe the second year? (1)

d) Rewrite the following sentence to include only the correct word from each choice.
This example shows how the removal of the primary/secondary consumers from a food chain/web enables a primary/secondary consumer to undergo a population increase/decrease and destroy most of the decomposer/producer. (2)

8 The relative humidity of the air in an enclosed space is affected by the chemical solution present (see Table 7.7).

percentage by weight of glycerol in solution	33	64	79	89	95
relative humidity (%) of air at 25°C	90	70	50	30	10

Table 7.7

In the choice chamber shown in Figure 7.48, two of the lower sectors were filled with 33% glycerol solution and the other two with 95% glycerol solution. The five holes in the top sector were sealed with rubber plugs and the apparatus left for 10 minutes to allow a humidity gradient to develop.

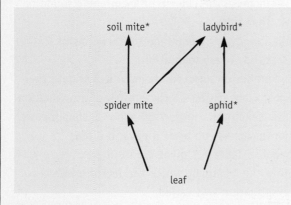

Figure 7.47

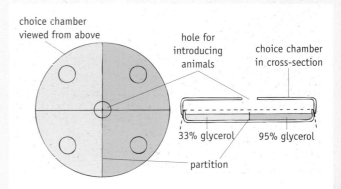

Figure 7.48

Ten woodlice (animals whose skins are permeable to water) were placed in a thin film of water in an enclosed dish in order to 'wet-adapt' them and they were then inserted into the choice chamber. Their distribution (Table 7.8) and rate of activity (* = more active) were observed at two-minute intervals.

time interval	33% glycerol	95% glycerol
1	2*	8
2	1*	9
3	2*	8
4	0	10
5	1*	9
6	0	10
7	1*	9
8	3*	7
9	2*	8
10	3*	7
11	4*	6
12	3*	7
13	5	5
14	6	4*
15	7	3*
16	6	4*
17	8	2*
18	9	1*
19	10	0
20	10	0

Table 7.8

a) What relationship exists between the relative humidity of air and the concentration of glycerol solution? (1)
b) Name TWO occasions in the experiment where some time must be allowed for the woodlice to become acclimatised before proceeding to the next stage. (2)
c) How many conditions of the one variable factor were studied? Briefly describe the state of the air in each of these conditions. (2)

d) (i) In which side did the woodlice tend to congregate initially?
   (ii) Where did the animals congregate finally?
   (iii) Account for this change in preference.
   (iv) Why is such behaviour of survival value under natural conditions? (4)

9 Figure 7.49 shows the positions (as black dots) of two flatworms in a Petri dish at 15-second intervals.

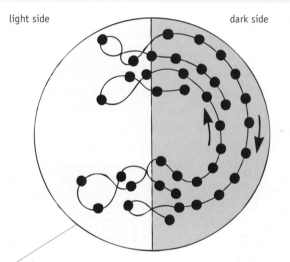

Figure 7.49

a) (i) On which side of the dish were the worms moving at a faster rate?
   (ii) Explain how you arrived at your answer. (3)
b) On which side does the worms' rate of turning increase? (1)
c) (i) On which side do the worms tend to congregate?
   (ii) State TWO features of their behaviour that increase the chance of them being found on that side.
   (iii) Suggest an advantage to flatworms of this behavioural adaptation. (4)

10 A student set up an experiment to investigate the effect of competition on the growth of a population of cress seedlings. Figure 7.50 shows a simplified version of his set-up after 5 days in a seed propagator. Table 7.9 gives his results.
a) State the means by which the factor under investigation was varied. (1)

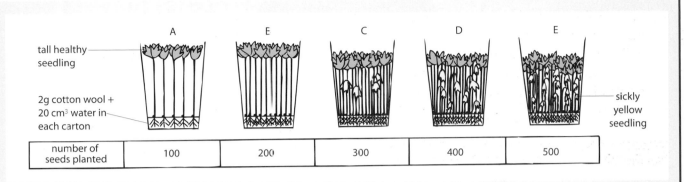

Figure 7.50

b) State THREE factors that the diagram shows were kept constant. (3)

c) (i) Copy and complete Table 7.9.

(ii) Why is it necessary to convert the results to percentages? (3)

number of seeds planted	number of healthy seedlings with green leaves	percentage number of healthy seedlings with green leaves
100	87	
200	178	
300	204	
400	228	
500	225	

Table 7.9

d) Plot a line graph of the number of seeds planted against the percentage number of healthy seedlings with green leaves. (3)

e) Make a generalisation from your graph about the effect of competition on the growth of a population of cress seedlings. (1)

f) Suggest ONE factor other than water that the cress seedlings in carton E could be competing for. (1)

g) If a plant is short of water, its stomata remain closed for a longer time than when water is plentiful. Explain why closed stomata eventually lead to stunted growth of the affected plant. (2)

h) Figure 7.51 is a partly completed diagram of an alternative method of investigating the effect of competition on cress seedlings. Make a simple diagram of plates B and C to show them set up and ready at the start of the experiment. (2)

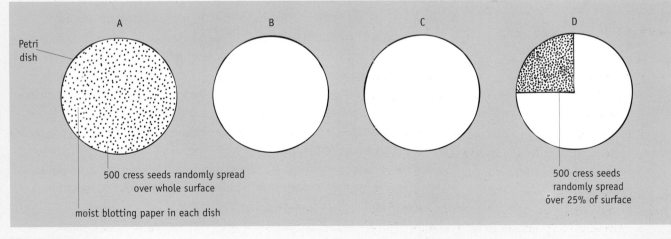

Figure 7.51

Word bank

adaptations	habitat
behavioural	lost
biodiversity	numbers
biomass	nutrients
chains	pollution
community	population
compete	prey
consumers	primary
decomposers	producers
destruction	pyramid
disrupted	requirements
ecosystem	shelter
energy	stable
environmental	survival
fertile	variety

Table 7.10 *Word bank for Chapters 6 and 7*

What You Should Know

(Chapters 6 and 7) (See Table 7.10 for word bank)

1 A group of living organisms of the same species is called a _____. The sum total of all the populations living in an ecosystem is called the _____. The place where an organism lives is called its _____. The term _____ refers to the balanced interaction between the members of a community and their habitat.

2 Green plants are described as _____ because they convert light energy to chemical energy during photosynthesis. This plant food is eaten by herbivorous animals (_____ consumers) who may in turn be the _____ of predatory carnivorous animals (secondary _____).

3 Organisms that obtain their energy by breaking down dead remains or waste materials are called _____.

4 During the feeding relationships that occur in an ecosystem, energy flows from organism to organism through food _____ and webs but much of this energy is _____ along the way.

5 Amongst the organisms in a food chain, the producers are the most numerous. The number decreases at each link and the final consumers are the least numerous. This relationship can be represented as a pyramid of _____.

6 In a food chain, the producers have the greatest biomass. This decreases at each link and the final consumers have the smallest _____. This relationship can be represented as a _____ of biomass.

7 In a food chain, some of the energy passed on at each link is incorporated in the tissues of the next organism. Most energy is transferred in this way from producer to primary consumer and least energy is transferred at the other end of the food chain. This relationship can be represented as a pyramid of _____.

8 A species is a group of interbreeding organisms whose offspring are _____. The total range of species present in an ecosystem's community is also known as its _____. The distribution of species is influenced by the _____ to niche and habitat that they possess.

9 A _____ ecosystem is found to contain a wide range of species (producers, consumers and decomposers) living and interacting in a balanced way and dependent on one another.

10 Unselective grazers maintain the _____ of plant species present in a grassland ecosystem. _____ reduces the variety of species present in an affected ecosystem.

11 Some human activities such as deforestation cause habitat _____ which leads to a loss of biodiversity.

12 When a food web is _____ by the removal of one of its members, this may lead to changes in population numbers of the remaining members.

13 Animals show _____ adaptations by responding to _____ stimuli in certain ways. These are of adaptive significance since they increase the animal's chance of _____.

14 Plants _____ for light and soil _____ if these are in limited supply. Animals compete for food, water and _____. Since the members of a species all have the same _____, competition between them is normally more intense than between members of different species.

8 Factors affecting variation in a species

Variation

Although the members of the same species are very similar to one another, they are not identical. This is because **variation** exists amongst the members of a species.

Types of variation

Discontinuous variation

A characteristic shows **discontinuous variation** if it can be used to divide up the members of a species into two or more **distinct groups**. For example ivy plants can be divided into two separate groups according to leaf type (green or variegated as shown in Figure 8.1). Similarly humans can be split into two separate groups depending on their ability or inability to roll the tongue (see Figure 8.2) and into four groups based on blood group types A, B, AB and O.

Figure 8.1 *Leaf type in ivy plants*

Figure 8.2 *Tongue-rolling ability and inability*

The data obtained from a survey of a characteristic that shows discontinuous variation is presented as a **bar graph** or **pie chart**. Figure 8.3 shows a bar graph of the blood groups of 100 people with each distinct group represented by a separate bar.

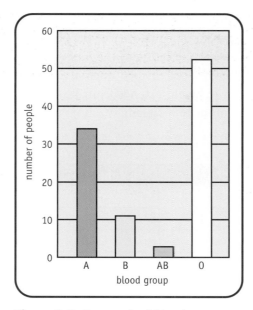

Figure 8.3 *Bar graph of blood groups*

Continuous variation

A characteristic shows **continuous variation** when it varies amongst the members of a species in a **smooth, continuous way** from one extreme to the other and does not fall into distinct groups. For example seed length in broad beans varies continuously as shown in Figure 8.4. Similarly shell length in mussels (see Figure 8.5) and height in humans vary continuously.

Figure 8.4 *Seed length in broad beans*

Figure 8.5 *Shell length in mussels*

Table 8.1 shows the heights of a sample group of 50 people arranged in increasing order. They do not fall into distinct groups. For convenience, the entire range of the characteristic is divided into small groups (subsets) of 5cm. For example eight people are found to fall into subset 155–159cm whereas only two people fall into subset 180–184cm.

Grouping the data into subsets allows a **histogram** to be drawn as shown in Figure 8.6. The majority of the people in the sample group have a height that is close to the centre of the range with fewer at the extremities. When a curve is drawn, a bell-shaped **normal distribution** is produced. In this example, the range in height extends from 140cm to 189cm and the most common height is the subset 160–164cm.

person	height (cm)	person	height (cm)
1	144	26	164
2	145	27	164
3	148	28	164
4	151	29	165
5	152	30	165
6	152	31	165
7	153	32	166
8	154	33	166
9	155	34	167
10	155	35	167
11	156	36	168
12	156	37	168
13	157	38	169
14	158	39	170
15	158	40	171
16	159	41	172
17	160	42	172
18	160	43	174
19	161	44	174
20	161	45	176
21	162	46	178
22	162	47	179
23	162	48	181
24	163	49	183
25	163	50	186

Table 8.1 *Continuous variation in human height*

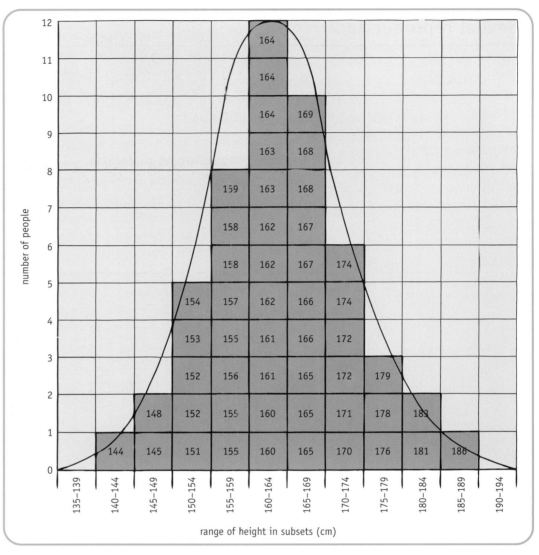

Figure 8.6 *Histogram of human height*

Variation continued

Further examples of continuous and discontinuous variation are given in Table 8.2.

continuous	discontinuous
body mass in humans	ear lobe type in humans (attached or unattached)
hand span in humans	fingerprint type in humans (loop, whorl, arch or compound)
resting heart rate in humans	hair type in humans (straight, curly or wavy)
length of index finger in humans	eye colour in fruit flies (red or white)
diameter of shell in limpets	wing length in fruit flies (long or short/vestigial)
body length in trout	coat colour in guinea pigs (black or white)
milk yield in cattle	flower colour in foxgloves (purple or white)
mass of seed in sunflower plants	seed shape in garden pea plants (round or wrinkled)
number of fruits on apple trees	height in garden pea plants (tall or dwarf)
length of petal in buttercup flowers	colour of grain in maize plants (purple or yellow)

Table 8.2 *Examples of continuous and discontinuous variation*

Sexual reproduction

Reproduction is the process by which the members of a species produce offspring. **Sexual** reproduction involves the fusion of two sex cells (gametes) during **fertilisation.**

Gamete production in mammals

The human reproductive organs are shown in Figures 8.7 and 8.8. The **testes** are the site of **sperm** production; the **ovaries** are the site of **egg** production.

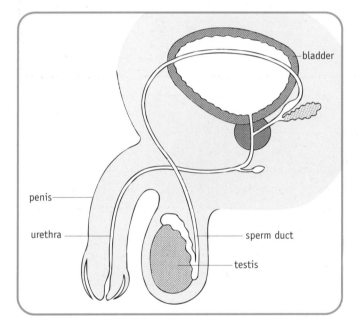

Figure 8.7 *Human male reproductive organs*

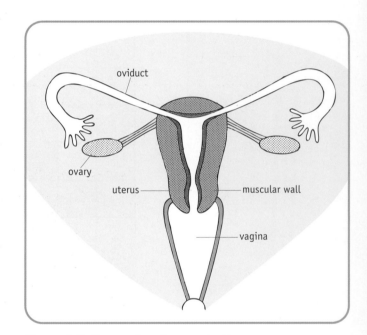

Figure 8.8 *Human female reproductive organs*

Fertilisation

The release of an egg from a mammalian ovary is called **ovulation**. Eggs are released at regular intervals into the oviducts (see Figure 8.9). The inner lining of an oviduct bears hair-like cilia which beat and gently move the egg along towards the uterus. This journey takes about 3 days in humans.

Following copulation (sexual intercourse in humans) and ejaculation (release of sperm by the male), sperm swim up the uterus and into the oviducts. It is here in an **oviduct** that fertilisation occurs. Although many sperm may meet an egg, only one sperm fertilises the egg. Its nucleus fuses with that of the egg forming a **zygote** which contains genetic material from **both** parents.

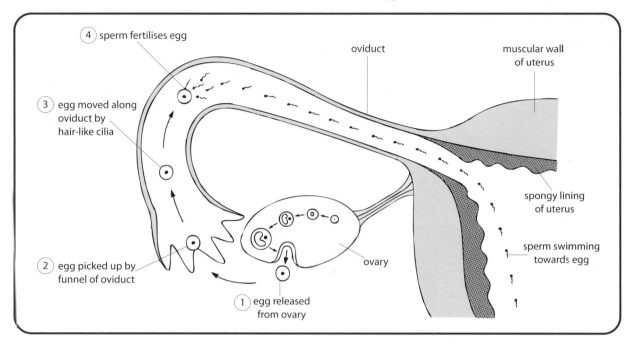

Figure 8.9 *Events leading to fertilisation*

Gamete formation in flowering plants

Flowers are the structures responsible for sexual reproduction in flowering plants. Usually the male and female reproductive organs are both present in the same flower (see Figure 8.10).

Pollen grains contain the plant's male sex cells and are produced in the **anthers. Ovules** contain the plant's female sex cells (egg cells) and are produced in the **ovary.**

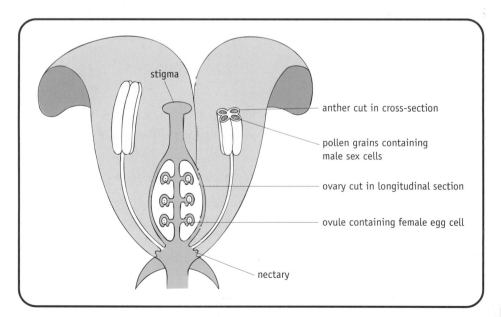

Figure 8.10 *Half section of a flower*

Fertilisation

Pollination is the **transfer** of pollen grains from an anther to a stigma and should not be confused with fertilisation, the process by which the nucleus of a male sex cell from a pollen grain **fuses** with the nucleus of an egg cell to form a zygote. Fertilisation takes place inside an ovule following the growth of a pollen tube which enables the male cell to reach the female cell (see Figure 8.11).

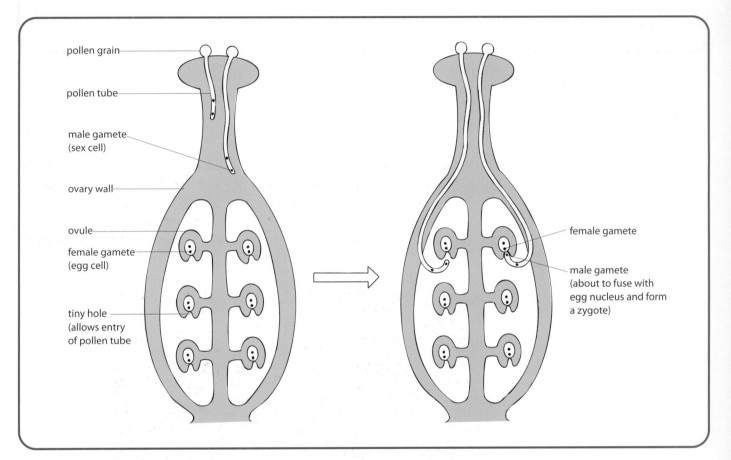

Figure 8.11 *Growth of a pollen tube*

The **zygote** formed becomes an embryo which later grows into a plant containing genetic material from both the male and the female parents.

Variation produced by random combination of gametes

Each new member of a species formed by sexual reproduction receives half of its genetic material from one parent and half from the other parent. It therefore resembles each parent in some ways but differs from each parent in other ways. It is genetically distinct from all other members of the species.

Since sexual reproduction brings about the **random combination** of two parental gametes at fertilisation, it is constantly producing **variation** within each species. The importance of this variation is discussed on page 176.

Testing Your Knowledge

1. a) Where are (i) sperm; (ii) eggs produced in the body of a mammal? (2)
   b) Identify the site of fertilisation in a mammal's body. (1)
   c) What is formed as a result of fertilisation? (1)
2. a) Which structures in a flower contain pollen grains? (1)
   b) Of which sex are the gametes present in the pollen grains? (1)
   c) Which structure in a flower contains egg cells in ovules? (1)
   d) How does a male sex cell reach and fertilise a female sex cell? (2)
3. Although human beings resemble their parents in many ways, they are never identical to either of them. Explain why. (1)

Genetics

The **variation** that exists among the members of a species is due largely to the information present in the genetic material in the nuclei of their cells. The study of this genetic material and its transmission from generation to generation is called **genetics**. The genetic information that gives rise to an organism's characteristics is contained in **genes** on **chromosomes** (see Figure 8.12).

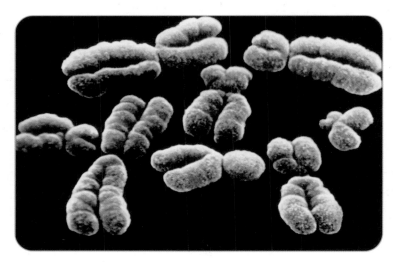

Figure 8.12 *Chromosomes*

Chromosome structure

Chromosomes are thread-like structures found inside the nucleus of every living cell. They contain **DNA** (deoxyribonucleic acid). A molecule of DNA consists of two strands. Each strand has a 'chain' of **base** molecules (see Figure 8.13). Normally the two strands are held together by weak bonds between pairs of bases. Each base's molecular structure is such that it can only fit one other type of base.

167

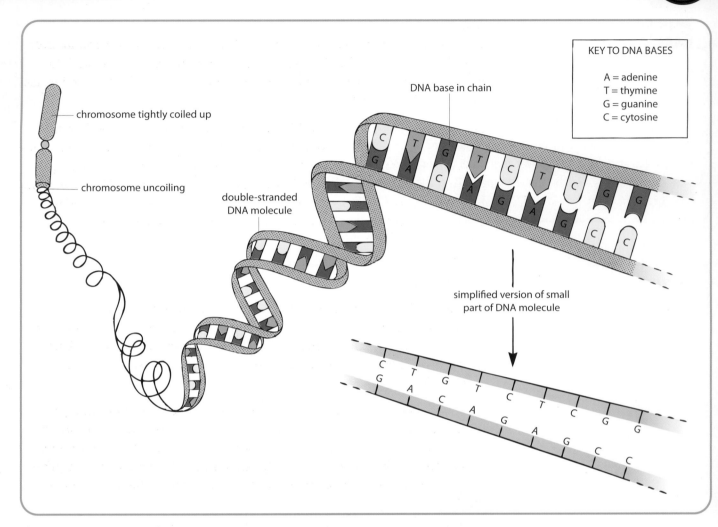

Figure 8.13 *Structure of DNA*

KEY TO DNA BASES

A = adenine
T = thymine
G = guanine
C = cytosine

Sequence of DNA bases

The DNA of one member of a species differs from that of another member by the order in which the bases occur in their chromosomes. It is this **sequence** of bases along the DNA strands that is unique to an organism. The sequence contains the **genetic instructions** that control the organism's inherited characteristics.

Proteins

Inherited characteristics are the result of many biochemical processes controlled by **enzymes**. In humans, for example, certain enzymes govern the biochemical pathways that lead to the formation of hair of a particular texture, eye irises of a certain colour, and so on. Every enzyme is made of **protein**.

In addition, many of the body's tissues, such as muscles, bones, tendons and ligaments, are composed largely of protein. The body also possesses protein in many other forms such as **hormones** (chemical messengers), **antibodies** (protective molecules) and **haemoglobin** (the protein-containing chemical that gives red blood cells their characteristic colour.

Amino acids

Each protein is built up from a large number of sub-units called **amino acids** of which there are about 20 different types. These are joined together into polypeptide chains each of which normally consists of hundreds of amino acid molecules linked together.

Depending on which amino acids are present in the chain, further bonds between certain amino acids form, making the chain coil up and often become folded in a characteristic way. The exact nature of the coiling and folding depends on the **sequence** of amino acids present in the chain. This determines the final structure of the protein and whether or not it is exactly suited to carrying out its function. So how is this critical order of amino acids in a protein determined?

Genetic code

A region of DNA on a chromosome is called a **gene** and it is normally hundreds or even thousands of DNA bases long. The information present in DNA takes the form of a molecular language called the **genetic code**. Each group of three bases along a DNA strand represents a '**codeword**' for an amino acid.

Each gene codes for a particular protein (or polypeptide) by making a molecular 'mirror image' of its DNA sequence and passing it out into the cytoplasm (see Figure 8.14). This message is then translated into protein. By this means the order in which the amino acids become joined together into protein is determined (indirectly) by the order of the bases on the DNA.

Thus DNA encodes the information for the particular sequence of amino acids in a protein, which in turn dictates the structure and function of that protein. Figure 8.15 shows the formation of a molecule of an enzyme whose folded structure contains the active site it needs to function as a biological catalyst.

Testing Your Knowledge

1 What are chromosomes and where are they located in a cell? (2)

2 a) What do the letters DNA stand for? (1)
  b) How many strands are present in a DNA molecule? (1)
  c) Name ONE of the main types of molecular component present in a DNA chain. (1)

3 a) Of what type of organic compound are enzymes composed? (1)
  b) What sub-units make up a protein? (1)
  c) What information determines the sequence of amino acids in a protein? (1)
  d) Why is the sequence of amino acids in a protein important? (1)

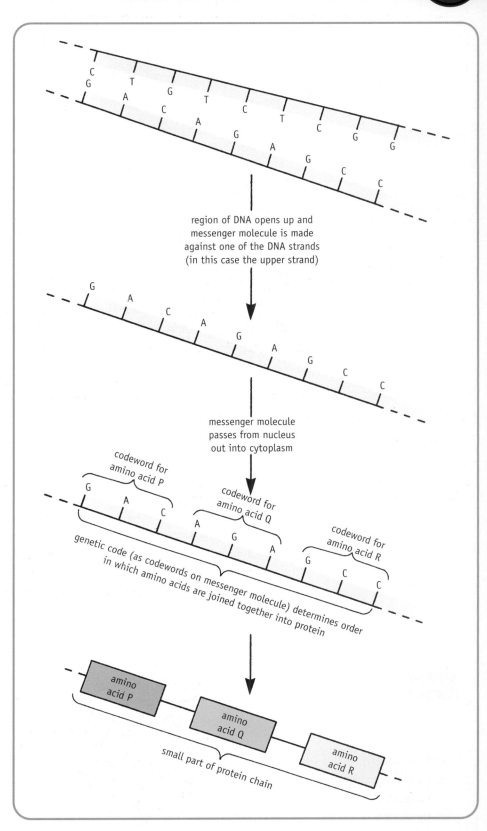

Figure 8.14 *Amino acid sequence determined by DNA*

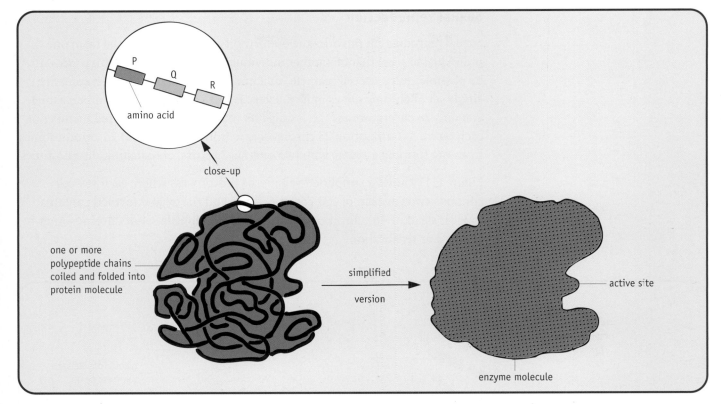

Figure 8.15 *Formation of an enzyme molecule*

Matching sets of chromosomes

Each body cell of a multicellular organism has two **matching sets** of chromosomes in its nucleus. For example, every normal human body cell (and zygote) contains 46 chromosomes that can be arranged as 23 pairs (see Figure 8.16).

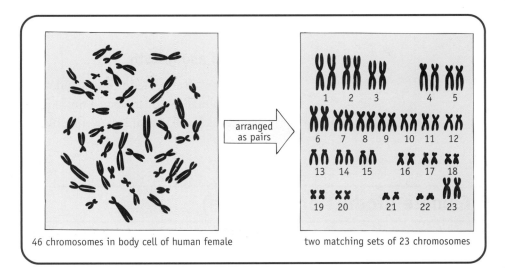

Figure 8.16

Sexual reproduction

Sexual reproduction provides the opportunity for genetic material from one individual to meet that of another individual at fertilisation. For this process to be possible, every sex cell (gamete) of a multicellular organism must contain a **single set** of chromosomes in its nucleus. For example, a normal human gamete contains 23 chromosomes (i.e. a single set made up of one unpaired member of each type). At fertilisation 23 chromosomes from a sperm meet 23 chromosomes in an egg forming a zygote with two sets, i.e. 23 **pairs**, of matching chromosomes.

Figure 8.17 shows a simplified version of this process where each sex cell contains two (instead of 23) chromosomes and the zygote formed contains four (instead of 46) chromosomes. A complete **double** set of chromosomes is restored at fertilisation.

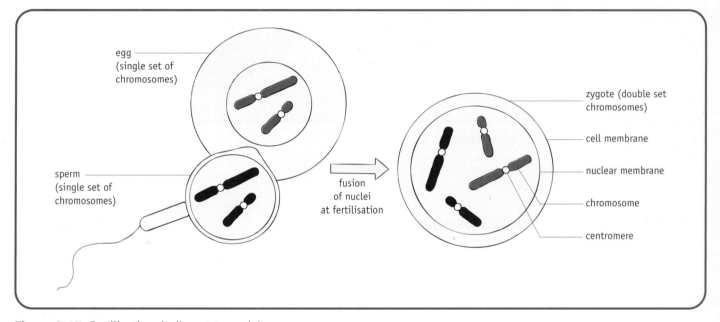

Figure 8.17 *Fertilisation (cells not to scale)*

Mitosis and growth

As a result of repeated **mitosis** and cell division, the zygote grows into a multicellular organism, e.g. a human being, consisting of billions of cells where each cell contains two matching sets of chromosomes in its nucleus.

Gamete production (meiosis)

At sexual maturity, certain body cells begin to produce gametes. For example, in humans, sperm are formed by cells in the testes and eggs are formed by cells in the ovaries. During the process of gamete formation (**meiosis**), a type of

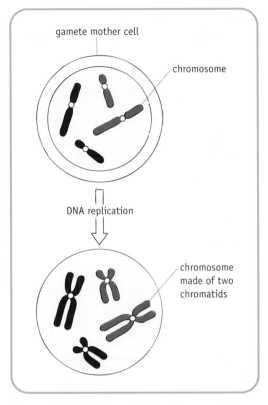

Figure 8.18 *DNA replication in a gamete mother cell*

body cell called a gamete mother cell divides to form four sex cells. Each gamete mother cell contains **two** sets of chromosomes and each sex cell formed contains **one** set of chromosomes.

DNA replication

Immediately before gamete production, each chromosome in a gamete mother cell uncoils and undergoes **DNA replication** forming two identical chromatids (see Figure 8.18).

Random assortment of chromosomes

During meiosis, each gamete receives a copy of one but not both members of each matching pair of chromosomes present in the gamete mother cell. Thus if the gamete mother cell contains four chromosomes as two pairs then each gamete receives two chromosomes, one from each pair.

Since the separation of the members of a pair of chromosomes occurs **independently** of and at **random** relative to the separation of the members of any other pair of chromosomes (see Figure 8.19 on the following page), a gamete mother cell containing two pairs of chromosomes has the potential to produce 2^2 (i.e. 4) different combinations in the gametes.

A gamete mother cell containing three pairs of chromosomes has the potential to produce 2^3 (i.e. 8) possible combinations in the gametes (see Practical

Activity below). The greater the number of chromosomes present, the greater the number of possible combinations. For example, a human egg mother cell with 23 homologous pairs has the potential to produce 2^{23} (i.e. 8 388 608) different combinations. This 'shuffling' of matching pairs of chromosomes is also described as **random assortment** of chromosomes.

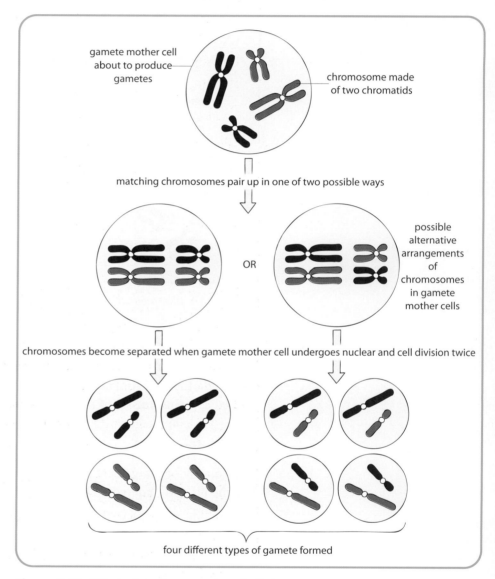

gamete mother cell about to produce gametes

chromosome made of two chromatids

matching chromosomes pair up in one of two possible ways

OR

possible alternative arrangements of chromosomes in gamete mother cells

chromosomes become separated when gamete mother cell undergoes nuclear and cell division twice

four different types of gamete formed

Figure 8.19 *Effect of random assortment of chromosomes*

Practical Activity

Using a model to show chromosome behaviour during gamete formation

You need

1 set of 'chromosomes' made of lengths of coloured cable (see Figure 8.20)
5 sheets of white paper
1 blue and 1 brown coloured pencil

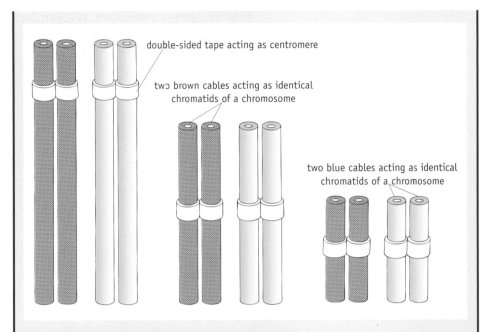

double-sided tape acting as centromere

two brown cables acting as identical chromatids of a chromosome

two blue cables acting as identical chromatids of a chromosome

Figure 8.20 *Apparatus for chromosome model*

What to do

1 Spread out the lengths of coloured cable on a sheet of paper and check that each chromosome's two identical chromatids are attached at the centromere. Imagine that they represent the chromosome complement of a cell about to form four gametes.

2 Match the chromosomes into pairs according to length so that each pair consists of four chromatids.

3 Show what would be formed as a result of meiosis by doing the following:
   (i)   lay out four sheets of paper to represent the four gametes
   (ii) take one pair of chromosomes (i.e. four chromatids) and tease them apart
   (iii) distribute these chromatids, one to each gamete, in a random manner
   (iv) repeat the process for the other two pairs of chromosomes.

4 Using coloured pencils, make a simple diagram of each type of gamete formed.

5 Reassemble the chromosomes by joining identical chromatids at their centromeres.

6 Repeat step 3 so that you form gametes where some or all are different from those that you obtained before.

7 Add simple drawings of any new types of gamete to your original gamete diagram.

8 Continue to repeat step 3 forming gametes different from before until you have exhausted all the possibilities.

9 Add simple drawings of the new types of gamete formed each time to your gamete diagram.

10 State the number of different types of gamete that can be formed by gamete mother cells that contain three pairs of matching chromosomes and then check your answer with Figure 8.21.

Significance of random assortment during meiosis

Random assortment ('shuffling') of matching pairs of chromosomes leads to **variation** among the sex cells formed during meiosis. In the example in Figure 8.19, four different types of gamete are formed. In the Practical Activity eight different types of gamete can be produced, as shown in Figure 8.21. Further variation is produced by the random combination of gametes that occurs during fertilisation (see page 166).

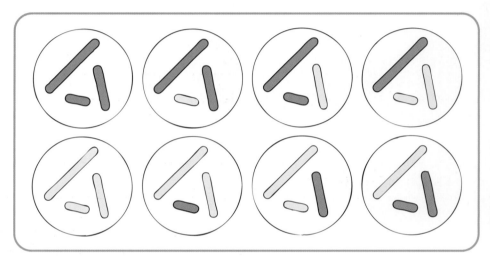

Figure 8.21 *Eight varieties of gamete*

Importance of variation

Such **variation** within a species is of great importance because it helps the species to adapt to a changing environment. Imagine, for example, that a new disease appears. If great genetic variation exists among the members of the species, then there is a good chance that some of them will be resistant and **survive**. If they were all identical and susceptible to the disease, the whole species would be wiped out.

Chromosome complement

Every species of plant and animal has a definite and characteristic number of chromosomes (the **chromosome complement**) present in each cell. This number varies from species to species. A few examples are given in Table 8.3.

species	chromosome complement
geranium	10
pea	14
barley	14
turnip	20
tomato	24
fruit fly	8
toad	36
human	46
chimpanzee	48
cow	60

Table 8.3 *Chromosome complements*

Sex chromosomes

Every normal body cell in a human being contains 46 chromosomes as 23 pairs. Of these, one pair makes up the **sex** chromosomes. In each body cell of a human female, these are the two equal-sized **X** chromosomes (see Figure 8.16). In each body cell of a human male, the sex chromosomes comprise a single **X** chromosome and a much smaller **Y** chromosome (see Figure 8.22).

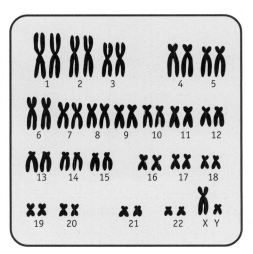

Figure 8.22 *Sex chromosomes*

Gametes and sex chromosomes

Figure 8.23 shows gamete formation in humans where only the sex chromosomes have been drawn. Each female gamete (egg) receives an X chromosome from its gamete mother cell. However, each male gamete (sperm) receives either an X or a Y chromosome from its gamete mother cell. Thus it is a 1 in 2 chance which type of sex chromosome a sperm receives during gamete formation.

Sex determination

Figure 8.24 shows how the sex of a child is determined by the type of sex chromosome carried by the sperm. If an egg (always containing an X chromosome) is fertilised by a sperm containing an X chromosome, a **female** child (**XX**) is formed. However, if an egg (X) is fertilised by a sperm containing a Y chromosome, a **male** child (**XY**) is the result.

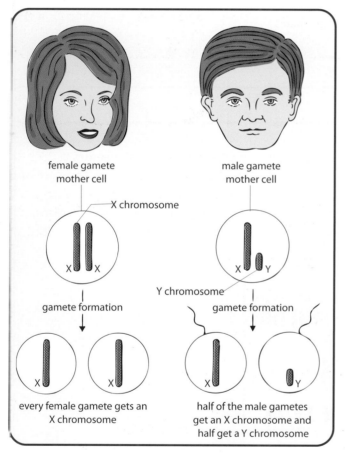

Figure 8.23 *Gametes and sex chromosomes*

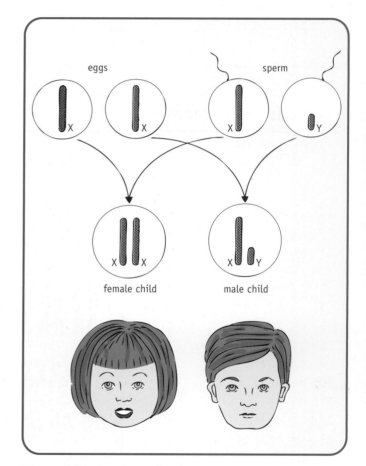

Figure 8.24 *Sex determination*

It is a 1 in 2 chance that an egg will be fertilised by an X sperm and a 1 in 2 chance that it will be fertilised by a Y sperm. This produces a **sex ratio** of **1 male:1 female** in the population as a whole.

Testing Your Knowledge

1 a) How many chromosomes are present in the nucleus of a normal body cell of a human being? (1)
   b) How many matching sets of chromosomes are present in a normal body cell of a human being? (1)

2 a) How many matching sets of chromosomes are present in a sperm mother cell? (1)
   b) How many sex cells are formed by a sperm mother cell during gamete production? (1)
   c) How many sets of chromosomes does each sperm receive? (1)

3 Rewrite the following sentences including only the correct word from each choice in brackets.
   During gamete formation, (sex/body) cells receive different combinations of the (single/paired) chromosomes present originally in gamete mother cells. This takes place as a result of (mitosis/random assortment) of chromosomes. Since the products of gamete formation are (identical/non-identical), this leads to increased (variation/uniformity) among the members of the species following (sexual/asexual) reproduction. (6)

4 a) In humans which type(s) of sex chromosome could be found in (i) an egg? (ii) a sperm? (3)
   b) With reference to sex chromosomes, explain how the sex of a child is determined. (2)

Applying Your Knowledge

1 Match the terms in list X with their descriptions in list Y.

List X	List Y
1) amino acid	a) type of variation that can be used to divide a species up into distinct groups
2) anther	b) sex chromosome found in human males only
3) base	c) male sex cell in animals
4) chromosome	d) sex chromosome found in human males and females
5) continuous	e) molecular component of protein
6) discontinuous	f) part of flowering plant containing pollen grains
7) egg	g) molecular component of a DNA chain
8) genetic code	h) organic compound made of amino acids
9) ovary	i) site of sperm production in animals
10) pollen grain	j) process by which an offspring of either one sex or the other is formed
11) protein	k) site of egg production
12) random assortment	l) thread-like structure found inside the nucleus of a cell
13) sex determination	m) molecular language determined by the sequence of bases in a DNA chain
14) sperm	n) structure present in anthers that contains male sex cells of a flowering plant
15) testis	o) fertilised egg
16) X chromosome	p) production of different combinations of matching pairs of chromosomes by 'shuffling' during meiosis
17) Y chromosome	q) female sex cell
18) zygote	r) type of variation that varies amongst the members of a species in an uninterrupted way

2 The following list gives the steps involved in the construction of a histogram to show the variation in mass that exists amongst the fish in a fish farm. Arrange the five steps into the correct sequence.

   a) Consider the range in mass obtained with respect to the lightest and heaviest.
   b) Prepare graph paper by putting range in mass as subsets on the x-axis and number of fish on the y-axis.

179

c) Weigh a large sample of fish.

d) Count the number of fish in each subset and draw the histogram.

e) Divide the total range in mass into several small equally spaced subsets of range.

3 Fifty pupils had their left index finger measured. The lengths are listed in Table 8.4.

pupil number	length of left index finger (mm)	pupil number	length of left index finger (mm)
1	48	26	68
2	51	27	68
3	52	28	69
4	53	29	70
5	55	30	70
6	56	31	71
7	57	32	71
8	58	33	71
9	58	34	72
10	60	35	72
11	60	36	73
12	61	37	74
13	62	38	74
14	62	39	75
15	62	40	75
16	63	41	76
17	63	42	78
18	65	43	78
19	65	44	79
20	65	45	81
21	66	46	83
22	66	47	84
23	66	48	85
24	67	49	88
25	67	50	91

Table 8.4

a) Divide the range into 10 subsets of 5mm (with the first subset being 45–49mm) and then present the information as a histogram.

b) How many pupils have a left index finger in the subset 70–74mm?

c) (i) What is the most common subset of the range of index finger length?

(ii) How many pupils have a left index finger that belongs in this subset?

d) How many pupils have a left index finger of 75mm or more?

4 Figure 8.25 shows a human sperm.

a) (i) Name parts A and B.

(ii) Enzymes are present in a sac in part A. Suggest the important role that they play. (3)

b) An average sperm is 60 micrometres in length. Express this as a decimal fraction of a millimetre (1 millimetre = 1000 micrometres). (1)

Figure 8.25

c) Sperm count is measured as millions of sperm per millilitre (ml) of semen. A man is considered to be fertile if his sperm count is 20 million or more. Table 8.1 shows the data for three patients at a fertility clinic.

patient	average volume of semen released (ml)	average total number of sperm in semen (millions)
X	3.0	63
Y	4.0	76
Z	4.5	117

Table 8.5

(i) Calculate the sperm count for patients X, Y and Z.

(ii) Which of these men is most likely to be infertile? (4)

d) Suggest why mammals produce many more sperm than eggs. (1)

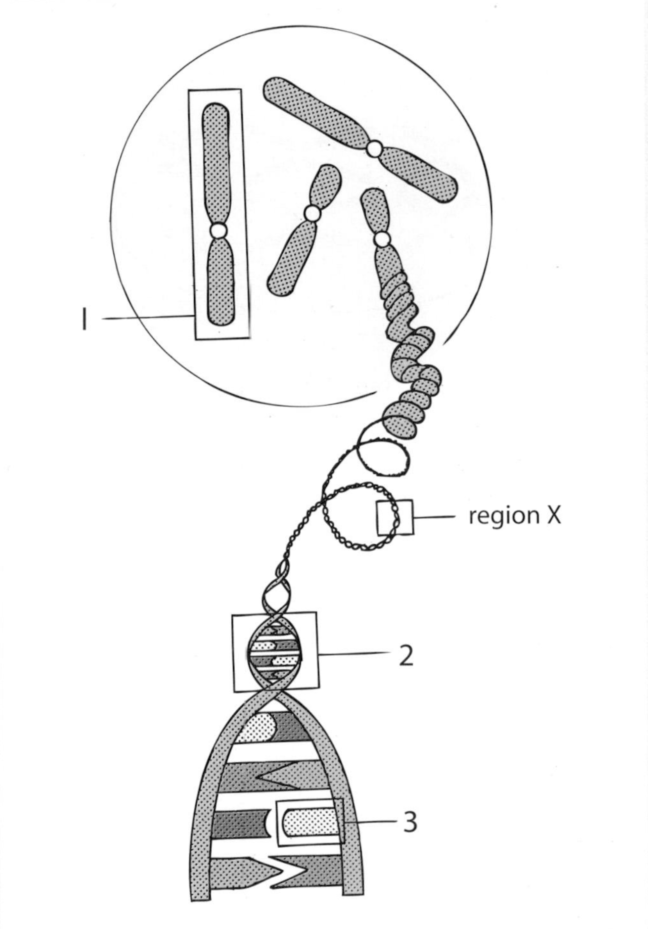

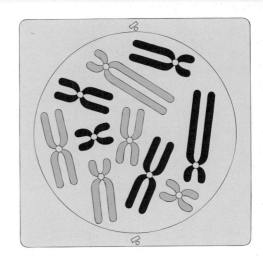

Figure 8.26

5 a) Match the contents of boxes 1, 2 and 3 in Figure
      8.26 with the following descriptions:
      (i)   a DNA base
      (ii) a chromosome
      (iii) a small portion of DNA's two strands held
            together.   (3)
   b) Suggest why region X in the diagram fails to
      represent one gene adequately.   (1)

6 The building-up of a molecule of protein whose
   structure is dictated by information held in a
   chromosome involves the stages listed below.
   Arrange them into the correct sequence.
   1) The messenger molecule passes out of the nucleus
      and into the cytoplasm of the cell.
   2) A region of DNA molecule uncoils and opens
      up.
   3) Amino acids are assembled into protein in a
      sequence determined by the order of the bases on
      the messenger.

4) The messenger molecule forms as a 'mirror image'
   of one of the DNA strands.   (1)

7 With reference to the cell shown in Figure 8.27,
   state the number of:
   (i)   chromosomes present
   (ii) chromatids present
   (iii) pairs of matching chromosomes present
   (iv) chromosomes that would be present in each
         gamete if this cell underwent gamete
         production.   (4)

Figure 8.27

8 Figure 8.28 represents the human life cycle.

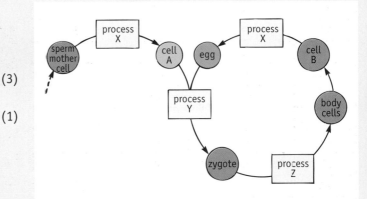

Figure 8.28

   a) Identify processes X, Y and Z.   (3)
   b) Identify cell types A and B.   (2)
   c) Name TWO types of cell that contain a single set
      of unmatched chromosomes.   (2)

d) Name THREE types of cell that contain a double
set of matching chromosomes. (2)

9 Figure 8.29 shows the formation of a zygote and
the start of its development into a multicellular
organism.

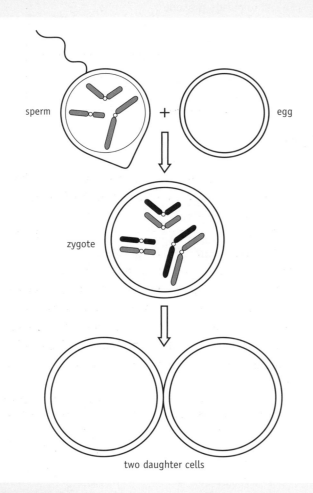

Figure 8.29

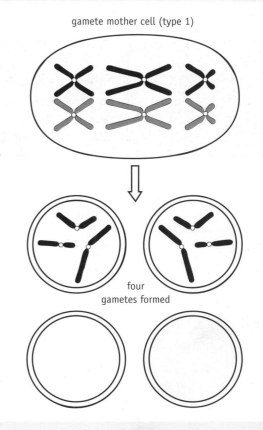

Figure 8.30

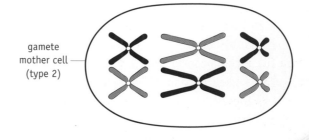

Figure 8.31

a) Draw a diagram of the egg nucleus to show its
complement of chromosomes. (2)

b) Draw a diagram of the nucleus of one of the
daughter cells to show its complement of
chromosomes. (2)

c) Figure 8.30 shows a gamete mother cell from the
same organism undergoing gamete production.
Draw a diagram to show the chromosome
complement of the two gametes that have been
left blank. (2)

d) Figure 8.31 shows another gamete mother cell
about to undergo gamete production. Draw the
four gametes that it would produce. (2)

10 Copy and complete Table 8.6.   (6)

chromosome number		number of different combinations of homologous chromosomes that can arise in gametes following random assortment
in body cell	in gamete	
4	2	$2^2 = 4$
	3	$2^3 = 8$
8		$2^4 =$
	5	$= 32$
12		$2^6 =$
		$2^{23} = 8\,388\,608$

Table 8.6

11 In human beings the chromosome complement of the body cells of males and females can be represented as shown in Figure 8.32. X and Y indicate the sex chromosomes and the number indicates the other chromosomes.

Figure 8.32

a) Using this format, draw simple diagrams to show:
   (i)   a male gamete mother cell
   (ii) the sperm formed during gamete production
   (iii) a female gamete mother cell
   (iv) the eggs formed during gamete production.   (4)

b) (i) Copy the Punnett square shown in Figure 8.33 and complete the sperm and egg boxes using your answers to a) parts (ii) and (iv).
   (ii) Complete the rest of the Punnett square to show the four zygotes that would be formed.   (4)

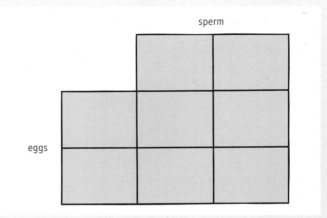

Figure 8.33

c) Which of the following correctly represents the ratio of the two sexes formed in the Punnett square?

   A 1:1   **B** 2:1   **C** 3:1   **D** 4:1
   (Choose ONE answer only)   (1)

12 The male and female grasshopper differ in chromosome complement. The male has only one X (sex) chromosome in his body cells whereas the female has two X chromosomes. Figure 8.34 shows how sex is determined in grasshoppers. Copy and complete the boxed part of the diagram.   (3)

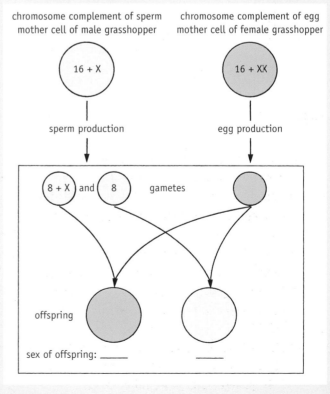

Figure 8.34

183

Word bank

amino	independently
anthers	ovaries
base	protein
chromosomes	random
continuous	separate
determined	sequence
different	single
discontinuous	sperm
DNA	testes
eggs	variation
female	Y
gamete	zygote

Table 8.7 *Word bank for Chapter 8*

What You Should Know

(Chapter 8) (See Table 8.7 for word bank)

1 Variation exists amongst the members of a species. When a characteristic can be used to divide the species into distinct groups, it is said to show _____ variation. When the characteristic varies in an uninterrupted way from one extreme to the other, it is said to show _____ variation.

2 In mammals, sperm are produced in _____ and _____ are produced in ovaries; in flowering plants, the _____ produce pollen which contain male gametes and the _____ produce ovules which contain the female gametes.

3 During fertilisation, the nuclei of two gametes fuse to form a _____. This process produces variation within the species since each zygote receives genetic information from two _____ parents following the random combination of two parental gametes.

4 Chromosomes contain deoxyribonucleic acid (_____) which consists of two strands each bearing a chain of _____ molecules. The _____ of bases present in a region of DNA represents a series of codewords. These pass on a message that determines the sequence of _____ acids in, and the structure of, a particular _____.

5 Each body cell of a multicellular organism has two matching sets of _____. During _____ formation (meiosis), matching chromosomes pair up and then _____ again. Each gamete formed receives one but not both members of a pair of chromosomes giving it a _____ set of chromosomes.

6 The separation of the members of a pair of chromosomes during meiosis occurs _____ of the separation of all other pairs. This _____ assortment of chromosomes leads to _____ in the gametes and in the offspring produced.

7 Each body cell of a human _____ contains two X chromosomes; each body cell of a human male contains one X and one _____ chromosome.

8 All human eggs receive an X chromosome; half of the sperm receive an X and half a Y chromosome. The sex of a human individual is _____ by the type of _____ that fertilises the egg.

9 Phenotype and genotype

Phenotype

Leaf shape in plants, coat colour in guinea pigs, wing type in fruit flies and grain colour in maize plants are a few examples of the many physical characteristics possessed by living organisms (see Figure 9.1). These physical characteristics make up the organism's **phenotype**. Such characteristics are determined by genetic information. The information is contained in the organism's **genotype** (see page 188) and it is passed on from one generation to the next during sexual reproduction.

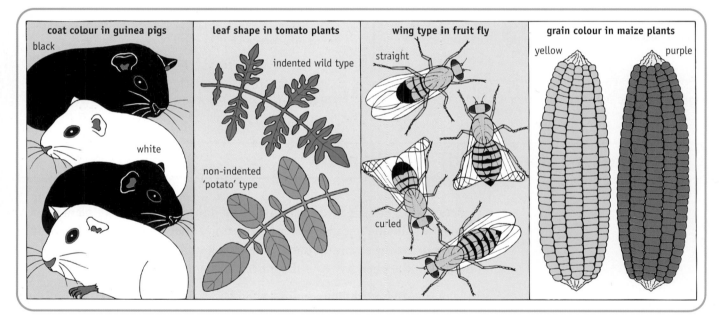

Figure 9.1 *Phenotypes*

True-breeding strains

Flower colour is a characteristic feature of pea plants that is controlled by genetic information. Lilac and white are two phenotypic expressions of flower colour in pea plants.

Figure 9.2 shows three generations of a strain of pea plants with lilac flowers; Figure 9.3 shows three generations of a strain of pea plants with white flowers. In each case the flower colour of the offspring is always identical to that of the parents and the members of the strain are described as being **true-breeding**.

Similarly, black guinea pigs that interbreed to form only black offspring and long-winged fruit flies that interbreed to produce only long-winged offspring are described as true-breeding strains.

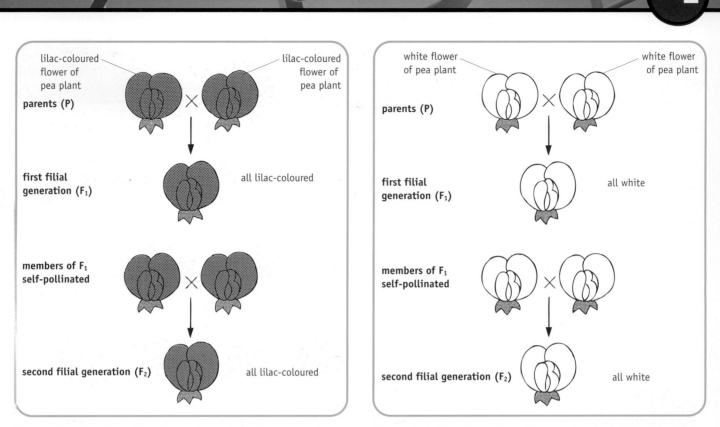

Figure 9.2 *True-breeding lilac-coloured strain of pea plants*

Figure 9.3 *True-breeding white strain of pea plants*

Single factor (monohybrid) inheritance

Geneticists begin their investigation of inheritance in a species of plant or animal by studying one characteristic at a time. They set up a cross between two true-breeding strains of the organism that differ from one another in only one way — the phenotypes of the inherited characteristic being studied.

In guinea pigs, for example, a cross could be set up between two true-breeding strains differing only in coat colour. In fruit flies a cross could be set up between two true-breeding strains differing only in wing type. In pea plants, the example that follows, a cross between two true-breeding strains differing only in flower colour is followed through two generations.

An experimental cross of this type that involves only one difference between the original parents is called a **monohybrid cross**.

Inheritance of flower colour in pea plants

Figure 9.4 summarises a monohybrid cross of flower colour in pea plants. The phenotypes of the members of the F_1 generation in a cross of this type are always found to be the same (i.e. uniform).

In this case they all bear lilac-coloured flowers. The white-flowered characteristic has disappeared in the F_1 because it has been masked by the lilac-flowered characteristic. Lilac flower colour is therefore said to be the **dominant** characteristic and white flower colour the **recessive** one.

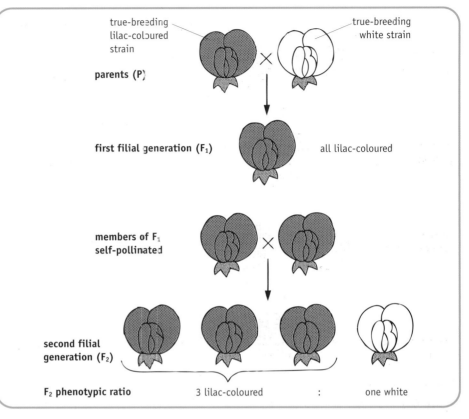

Figure 9.4 *Monohybrid cross in pea plants*

The F_2 generation does not contain any new 'in-between' forms of flower colour. Both lilac and white flower colour appear in the F_2 generation in their original form, unaffected by their union in the F_1 generation.

All monohybrid crosses of this type produce a 3:1 phenotypic ratio in the F_2 generation.

Testing Your Knowledge

The following crosses refer to a genetics experiment using pea plants.

| parents (a) | plant with round seeds | × | plant with wrinkled seeds |

| first filial generation (b) | | all round | |

| members of (c) self-pollinated | round | × | round |

| second filial generation (d) | | round and wrinkled | |

1 Before being used in this experiment, the parent plants were tested to ensure that they were true-breeding. How would this be done? (2)

2 Supply the symbols that would normally be used at positions **(a)**, **(b)**, **(c)** and **(d)** to denote the different generations. (2)

3 Which seed shape characteristic is recessive and which is dominant? Explain how you arrived at your answer. (2)

4 What can always be said about the phenotypes of the members of the F_1 generation resulting from a cross between two true-breeding parents? (1)

5 Predict the ratio of the two phenotypes that would occur in the F_2 generation. (1)

6 What term is used to refer to a cross of this type where the original parents differ from one another in one way only? (1)

Genes and genotype

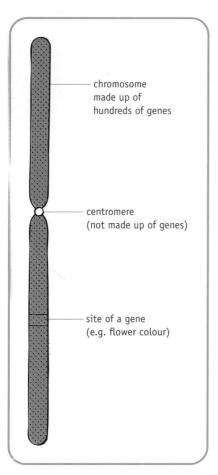

chromosome made up of hundreds of genes

centromere (not made up of genes)

site of a gene (e.g. flower colour)

Figure 9.5 *Site of a gene*

Chromosomes (see Chapter 8) are made up of smaller units called **genes**. Each gene is a unit of heredity that controls an inherited characteristic (e.g. wing type, hair texture, flower colour, leaf shape, etc.) Each gene occupies a specific site on a chromosome (see Figure 9.5). The complete set of genes possessed by an organism is called its **genotype**.

Alleles

At least two forms of each gene normally exist among the members of a species. For example, the gene controlling wing type in fruit flies may be the dominant form that produces straight wings or the recessive form that produces curled wings; the gene controlling flower colour in pea plants may be the dominant form that produces lilac-coloured flowers or the recessive form that produces white flowers. These different forms of a gene are called **alleles**.

Every normal body cell in an organism carries two matching sets of chromosomes, one originating from each parent. Thus every body cell has two alleles of each gene, one from each parent. These two alleles may be the same or different depending on which alleles an organism inherits from its parents.

Each gamete has only one set of chromosomes and therefore carries only one allele of each gene.

Symbols

For convenience the dominant and recessive alleles of a gene are often represented by symbols. In pea plants, for example, the two alleles of the gene for flower colour can be represented by the letters L, for the dominant allele (lilac), and l, for the recessive allele (white).

Since every **body cell** has **two** alleles of each gene, one from each parent, an organism's genotype is always represented by **two letters** per gene. A **gamete** only carries **one** allele of each gene and is always represented by **one letter** per gene.

Symbolic representation of pea plant cross

A pea plant's body cells each contain two matching sets of chromosomes as seven pairs. Each gamete contains a single set of seven chromosomes. The cross in Figure 9.4 can be represented at a chromosomal and gene level as shown in Figure 9.6. For the sake of simplicity only the chromosomes carrying the alleles of the gene for flower colour have been drawn.

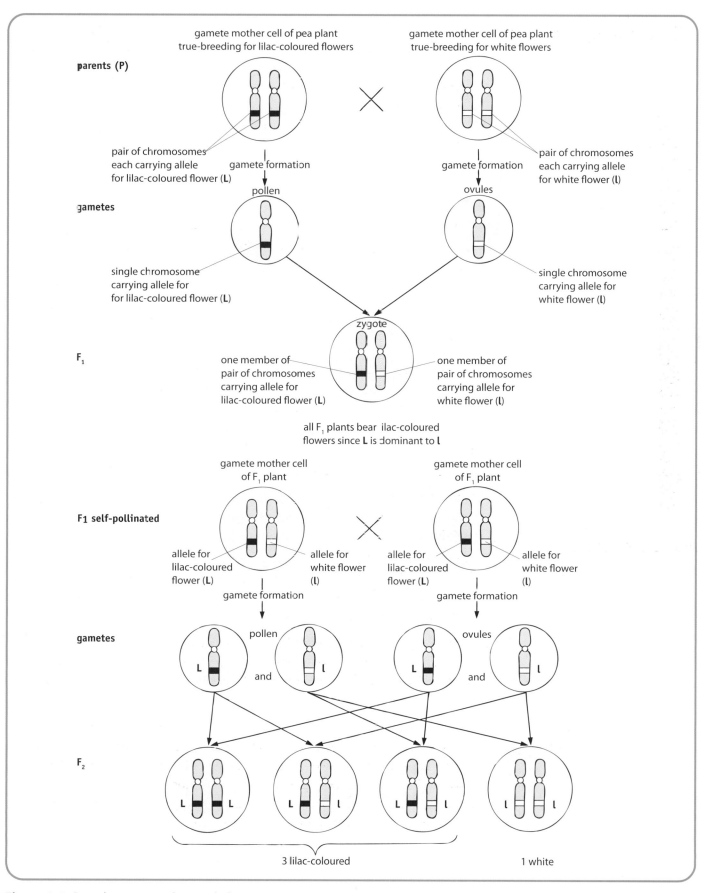

Figure 9.6 *Pea plant cross using symbols*

Normally the representation of a cross is further simplified using letter symbols as follows:

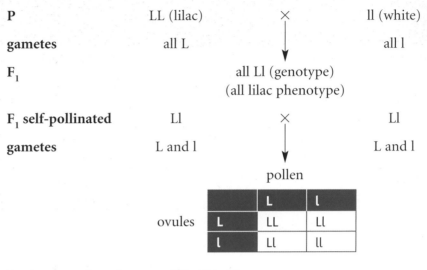

| | P | LL (lilac) | × | ll (white) |

P   LL (lilac)   ×   ll (white)

gametes   all L   all l

F_1 all Ll (genotype)
                     (all lilac phenotype)

F_1 self-pollinated   Ll   ×   Ll

gametes   L and l   L and l

pollen

ovules		**L**	**l**
L		LL	Ll
l		Ll	ll

F_2 genotypic ratio   =   **1**LL: **2**Ll: **1**ll
     phenotypic ratio   =   **3** lilac: **1** white

No genotypic or phenotypic ratio occurs in the F_1 generation in this type of cross because all members of the F_1 generation share the same genotype and phenotype.

Homozygous and heterozygous

When an organism possesses two identical alleles of a gene (e.g. L and L or l and l) its genotype is said to be **homozygous** and it is true-breeding. When an organism has two different alleles of a gene (e.g. L and l) its genotype is said to be **heterozygous** and it is not true-breeding.

Same phenotype, different genotype

In the above example, if a pea plant's phenotype is white flower then its genotype must be ll. However, if a pea plant's phenotype is lilac flower, its genotype may be LL or Ll. In other words the two **different genotypes** LL and Ll have the **same phenotype** (lilac).

Number of offspring — observed versus predicted figures

Monohybrid crosses of the type shown in Figure 9.4 always produce a 3:1 phenotypic ratio in the F_2 generation. However, there is often a difference between the **observed** and the **predicted** numbers of the different types of offspring, as shown in Table 9.1 which refers to pea plants.

In each of these monohybrid crosses, the F_2 offspring do not show the predicted ratio of 3:1 exactly although each ratio is very close to it. Continuing with flower colour as our example, an exact 3:1 phenotypic ratio would have occurred in the F_2 generation if during the Ll × Ll cross (see above) exactly half of the L ovules had been fertilised by L pollen grains and the other half of the L ovules by l pollen, while at the same time exactly half of the l ovules had been fertilised by L pollen and the other half by l pollen.

pea plant gene being studied in monohybrid cross	phenotypes of original true-breeding parents	total number of offspring in F_2	predicted F_2 numbers based on exact 3:1 ratio	observed F_2 numbers	actual F_2 phenotypic ratio
colour of flower	lilac and white	928	696 purple and 232 white	704 purple and 224 white	3.14:1
height of plant	tall and dwarf	1064	798 tall and 266 dwarf	787 tall and 277 dwarf	2.84:1
shape of seed	round and wrinkled	7324	5493 round and 1831 wrinkled	5474 round and 1850 wrinkled	2.96:1
colour of seed	yellow and green	8024	6018 yellow and 2006 green	6022 yellow and 2002 green	3.01:1

Table 9.1 *Predicted versus observed numbers*

However, this rarely happens in nature because fertilisation is a **random** process involving the element of **chance**. This principle can be illustrated by carrying out the following activity.

Practical Activity

Using a bead model to illustrate a monohybrid cross

Information

The model represents **self-pollination** of the F_1 generation of pea plants in the cross shown in Figure 9.6.

You need

2 beakers each containing 100 lilac and 100 white beads
1 beaker marked LL
1 beaker marked Ll
1 beaker marked ll

What to do

1 Follow the procedure shown in Figure 9.7.
2 Continue to form and classify 'zygotes' until no 'gamete' beads are left.
3 Count the number of LL, Ll and ll zygotes, take a note of your results and then express them as an F_2 genotypic ratio.
4 Express your results as an F_2 phenotypic ratio as follows lilac:white = _:1
5 On most occasions the results obtained from this activity are found to give a ratio that is close to, but not exactly, 3:1. Explain why.
6 If other students have carried out the same activity, pool your results and then repeat step 4 above.
7 You should find that the greater the number of zygotes included in your results, the closer the totals come to showing an exact 3:1 phenotypic ratio. Explain why.
8 Explain why your eyes should be kept closed during the bead selection part of the activity.

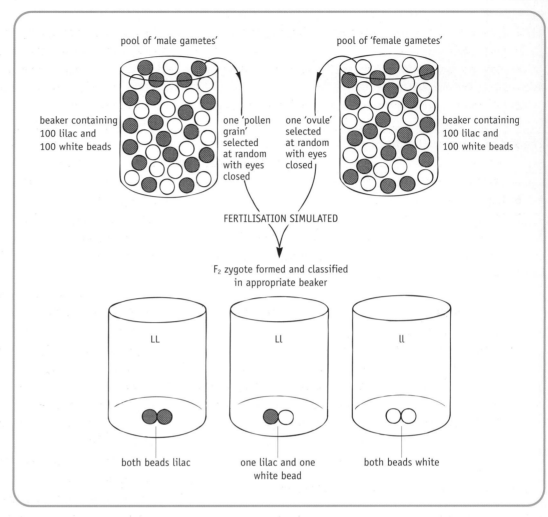

pool of 'male gametes'

pool of 'female gametes'

beaker containing
100 lilac and
100 white beads

one 'pollen
grain'
selected
at random
with eyes
closed

one 'ovule'
selected
at random
with eyes
closed

beaker containing
100 lilac and
100 white beads

FERTILISATION SIMULATED

F₂ zygote formed and classified
in appropriate beaker

LL

Ll

ll

both beads lilac

one lilac and one
white bead

both beads white

Figure 9.7 *Bead model*

Practical Activity and Report

Proportions of phenotypes of F₁ and F₂ offspring in tobacco plants

Information

- In tobacco plants, the seed leaves (cotyledons) emerge from the germinating seeds as the first two leaves of the plant.

- The tobacco seeds that you are going to plant belong to the F₁ and F₂ generations resulting from an original cross between two true-breeding strains of tobacco plant. In one parental strain (wild type) the seedlings always produced **green** seed leaves; in the other strain (albino) the seedlings always produced **white** seed leaves. The F₁ was self-pollinated to give the F₂.

- Tobacco seeds are tiny and their dispersal during sowing is aided by first mixing them with an equal mass of fine sand.

You need

10 mg tobacco seed marked F₁ (approximately 100 seeds)

10 mg tobacco seed marked F₂ (approximately 100 seeds)

2 envelopes each containing 10 mg fine sand

2 Petri dishes each containing sterile agar jelly

1 marker pen

1 pair of fine forceps

access to a seed propagator at 24°C (which allows entry of light)

What to do

1 Read all the instructions in this section and prepare your table of results before carrying out the experiment.

2 Mark the base of one of the Petri dishes F_1 and add your initials.

3 Mark the base of the other Petri dish F_2 and add your initials.

4 Mix the F_1 seeds with the grains of sand by shaking them together in the envelope.

5 Gently sprinkle the contents of the envelope as evenly as possible over the surface of the agar in the Petri dish marked F_1.

6 Repeat steps 4 and 5 for the F_2 seeds.

7 Place the Petri dishes in the seed propagator.

8 Inspect the dishes regularly and remove them from the propagator when the seeds have germinated and each seedling bears two seed leaves (at 7 days approximately).

9 Count the number of green and the number of white seedlings produced in the F_1 and F_2 generations by using the forceps to pick out each plant. Keep a score as you go along and record your results in your table which should refer to the numbers of green and white offspring present in the F_1 and F_2 generations.

10 If other students have carried out the same experiment, pool the results.

Reporting

Write up your report by doing the following:

1 Copy the title given at the start of this activity.

2 Put the subheading 'Aim' and state the aim of the experiment.

3 a) Put the subheading 'Method'.

b) Using the impersonal passive voice, briefly describe the experimental procedure that you followed and state how you obtained your results.

4 Put the subheading 'Results' and draw a final version of your table.

5 Put a heading 'Analysis of Results'. Extend your table to include a column giving your results expressed as a ratio and then answer the following questions using information extracted from your table:

a) Which gene in tobacco plants was investigated in this experiment?

b) What were the two alleles of this gene?

c) Which allele was dominant and which one was recessive? Explain how you arrived at your answer.

6 Put a subheading 'Conclusions' and answer the following questions:

a) What percentage of the F_1 generation were (i) green; (ii) white?

b) To which whole number ratio are your F_2 results closest?

c) Give the two alleles appropriate letter symbols and present the full cross through to the F_2 in diagrammatic form beginning with the true-breeding parents and including a Punnett square in your answer.

7 Put a final subheading 'Evaluation of Experimental Procedure' and then answer the following:

a) Why is even dispersal of seeds with the aid of sand a valuable piece of experimental procedure?

b) Why are the results for as many as 100 (or more) seedlings used when attempting to calculate the F_1 and F_2 proportions?

c) Why is it better to pool the class results than to depend on one group's set of results when calculating the phenotypic ratios?

d) Why is an exact 3:1 phenotypic ratio rarely obtained?

e) How could the reliability of the results be checked?

f) Feel free to comment on any of the following if you have an additional point that you wish to make:

(i) possible sources of error

(ii) further improvements that you would include in a repeat of the experiment

(iii) limitations of the equipment.

Testing Your Knowledge

1 a) What name is given to the basic unit of inheritance that controls a characteristic that is passed on from generation to generation? (1)

b) Of what structures found in a cell's nucleus do such units form a part? (1)

2 State the meaning of the terms genotype and phenotype. (2)

3 a) What term is used to refer to the different forms of a gene? (1)

b) How many forms of a gene does a zygote receive from each parent? (1)

4 Explain the difference between the terms homozygous genotype and heterozygous genotype with reference to a named example. (2)

5 a) Return to Table 9.1 and choose letter symbols to represent the alleles of the gene for seed colour. (1)

b) Using your symbols, present this monohybrid cross through to the F_2 generation. (4)

c) Identify TWO members of the F_2 generation that have the same phenotype but different genotypes. (1)

d) Explain why the actual observed F_2 results differ slightly from the predicted figures. (2)

Human inheritance

Unlike pea plants and fruit flies, human beings do not breed to suit the geneticist. In addition, they produce too few offspring to allow reliable conclusions to be drawn about the phenotypic ratios produced. Nevertheless the laws of genetics still apply to humans and particular traits can be traced through several generations of a family by constructing a **family tree**.

Hair colour

In humans the allele for red hair colour (h) is recessive to the dominant allele for non-red hair (H). If a person such as Sandy in Figure 9.8 has red hair yet neither of his parents has red hair then it can be concluded that Sandy must have the **homozygous** genotype (hh) and that both of his parents must have the **heterozygous** genotype (Hh).

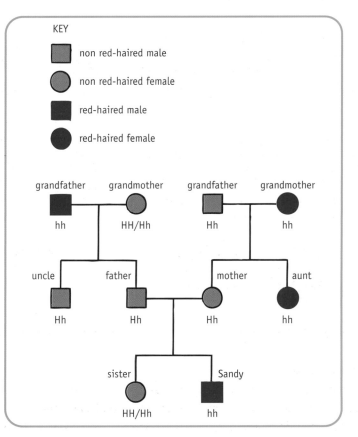

Figure 9.8 *Family tree for hair colour*

Since Sandy's aunt and maternal grandmother are red-haired they must be hh and his maternal grandfather must be Hh and so on. Piecing together such information for several generations of a family enables a geneticist to construct a family tree as shown in Figure 9.8.

Element of chance
Each time Sandy's maternal grandparents produced a child, it was a 1 in 2 chance that s/he would be non red-haired (Hh) and a 1 in 2 chance that s/he would be red-haired (hh).

Every time Sandy's parents (Hh and Hh) produce a child, it is a 1 in 4 chance that s/he will be non red-haired (HH), a 1 in 2 chance that s/he will be non red-haired (Hh) and a 1 in 4 chance that s/he will be red-haired (hh).

Activity

Selecting and presenting information showing that characteristics are inherited from both parents

Information
- In humans, the allele for tongue-rolling ability (R) is dominant to that for inability to roll the tongue (r).
- In humans, the allele for wavy hair (H) is dominant to the allele for straight hair (h).

What to do
1 Study the family trees shown in Figures 9.9 and 9.10.

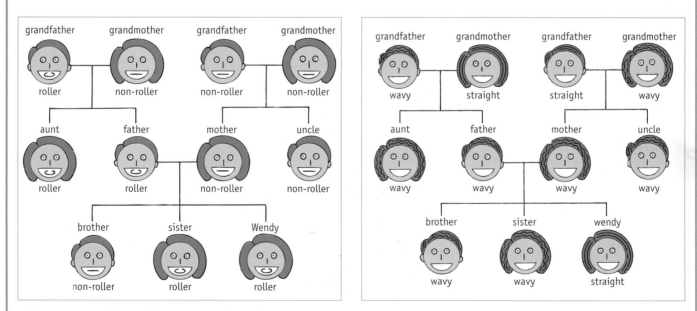

Figure 9.9 *Family tree for tongue-rolling ability*   **Figure 9.10** *Family tree for hair type*

2 Answer the following questions.
   a) From which parent did Wendy inherit her tongue-rolling ability?
   b) From whom did Wendy receive the genetic information that gave her straight hair?
   c) Using the format shown in Figure 9.8, copy each family tree and write in the genotypes.

Co-dominance

In the examples of inheritance considered so far in this chapter, each allele of the gene controlling a characteristic is either completely dominant or recessive to the other allele. As a result the heterozygous genotype (e.g. Pp, Rr, Hh) has the same phenotype as the dominant homozygous genotype (e.g. PP, RR, HH).

However, there are exceptions to this rule. Sometimes two alleles of a gene are found to be **co-dominant**. This means that neither allele is dominant to the other and both alleles are expressed in the phenotype of an organism with the heterozygous genotype.

Blood groups

genotype	blood group phenotype
AA	A
AO	A
BB	B
BO	B
AB	AB
OO	O

Table 9.2 *Blood groups*

Blood group is determined by three alleles A, B and O. A and B are **co-dominant** to one another and both are **completely dominant** to O, the recessive allele. Taking the alleles two at a time, there are six ways in which they can be combined to give genotypes. The six different genotypes are expressed as four different phenotypes (see Table 9.2).

A person with blood group A may have genotype AA or AO (since allele A is completely dominant to allele O). A person with blood group B may have genotype BB or BO (since allele B is completely dominant to allele O).

However, a person with blood group AB must have genotype AB since both alleles are co-dominant and fully expressed in the phenotype. A person with blood group O must have the genotype OO since neither allele A nor B is present to be expressed in their phenotype.

Polygenic inheritance

Some characteristics are controlled by the alleles of a single gene. They are expressed as clear-cut phenotypic groups showing discontinuous variation. In humans the ability or inability to roll the tongue and in pea plants the possession of lilac or white flowers are two examples of such **single gene inheritance**.

Other characteristics are controlled by the alleles of several genes. These alleles interact with one another and each may or may not add its contribution to the overall phenotypic expression of the characteristic. This results in the characteristic being expressed among the members of the species as a **range of phenotypes**. An example is human height (see Figure 9.11). It varies from very small at one extreme to very tall at the other extreme without falling into distinct groups. A characteristic that is controlled in this way by more than one gene is said to show **polygenic inheritance**.

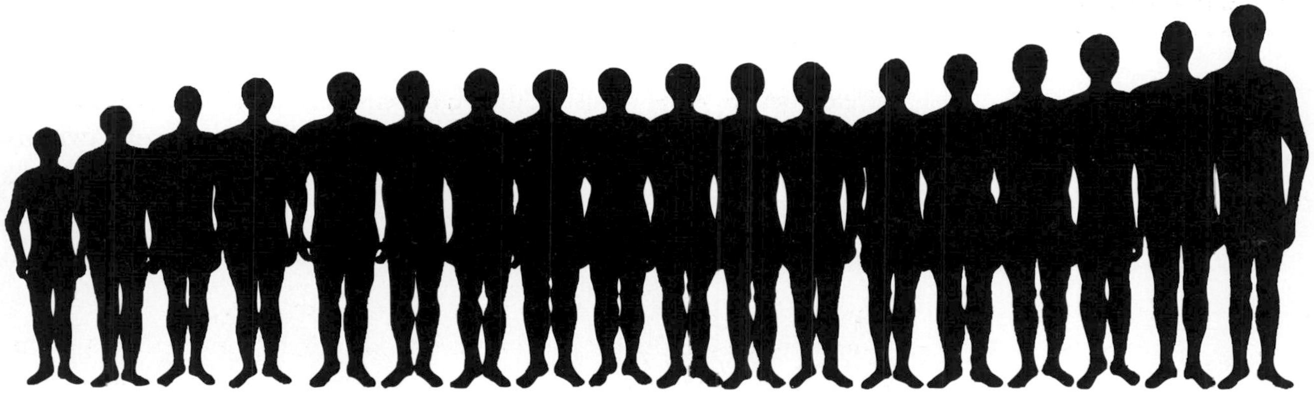

Figure 9.11 *Polygenic inheritance of height in humans*

Skin colour in humans is a further example of polygenic inheritance. Melanin is the dark pigment that gives skin its colour. Each of the genes involved has a dominant allele which promotes melanin formation and a recessive allele unable to promote melanin formation. Therefore within the human species a wide range of skin colour exists depending upon which combination of dominant and recessive alleles the person happens to have inherited.

In plants, seed length and mass are examples of polygenic inheritance giving a range of phenotypes which vary continuously as shown in Figure 8.4 (on page 161) for broad bean.

Environmental impact on phenotype

An organism's final appearance (**phenotype**) is the result of the interaction between the information held in its **genotype** and the effect of the **environment** acting on it during growth and development as in the equation:

$$genotype + environment \rightarrow phenotype$$

The phenotypic expression of some characteristics remains unaffected by environmental factors. Tongue-rolling ability and blood group in humans, for example, are determined solely by genotype.

The phenotype of other characteristics is influenced in part by environmental factors. If, for example, a person inherits the genetic information to become tall but consumes a poor diet during childhood then s/he will not reach full potential height.

Same genotype, different phenotype

In order to investigate the impact of environmental factors on phenotypic expression, scientists study organisms that possess the same genotype.

Examples include **clones** of plants produced by asexual reproduction and identical human **twins**, each derived from the same zygote.

Since such individuals are genetically identical, any phenotypic differences that develop between them must be due to the effect of environmental factors acting on them during growth. Figure 9.12 shows the outcome of a **twin study** in humans.

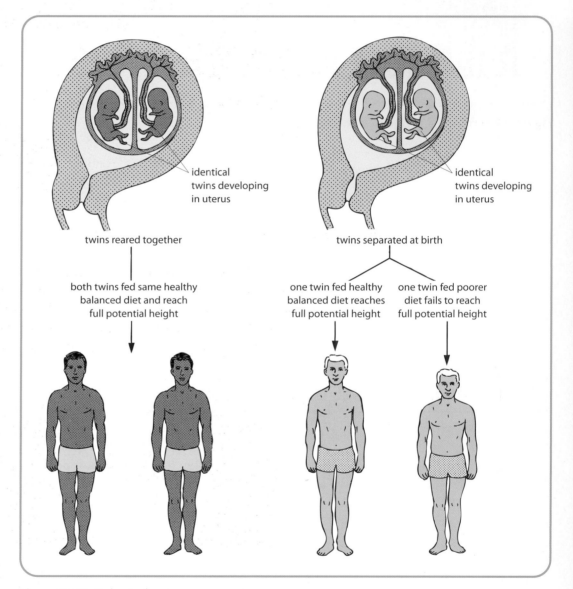

identical twins developing in uterus

identical twins developing in uterus

twins reared together

twins separated at birth

both twins fed same healthy balanced diet and reach full potential height

one twin fed healthy balanced diet reaches full potential height

one twin fed poorer diet fails to reach full potential height

Figure 9.12 *Twin study*

The arrowhead plant in Figure 9.13 reproduces asexually by means of bud-like 'plantlets'. The diagram shows the results of an investigation into the effect of environmental factors on the phenotypes of two arrowhead plants with the same genotype. Considerable phenotypic variation between them is found to occur.

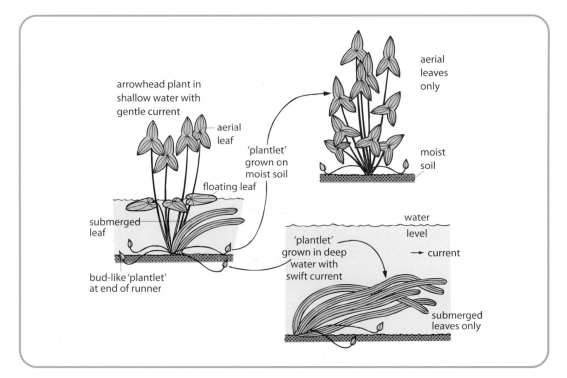

Figure 9.13 *Same genotype, different phenotypes*

Figure 9.14 shows the outcome of the next part of the investigation. When plantlets from these phenotypically different plants are grown in the same environmental conditions, they develop into plants with the same phenotype! This shows that the differences that existed between them due to the effect of environmental factors have **not** been passed on to the new offspring.

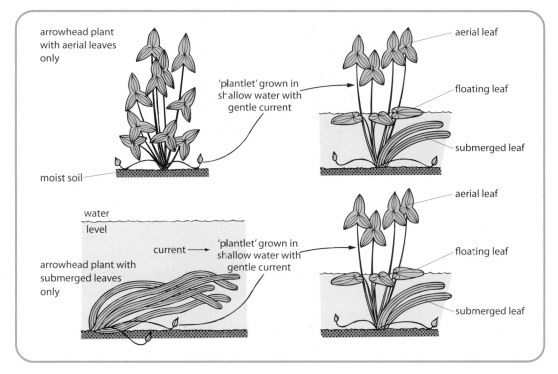

Figure 9.14 *Same genotype, same phenotype*

Non-transmission of acquired characteristics

Similarly, in other living things differences in phenotype caused by environmental factors ('**acquired characteristics**') are not transmitted from one generation to the next. A woman with a deep suntan maintained by a sunray lamp does not pass this on to her children; a bodybuilder does not father children with better developed biceps muscles than those of a non-bodybuilder; dogs whose tails have been docked do not produce tail-less puppies.

Since these differences are due to environmental factors and not to information held in the organism's genotype, they have little or no significance in the evolution of the species. They are not passed from generation to generation and play no part in natural selection.

Natural selection

Every species of living organism has an enormous **reproductive potential** (see Table 9.3). Its members produce far more offspring than the environment can support and this leads to a **struggle for survival**. A simplified example involving rabbits is shown in Figure 9.15.

animal species	average number of offspring per year
fox	5
red grouse	8
rabbit	24
mouse	30
trout	800
cod	4 000 000
oyster	16 000 000

Table 9.3 *Reproductive potential*

Many offspring die before reaching reproductive age as a result of factors such as competition, lack of food, overcrowding, inability to escape predators or lack of resistance to disease.

The members of a species are not identical to one another but show phenotypic **variation** in all characteristics. Much of this variation is inherited. It is this inherited variation that is of significance in the following sequence of events.

Survival of the fittest

Those offspring whose phenotypes are better adapted to their environment have a better chance of surviving, reaching reproductive age and passing on

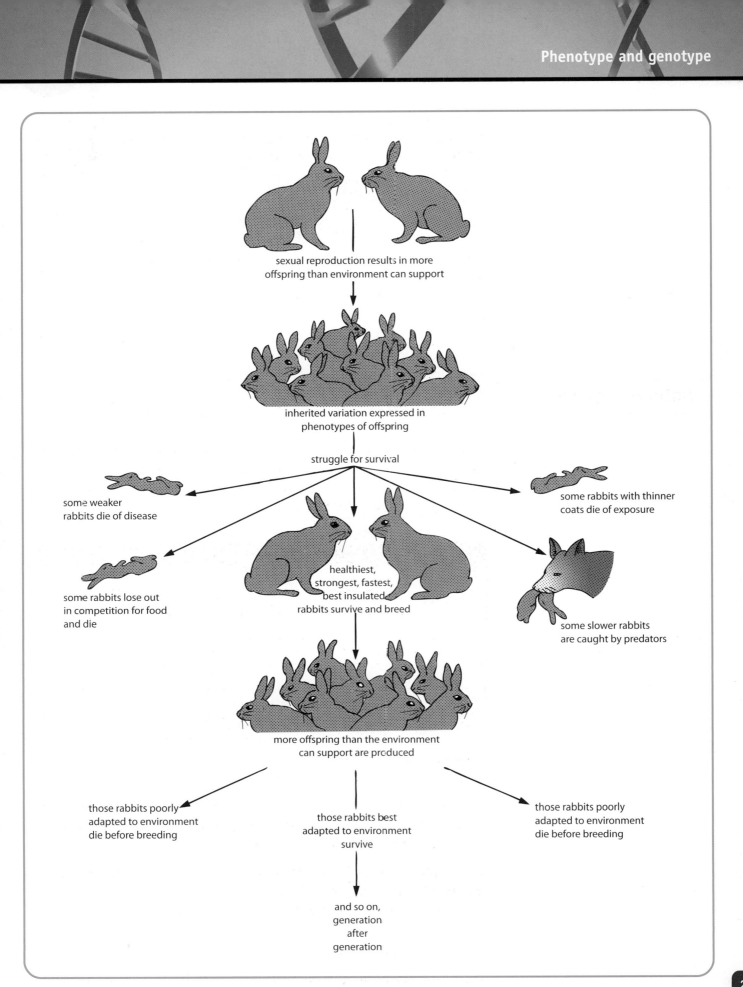

Figure 9.15 *Natural selection in rabbits*

the genes for the favourable characteristics to their offspring. Those offspring whose phenotypes are less well suited to their environment die before reaching reproductive age and fail to pass on their genes.

This process is repeated generation after generation and the organisms with the phenotypes best suited to the environment (i.e. the fittest) are 'selected' and predominate in the population. The poorer members are weeded out and perish. This process is known as **natural selection**. It was first described by Charles Darwin in 1858 as the main factor producing evolutionary change in species.

Natural selection in action

Peppered moth

One form of the peppered moth is light brown with dark speckles; the other is completely dark in colour. They differ by only one allele of the gene controlling the formation of **dark pigment** (melanin). Both forms of the moth fly by night and rest on the bark of trees during the day. Prior to the industrial revolution in the 1800s, the light form of peppered moth was common throughout Britain and the darker form which occasionally arose by mutation, was very rare indeed.

Surveys in the 1950s (see Figure 9.16) showed that the pale form was most abundant in non-industrial areas whereas the dark form was abundant in areas suffering from heavy industrial air pollution. Experiments and direct observations strongly support the following explanation of these findings.

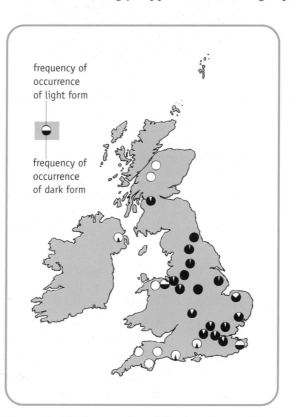

frequency of occurrence of light form

frequency of occurrence of dark form

Figure 9.16 *Frequencies of the two forms of peppered moth*

In non polluted areas, the tree trunks are covered with pale-coloured lichens and the light-coloured moth is well **camouflaged** against this pale background (see Figure 9.17). However, the darker form is easily seen and eaten by predators such as thrushes.

In polluted areas, toxic gases kill the lichens and soot particles darken the tree trunks. As a result the light-coloured moth is easily seen whereas the dark one is well hidden and favoured by **natural selection**.

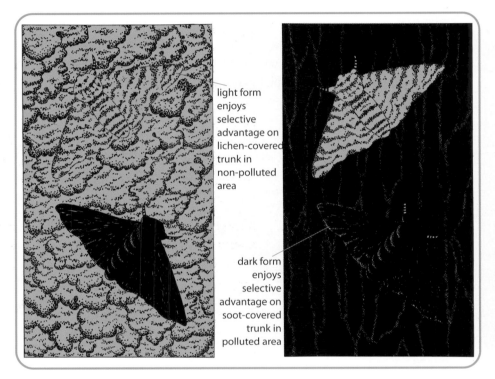

light form enjoys selective advantage on lichen-covered trunk in non-polluted area

dark form enjoys selective advantage on soot-covered trunk in polluted area

Figure 9.17 *Natural selection in peppered moth*

Nowadays, in polluted areas currently undergoing cleaning up campaigns in response to the Clean Air Act, the pale form is being naturally selected at the expense of the darker form which is losing its selective advantage. As the environment continues to change and more pale moths survive and breed, the situation that existed before the industrial revolution may return.

Evolution

Most scientists are of the opinion that the Earth's present biodiversity is the result of millions and millions of years of **evolution** involving natural selection. If humankind allows this process to continue normally, still more varied and complex forms of life will evolve in the future. However, this depends on humans reducing or bringing to a halt those activities that lead to habitat destruction (see Chapter 7).

Testing Your Knowledge

1 a) Give the simple equation that summarises the factors that determine an organism's phenotype. (1)

b) What information do scientists hope to obtain when they study identical twins who have been raised in different environments? (1)

2 Figure 9.18 shows an experiment using *Bryophyllum* plants.

a) Are the phenotypic differences between plants A and B due to genotype or environment? (1)

b) Would these differences be passed on to succeeding generations? Explain your answer. (2)

3 a) Table 9.3 shows that an oyster has the potential to produce 16 million offspring annually. Give THREE reasons why the seas are not over-populated with oysters. (3)

b) In the struggle for survival which oysters will

(i) survive and pass on their genes to succeeding generations?

(ii) die before reaching reproductive age? (2)

c) What name is given to this 'weeding out' process that promotes survival of the fittest? (1)

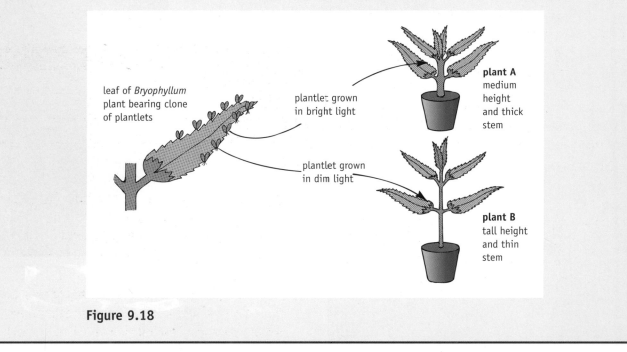

leaf of *Bryophyllum* plant bearing clone of plantlets

plantlet grown in bright light

plantlet grown in dim light

plant A medium height and thick stem

plant B tall height and thin stem

Figure 9.18

Applying Your Knowledge

1 Match the terms in list X with their descriptions in list Y.

List X
1) alleles
2) co-dominant
3) dominant
4) gene
5) genotype
6) heterozygous

List Y
a) basic unit of inheritance, many of which make up a chromosome
b) describes a genotype that contains two different alleles of a particular gene
c) describes a genotype that contains two identical alleles of a particular gene
d) describes the member of a pair of alleles that always shows its effect and masks the presence of the recessive allele
e) describes the member of a pair of alleles that is always masked by the dominant allele
f) describes a pair of alleles where neither is dominant and both are expressed in the phenotype

7) homozygous

8) monohybrid

9) natural selection

10) phenotype

11) polygenic

12) recessive

13) twin study

g) the physical characteristics of an organism

h) the set of genes possessed by an organism

i) close observation of identical twins to investigate the effect of the environment on phenotype

j) describes a cross between two true-breeding parents that differ from one another in one way

k) alternative forms of a gene that are responsible for different expressions of an inherited characteristic

l) process by which individuals best adapted to the environment survive and pass their genes on to succeeding generations

m) type of inheritance where a characteristic is controlled by several genes

2 In guinea pigs, black coat colour (B) is dominant to white coat colour (b).

a) In diagrammatic form, follow to the F_2 generation a cross between a white male and a true-breeding black female where the members of the F_1 generation are interbred. (4)

b) State TWO genotypes in the F_2 generation that have the same phenotype. (1)

3 In pea plants the gene for pod type has two alleles, inflated and constricted, as shown in the cross in Figure 9.19.

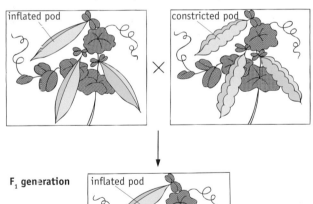

parents (both true-breeding)

inflated pod   constricted pod

×

F_1 generation   inflated pod

Figure 9.19

a) Which allele is dominant? Explain your answer. (2)

b) Using letters of your own choice, give the genotype of the F_1 generation. (1)

c) A further generation of pea plants was produced by allowing the F_1 generation to self-pollinate. Using your chosen symbols, copy and complete Table 9.4 to show the outcome of this cross. (4)

		genotypes of pollen	
genotypes of ovules			

Table 9.4

d) The F_2 generation consisted of 1180 plants. In theory how many would be

(i) homozygous for inflated pod?

(ii) heterozygous for inflated pod?

(iii) constricted pod? (3)

4 In tomato plants, fruit colour may be red or yellow. Table 9.5 give details of crosses between certain tomato plants and the offspring produced in each case.

cross	parents	number and phenotype(s) of offspring
1	plant A (red) × plant B (red)	258 red
2	plant B (red) × plant C (red)	197 red and 65 yellow
3	plant D (red) × plant E (yellow)	128 red and 134 yellow
4	plant E (yellow) × plant F (yellow)	261 yellow

Table 9.5

Using letters of your choice, identify the genotypes of tomato plants A, B, C, D, E, and F. (6)

5 In fruit flies, straight wing (S) is dominant to curled wing (s). Figure 9.20 shows the results of an investigation carried out to examine the phenotypes arising from a monohybrid cross involving the wing type gene.

	straight wing	curled wing
parents of F_1	6 true-breeding males	6 true-breeding females
F_1	168 flies of both sexes	0
parents of F_2	6 males from F1	6 true-breeding females
F_2	81 flies of both sexes	87 flies of both sexes

Figure 9.20

a) Present the information in the standard diagrammatic form. (4)
b) Explain why no curled-winged flies were produced in the F_1 generation. (1)
c) (i) What is the expected ratio of straight to curled in the F_2 generation?
   (ii) What is the expected number of straight to curled in the F_2 generation?
   (iii) Why do the actual results vary slightly from the expected ones? (4)

6 In humans, a patch of white hair at the front of the head is called white forelock. Possession of white forelock (F) is dominant to lack of white forelock (f). Read the details that follow about a family.
Mandy's mother and maternal grandparents are all homozygous for white forelock. Although Mandy's father lacks white forelock, both of his parents have white forelock.

a) Draw a diagram to show Mandy's family tree and include a key to explain the meanings of the symbols you have chosen. (4)
b) Add the genotypes to the family tree. (4)
c) What is Mandy's phenotype with respect to this inherited characteristic? (1)

7 Cystic fibrosis is an inherited disorder of the human body. The sufferer (1 in 2500 people in the UK) produces abnormally thick mucus that blocks the tiny air passages of the lungs. The condition is due to a recessive allele that is carried by 1 in every 25

people. Figure 9.21 shows a family tree involving this inherited disorder.

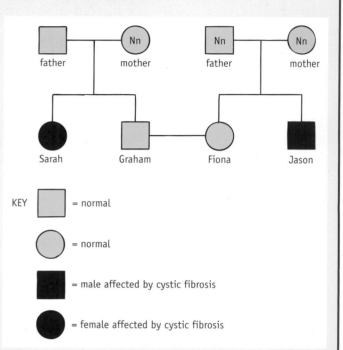

KEY

◻ = normal

◯ = normal

◼ = male affected by cystic fibrosis

● = female affected by cystic fibrosis

Figure 9.21

a) State the genotype of:
   (i) Jason
   (ii) Sarah
   (iii) the father of Sarah and Graham. (3)
b) State the possible genotypes of:
   (i) Graham
   (ii) Fiona. (2)
c) If Graham and Fiona both have homozygous genotypes, what is the chance of each of their children being a sufferer of cystic fibrosis? (1)
d) If Graham and Fiona both have heterozygous genotypes, what is the chance of each of their children being a sufferer of cystic fibrosis? (1)

8 A man with blood group A whose father was blood group O marries a woman with blood group AB.
a) Draw a diagram to show all the possible genotypes that could occur among their children. (3)
b) Which phenotype could not occur among their offspring? (1)

9 In the family tree shown in Figure 9.20 a circle represents a woman and a square a man. The upper half of each circle or square contains the person's blood group phenotype and the lower half the person's blood group genotype. Some letters have been omitted. Copy and complete Figure 9.22. (5)

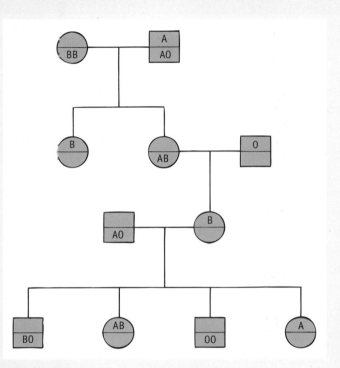

Figure 9.22

10 When grown in soil rich in iron, *Hydrangea* plants develop blue flowers; when grown in soil deficient in iron, they develop pink flowers.
  a) Predict the effect on flowering of transferring cuttings from pink-flowered plants to (i) iron-rich soil and (ii) iron-deficient soil. (2)
  b) Which ONE of the following correctly applies to the plants that develop from the cuttings grown in the two soil types referred to in part a)?
  A They have the same genotype and the same phenotype.
  B They have the same genotype but different phenotypes.
  C They have different genotypes but the same phenotype.
  D They have different genotypes and different phenotypes. (1)
11 On average, a pair of rabbits produce a litter of six offspring four times a year. Since the offspring mature quickly, millions of rabbits could be produced within a few generations.

Using the terms *over-production, competition, natural selection* and *variation* in your answer, explain why the world is not over-populated by rabbits. (4)

12 Figure 9.23 shows the results of growing cuttings from three genetically different varieties of a species of plant in each of three locations which varied in height above sea level. The numbers in the nine boxes refer to growth score (0 = least growth, 5 = most growth).
  a) In terms of growth score, which variety of plant showed (i) the most overall success? (ii) the least overall success? (2)
  b) Which environment in general was (i) least favourable for plant growth? (ii) most favourable for plant growth? (2)
  c) In which direction must the table be read (horizontally or vertically) to find out if phenotype varies as a result of differences in genotype? (1)
  d) In which direction must the table be read (horizontally or vertically) to find out if phenotype varies due to environmental factors? (1)
  e) What determines a plant's phenotype in this investigation? (1)

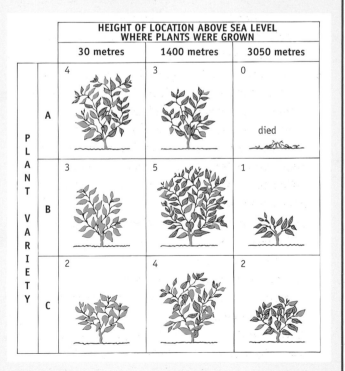

Figure 9.23

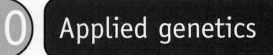

10 Applied genetics

Selective breeding

For thousands of years, people have cultivated plants and domesticated animals for the benefit of humankind. Almost all of the species of plants and animals that we eat today have been derived from their wild ancestors by **selective breeding**.

Variation exists among the members of a species. Selective breeding is the process by which the breeder attempts to **improve** the stock of a useful plant or animal species by selecting those individuals that have a version of a certain characteristic closest to the one required by humans. These animals or plants are used to breed the next generation. Other individuals lacking the desirable characteristic are prevented from breeding. When this process is repeated over many generations, it leads to the **enhancement** of the desired characteristic.

Enhancement of animal characteristics

Many characteristics of domesticated animals have been enhanced through selective breeding. Dairy cattle, for example, have been selected over many generations for milk yield and the butterfat content of milk (see Figure 10.1).

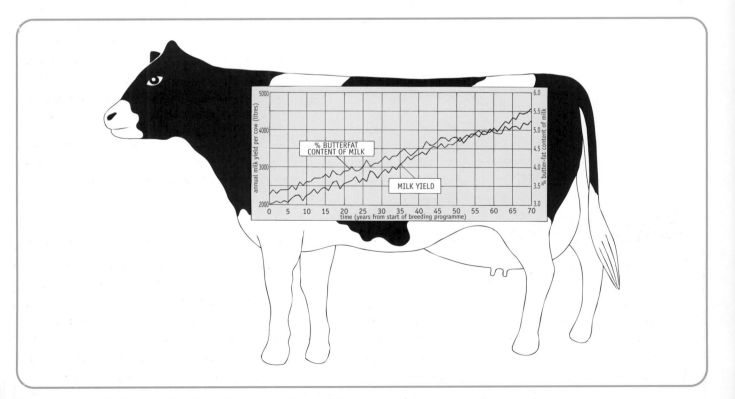

Figure 10.1 *Selective breeding for milk yield and butterfat content*

Other varieties of cattle have been selected for their meat production. The quantity of meat produced by pigs and the quantity of meat and quality of woolly fleece in sheep (see Figure 10.2) have also been greatly enhanced by selective breeding over many generations.

Figure 10.2 *Selective breeding for woolly fleece in sheep*

The birds on a modern poultry farm are found to grow more rapidly and require only half the food per kilogram gain in weight compared with their ancestors of 50 years ago. Again this is the result of selective breeding.

Enhancement of plant characteristics

In 1895 a group of American biologists began a breeding experiment using a variety of maize (sweet corn) whose seed grains varied slightly in oil and protein content. One part of the experiment involved selecting only those plants that produced seed grains with the highest **oil** content and using them to breed **strain O**; another part of the investigation involved selecting those plants that produced seed grains with the highest **protein** content and using them to breed **strain P**.

The selection procedure was repeated for 50 generations as shown in Figure 10.3. In strain O the oil-producing characteristic and in strain P the protein-producing characteristic were found to have been enhanced by selective breeding.

Testing Your Knowledge

1 a) Within a population of animals or plants, which organisms are normally used for selective breeding? (1)
   b) Which individuals are prevented from breeding? (1)
   c) If the process of selective breeding is successful, what effect does it have on the desired characteristic? (1)

2 a) (i) Name THREE types of farm animal that have been subjected to selective breeding programmes.

   (ii) For each of these give ONE example of a characteristic that has been improved for the benefit of humankind. (3)
   b) Give TWO ways in which maize plants have been enhanced by selective breeding. (2)

3 Predict what wild cattle would have been like today if their milk and meat had not been of use to human beings. (2)

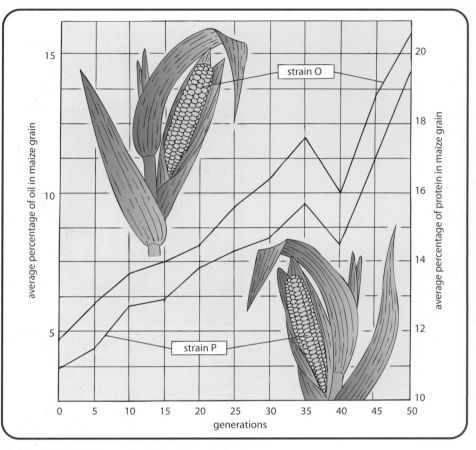

Figure 10.3 *Selective breeding in maize*

Genetic engineering

Chromosomal material of bacterium

A bacterium has one **chromosome** in the form of a complete circle and one or more small circular **plasmids** (see Figures 10.4 and 10.5). The chromosome and plasmid(s) are made up of **genes**. Each gene carries the information necessary to make a certain **protein** (e.g. an enzyme). Each enzyme in turn controls a particular chemical reaction. The normal activity of a bacterial cell depends therefore on its chromosomal material.

Genetic engineering

In recent years scientists have developed the technology that allows them to transfer pieces of DNA (consisting of one or more genes) from one organism (e.g. human) to another organism (e.g. bacterium). This process is called **genetic engineering**. The organism whose genetic make-up has been altered is said to have been **reprogrammed.** The reprogrammed organism acts as a chemical 'factory' and manufactures a substance that is new to it and useful to humankind.

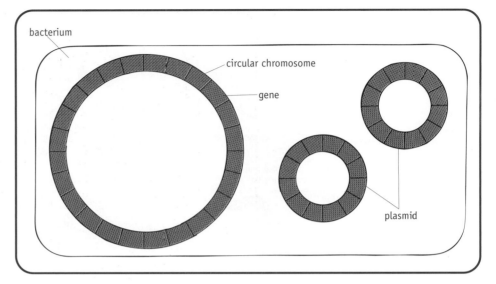

Figure 10.4 *Genetic material of bacterium*

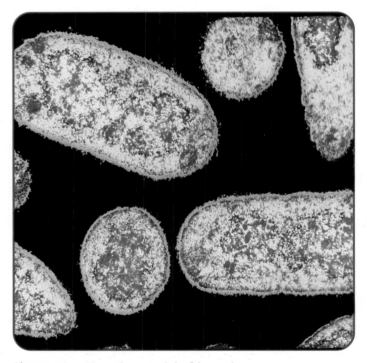

Figure 10.5 *Genetic material of bacterium*

Insulin gene

Figure 10.6 (on the following page) shows a simplified version of genetic engineering where the desired product is insulin. The process is made up of the following stages:

1 identification of the required gene for human insulin
2 cutting of the source chromosome using a special enzyme that acts as chemical 'scissors' and releases the gene
3 extraction of a plasmid from a bacterial cell

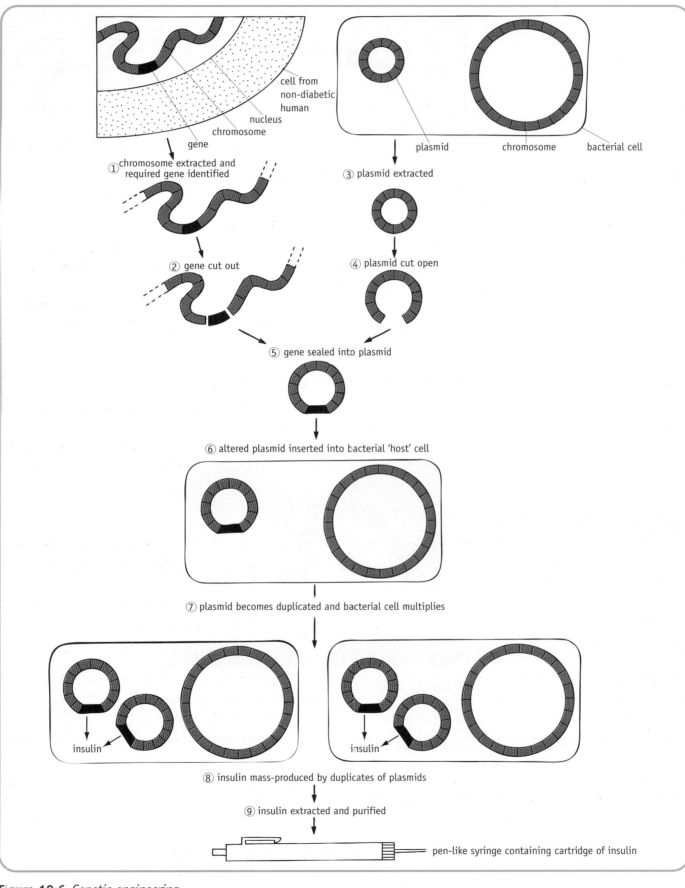

cell from non-diabetic human

nucleus

chromosome

gene

① chromosome extracted and required gene identified

② gene cut out

plasmid   chromosome   bacterial cell

③ plasmid extracted

④ plasmid cut open

⑤ gene sealed into plasmid

⑥ altered plasmid inserted into bacterial 'host' cell

⑦ plasmid becomes duplicated and bacterial cell multiplies

insulin   insulin

⑧ insulin mass-produced by duplicates of plasmids

⑨ insulin extracted and purified

pen-like syringe containing cartridge of insulin

Figure 10.6 *Genetic engineering*

4 opening up of the plasmid using the same special enzyme that acts as chemical 'scissors'

5 insertion of the gene into the plasmid ring using a further enzyme that 'seals' the gene into place

6 insertion of the altered plasmid into the bacterial 'host' cell

7 duplication of the altered plasmid and growth and multiplication of the bacterial cell

8 synthesis and mass-production of insulin by duplicate plasmids that express the 'foreign' gene

9 extraction and purification of insulin.

Advantages of genetic engineering

- Genetic engineering employs reprogrammed bacteria and yeasts. Given suitable conditions, these micro-organisms multiply at a **rapid rate** and make **large quantities** of the desired product.
- Genetic engineering has made possible the production of a **wide range** of substances that are of great medical and commercial value (see pages 214–5).

Disadvantages of genetic engineering

- The initial development of a strain of micro-organism reprogrammed exactly to requirements is very **expensive**.
- There exists the possibility of genetically engineered bacteria being released into the environment and having some **harmful effect** on the ecosystem.

Testing Your Knowledge

1 An agent that acts as a carrier between two species is often called a *vector*. Which part of a bacterium's chromosomal material could this term be used to describe? (1)

2 a) Give a brief definition of the term *genetic engineering*. (2)

   b) Arrange the following steps involved in the process of genetic engineering into the correct order:

1 insertion of gene into plasmid and plasmid into bacterial cell

2 identification and removal of the required gene from organism (e.g. human)

3 growth and multiplication of bacterial cell, forming product

4 extraction and opening up of bacterial plasmid. (1)

3 Give one possible advantage and one possible disadvantage of genetic engineering. (2)

Genetic engineering versus selective breeding

Genetic engineering and selective breeding are two methods by which scientists can **alter** the **genotype** of another species for the benefit of human beings. Each method involves producing **new combinations** of genetic material that would be unlikely to exist normally in nature. The two methods are compared in Table 10.1.

selective breeding (traditional method)	genetic engineering (modern method)
organism's genotype can only be altered indirectly and this means that production of the ideal organism is not necessarily guaranteed	organism's genotype can be altered directly by manipulating its chromosomal material to produce the ideal organism
production of new, improved varieties involves many years of careful selection	once the required gene has been identified it can immediately be transferred to a micro-organism to form a new variety which can then be put to work
usually only allows humans to develop an improved version of an existing variety of organism	allows humans to produce an organism with a new genotype able to make useful products previously only made by another species

Table 10.1 *Comparison of two methods of altering a species' genotype*

Applications of products of genetic engineering

Medical value

Many gene products have important **medical applications**, as shown in Table 10.2. In each of these examples a particular gene has successfully been inserted into a micro-organism which is then allowed to multiply and make a product of medical value.

Gene products carry no risk of viral contamination. In the past Factor VIII extracted from donated blood was sometimes contaminated with HIV which

product of genetic engineering	normal source and function of substance	medical application of gene product made by micro-organism
insulin	made by pancreas cells; controls level of glucose in blood	given to people who do not make enough insulin naturally, thereby relieving the symptoms of *diabetes mellitus* (see Figure 10.7 for insulin pen)
Factor VIII	chemical present in blood; required for clotting of blood at wounds	given by injection to sufferers of haemophilia who lack Factor VIII and whose blood fails to clot
human growth hormone	made by cells in pituitary gland; essential during childhood and adolescence to control normal growth and development	given by regular injection to children who do not make enough of their own; prevents reduced growth and dwarfism

Table 10.2 *Medical applications of gene products*

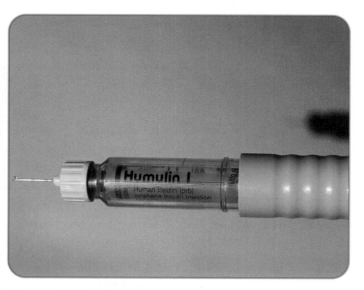

Figure 10.7 *Insulin pen*

then infected the haemophiliac recipient. Blood products given to haemophiliacs are now heat-treated to prevent transmission of viruses.

Scientists have also managed to reprogramme yeast cells to produce human serum albumen (used in blood replacements), epidermal growth factor (which speeds up wound healing) and hepatitis B antigens (used in the manufacture of hepatitis B vaccine).

Commercial applications

Bacteria

Many reprogrammed microbes produce enzymes that are added to detergents to digest difficult stains. Other varieties of reprogrammed bacteria are able to make ethylene glycol (antifreeze) and other chemicals used in the plastics industry.

Yeast

Brewing

Genetic engineers have produced a form of yeast that makes lager with a high alcohol content and a low carbohydrate content. This product is expected to appeal to weight-watchers.

Cheese-making

Rennin is an enzyme secreted by the stomachs of young mammals. It causes milk proteins to coagulate. This curdling of milk is an essential stage in cheese-making. In the past rennin for cheesemaking was extracted from the stomach lining of calves. Genetic engineers have recently enabled the rennin gene to be transferred from calves to yeast cells, creating a strain that can mass-produce the rennin needed by the cheese-manufacturing industry.

Transgenic multicellular organisms

Much of the earlier work in genetic engineering concentrated on the activities of reprogrammed micro-organisms. However, new technology is now rapidly advancing into the realm of transgenic multicellular organisms.

Agriculture

Agrobacterium tumefaciens (see Figure 10.8) is a primitive bacterium present in soil that invades wounded plant tissue. The bacterium injects a plasmid into a plant cell and some of the plasmid's genetic material becomes incorporated into the plant's DNA. *Agrobacterium* is therefore a 'natural genetic engineer'. It provides scientists with opportunities to introduce into crop plants desirable genes from other organisms that have first been inserted into one of *Agrobacterium*'s plasmids.

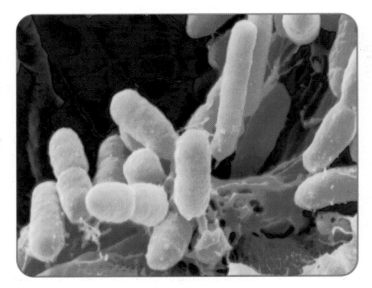

Figure 10.8 *Agrobacterium tumefaciens*

Figure 10.9 shows a simplified version of this process. It has already been successfully performed on several types of crop plants. These altered plants are called **transgenic** varieties (see Table 10.3).

transgenic crop plant	role of inserted gene	beneficial effect
apple	part blockage of production of chemicals that promote ripening	fruit's shelf-life is extended
pea	production of insecticide protein	leaves resist attack by caterpillars
soya	production of protein that gives resistance to weedkiller	crop survives but weeds die when weedkiller is applied
strawberry	production of antifreeze chemical	fruit is protected against frost damage
tomato	blockage of production of an enzyme	fruit is prevented from becoming soft and mushy

Table 10.3 *Transgenic crops*

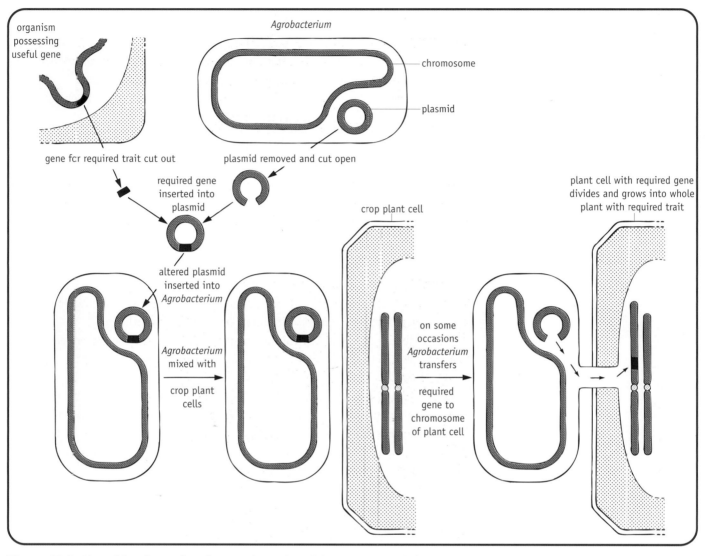

organism possessing useful gene

Agrobacterium

chromosome

plasmid

gene fcr required trait cut out

plasmid removed and cut open

required gene inserted into plasmid

plant cell with required gene divides and grows into whole plant with required trait

crop plant cell

altered plasmid inserted into *Agrobacterium*

Agrobacterium mixed with crop plant cells

on some occasions *Agrobacterium* transfers required gene to chromosome of plant cell

Figure 10.9 *Use of* Agrobacterium *in genetic engineering*

Future applications

Already some animal genes have been successfully inserted into plants. Using *Agrobacterium tumefaciens* as the vector, scientists have produced a transgenic variety of tobacco plant that contains and expresses the gene for human haemoglobin.

Such use of transgenic plants to produce blood products in the future would reduce our dependency on donated blood and eliminate the risk of microbial contamination resulting from a transfusion.

It is also possible that farmers will grow **transgenic crops** for their insulin, human growth factor or blood-clotting agent. It is certainly the case that we can look forward to an infinite variety of organisms being developed with the potential to produce a vast range of products useful to humankind.

Some of these applications are controversial and require extensive public debate. What is possible scientifically is not necessarily acceptable ethically to society. It is of critical importance that scientists proceed in this area with the utmost caution.

Testing Your Knowledge

1 State TWO advantages of genetic engineering over selective breeding. (2)

2 Describe THREE medical applications involving gene products. (3)

3 a) Describe the role of *Agrobacterium tumefaciens* in the production of a transgenic crop plant.

   b) Give TWO examples of such plants and their economic benefit to human beings. (6)

Practical Activity

Selecting, presenting and discussing issues arising from genetic engineering

INFORMATION

- Public attitudes to genetically engineered food vary from wholehearted enthusiasm at one extreme to intense hostility at the other.
- Some people are happy to consume genetically modified food provided that it has met with all the strict criteria laid down by the government's Advisory Committee on Genetic Manipulation.
- At the other end of the opinion spectrum, some people oppose all forms of genetic engineering claiming that humankind is running grave risks by mixing genes from unrelated species. They suggest that these genes could spread in time to other organisms with unknown long-term results.
- You are now invited to join this debate. Figure 10.10 shows six statements made by members of a pressure group expressing their concern about genetic engineering. It also shows the six replies given by scientific experts in this new branch of science.

WHAT TO DO

1 Obtain an enlarged photocopy of Figure 10.10 from your teacher.
2 Cut out the 12 people, each accompanied by his or her 'speech bubble'.
3 Match each statement made by a member of the pressure group with the reply made by one of the experts.
4 Draw up and complete a copy of Table 10.4.

statement made by member of pressure group	reply given by expert	person with whom I agree (P = pressure group E = expert N = neither)	reason for my choice in column 3

Table 10.4 *Genetic engineering issues*

Figure 10.10 *Debate on genetic engineering*

Applying Your Knowledge

1 Match the terms in list X with their descriptions in list Y.

List X
1) DNA
2) enhancement
3) factor VIII
4) genetic engineering
5) insulin
6) plasmid
7) reprogrammed
8) selective breeding

List Y
a) hormone made by human pancreas needed to control the glucose level of blood
b) small circular structure in bacterium used to transfer genes from one species to another
c) process by which humans choose, as the parents of the next generation, animals or plants with characteristics useful to humankind
d) term used to describe a micro-organism after the insertion into it of DNA from another species
e) improvement of a plant or animal characteristic by selective breeding
f) blood-clotting chemical that haemophiliacs lack
g) genetic material that can be transferred from one species to another by genetic engineering
h) process by which a piece of DNA is removed from one species and inserted into another

2 a) Table 10.5 gives data from two selective breeding programmes carried out over the same five-year period. Draw TWO conclusions from the information. (2)

   b) A breed of pig has been produced that reaches bacon weight in 150 days instead of the normal 185 days. In addition, this is achieved on 20% less food. Can this improvement process be continued indefinitely? Explain your answer. (2)

3 The symbols identified in Figure 10.11 are used in Figure 10.12 to represent the parents and offspring of a cross carried out during a selective breeding programme in potato plants. The overall aim of the programme was to produce a variety that develops many large tubers and whose leaves are fully resistant to disease.

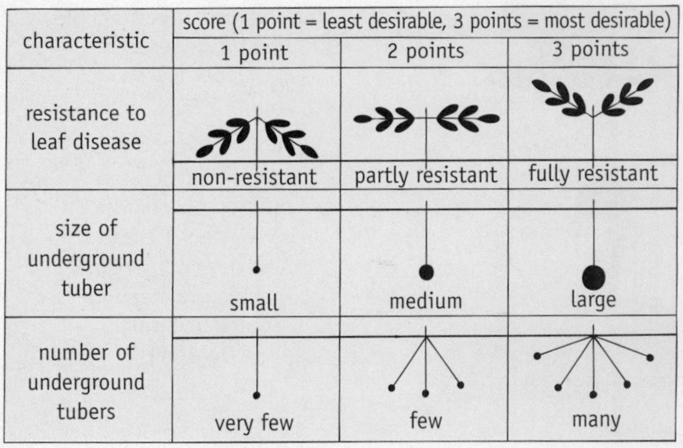

characteristic	score (1 point = least desirable, 3 points = most desirable)		
	1 point	2 points	3 points
resistance to leaf disease	non-resistant	partly resistant	fully resistant
size of underground tuber	small	medium	large
number of underground tubers	very few	few	many

Figure 10.11

Figure 10.12 shows the nine best offspring from the cross.

year	Ayrshire cattle		Jersey cattle	
	average milk yield per cow (l)	average butter fat content (%)	average milk yield (per cow (l)	average butter fat content (%)
1	3849	3.68	3093	4.76
2	3894	3.74	3134	4.82
3	3931	3.79	3169	4.87
4	3972	3.83	3207	4.95
5	4007	3.87	3239	5.01

Table 10.5

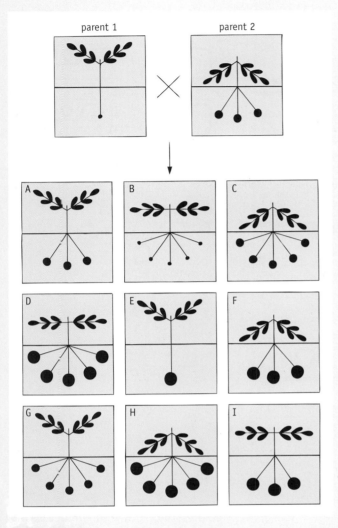

Figure 10.12

a) Calculate the number of points scored by each of the offspring. (9)
b) How many of these offspring scored more points than the parents? (1)
c) Which TWO offspring would you select as the parents of the next cross? (2)
d) (i) With reference only to the characteristics given in this question, draw a diagram of the plant that the farmer is hoping to achieve eventually by selective breeding.
   (ii) How many points would this plant score? (4)

4 Figure 10.13 (on page 222) shows eight of the stages involved in the process of genetic engineering. They are in a mixed-up order.

a) Arrange the stages into the correct order by listing their letters, starting with B. (1)

b) Against each letter in your list, write a sentence to describe what has taken place at that stage of the process. (8)

5 Read the passage and answer the questions that follow it.

The cassava plant (Figure 10.14) is native to South America. Its starchy root is used to make tapioca, a staple foodstuff consumed by around 500 million people in Asia and Africa. In the past, cultivation of cassava plants was severely hampered by insect pests and viral and fungal diseases. Attempts to cross cassava with its wild relatives in order to produce resistant strains proved unsuccessful.

Genetic engineers are now solving this problem by adding genes from other species to cassava plants. A gene from a bacterium that makes a natural pesticide against mealy bug and a gene for resistance to a viral disease have already been added using the bacterium *Agrobacterium tumefaciens*, the natural genetic engineer. These genes are unlikely, however, to provide a permanent solution to the problem since new strains of the pest will probably emerge in the future. Genetic engineers will then need to produce another new variety of cassava.

Scientists are at present also attempting to produce genetically modified cassava which will make starch suitable for use in the manufacture of biodegradable plastics. This industry would be based on a renewable resource rather than fossil fuels and could provide employment and profit for people in developing countries. However, if land previously used to grow cassava for food is used by a multinational chemical company to grow the new variety for plastics then local people could lose out since less food would be grown and the profits from the plastics could bypass them.

a) What is tapioca? (1)
b) What method did scientists use to try to create crops resistant to disease before genetic engineering began? (1)
c) (i) Imagine that genetic engineers are trying to create a new variety of cassava containing a gene from pea plants that gives resistance to greenfly. Describe what they must do to *Agrobacterium*

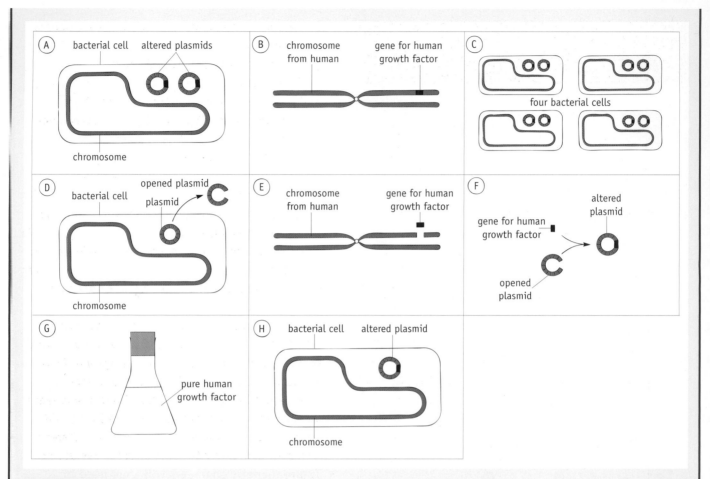

Figure 10.13

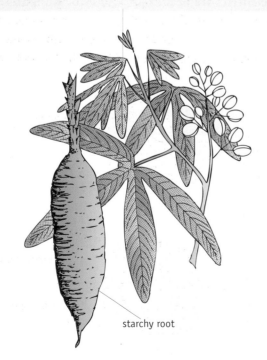

Figure 10.14

tumefaciens before bringing it into contact with cassava plants.

(ii) Why might the new cassava plant's resistance to greenfly be short-lived? (3)

d) State the case for and against developing genetically modified cassava that would make starch suitable for the manufacture of biodegradable plastics. (4)

6 When a tomato fruit is fully grown but still green and unripe, it begins to produce ethylene. Ethylene is a growth substance that promotes the ripening of fruit. Genetic engineers have developed a variety of tomato plant that makes very little ethylene by inserting a gene that almost completely blocks ethylene production. Table 10.6 gives the results from one of their experiments. Figure 10.15 shows a typical fruit from each variety of tomato plant.

a) Draw a line graph of the results (2)

b) Match the terms *control* and *genetically modified variety* with the two lettered varieties in the diagram. (1)

time from green fruit reaching full size (days)	ethylene production (units)	
	genetically modified variety	control
0	1	1
1	4	5
2	7	25
3	5	70
4	6	71
5	8	68
6	7	70
7	6	52
8	7	41
9	7	32
10	8	34
11	6	31
12	7	32
13	7	29
14	8	32
15	7	31
16	6	28
17	8	25
18	7	26
19	7	23
20	6	21
21	6	19

Table 10.6

c) State how many units of ethylene were produced on day 6 by
(i) the genetically modified variety
(ii) the control. (1)
d) Calculate the percentage decrease in ethylene production at day 6 caused by the blocked gene in the genetically modified variety. (1)
e) Suggest why it is important that ethylene production is not completely blocked in the modified variety. (1)
f) Suggest the benefit gained from this application of genetic engineering. (1)
g) Which variety of tomato would you prefer to eat? Explain why. (1)

7 Read the passage and answer the questions that follow it.

Soya beans produced by the soya plant (Figure 10.16) are rich in protein and oil and play a vital role in the nutrition of human beings and farm animals. Weeds can easily devastate a crop of soya plants by depriving them of water, nutrients and sunlight. Until recently farmers in the USA often needed to apply as many as five different herbicides to guarantee a successful harvest of soya.

In the 1990s scientists developed a genetically modified variety of soya by inserting a bacterial gene into the plant, which makes it resistant to the herbicide 'Roundup'. 'Roundup' kills weeds by blocking the production of a particular protein

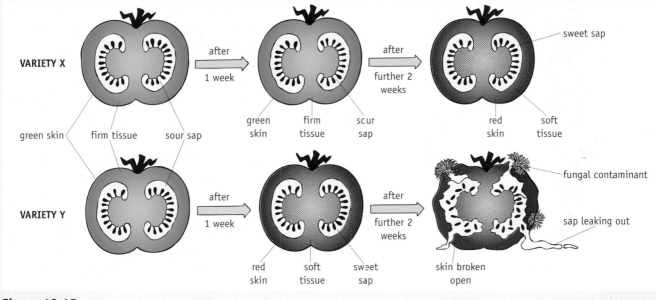

Figure 10.15

soya bean in pod

Figure 10.16

needed by the weeds for their growth. The modified soya plant produces a different protein and its metabolism remains unaffected by 'Roundup'.

In the past, 1.13 kg of herbicide per hectare was needed to control weeds; with 'Roundup' only 0.54 kg per hectare is needed, reducing the cost by more than 50%. Some pesticides persist in the ecosystem and accumulate in living organisms but not 'Roundup'. It is rapidly broken down by soil micro-organisms to carbon dioxide, ammonia and phosphate.

Although the genetically modified soya beans contain one additional protein, the beans are processed in such a way that the plant's genetic material and the 'bacterial' protein are broken down to their basic components. Flour from the new soya beans is indistinguishable from the traditional type and its manufacturers claim that the nutritional value and quality of foods remains unaffected.

In the 1990s, the beans from the GM plants only made up about 2% of the total soya crop and they were thoroughly mixed with the traditional variety prior to food manufacture. Since the GM portion was so small, the label on a product containing soya bean did not state whether or not it contained the GM type. The percentage of GM soya in foods has gradually increased over the years but labels still lack detailed information.

a) (i) What is meant by the term *herbicide?*
   (ii) By what means does the herbicide 'Roundup' normally bring about its effect on plants? (2)

b) (i) How could you demonstrate experimentally that a certain type of bacterium contains a gene that gives it resistance to the herbicide 'Roundup'?
   (ii) By what means did scientists create a strain of soya plant resistant to 'Roundup' ?
   (iii) Why do these modified soya plants not die when sprayed with 'Roundup'? (4)

c) (i) Explain why it is of economic benefit to farmers to use 'Roundup' on their soya crops.
   (ii) Quote the sentence in the passage that confirms that 'Roundup' is biodegradable. (2)

d) In November 1996, Greenpeace blockaded the Port of Liverpool in an attempt to prevent the first consignment of genetically engineered soya beans from the USA entering Britain.
   (i) Suggest why they objected to the new soya beans.
   (ii) Do you approve or disapprove of the action that Greenpeace took in this case? Explain your answer. (2)

e) (i) What explanation is given by the manufacturers of soya bean products for the lack of labelling of genetically modified ingredients?
   (ii) In 1996 a series of surveys in Europe showed that the majority of people wished mandatory labelling of genetically modified foodstuffs. Do you support this view? Justify your answer. (2)

Word bank

adapted	heterozygous
alleles	homozygous
breeding	inherited
directly	inserted
dominant	insulin
engineering	monohybrid
environment	natural
factories	parents
favourable	phenotype
fertilisation	ratio
generation	recessive
genes	reproductive
genetic	reprogrammed
genotype	

Table 10.7 *Word bank for Chapters 9 and 10*

What You Should Know

(Chapters 9 and 10) (See Table 10.7 for word bank)

1 An organism's physical characteristics are known collectively as its _____; an organism's inherited characteristics are determined by _____ information received from both of its _____.

2 Each inherited characteristic is controlled by one or more units of heredity called _____; each gene is part of a chromosome.

3 Each gene normally has two or more different forms called _____. An allele that always shows its effect and masks the presence of the other form is said to be _____; an allele that is masked by the dominant form is said to be _____.

4 Each gamete carries only one allele of a gene; each zygote formed at _____ receives two alleles for each gene, one from each parent.

5 The complete set of genes possessed by an organism is called its _____. The genotype of an organism with identical alleles of a gene is described as _____; the genotype of an organism with two different allelles of a gene is described as _____.

6 An experimental cross involving one difference between two true-breeding parents is called a _____ cross. All members of the F_1 generation resemble the parent with the dominant allele. Selfing of the F_1 produces an F_2 generation with a phenotypic _____ of 3:1.

7 An organism's phenotype is the result of the interaction between its genotype and the _____. Differences in phenotype caused by differences in genotype are _____ and passed on from generation to generation. Differences in phenotype caused by environmental factors are not passed on to the next _____.

8 Those members of a species that are better _____ to the environment survive to _____ age and pass on the _____ characteristics to succeeding generations; those members less well suited fail to do so. This survival of the 'fittest' in each generation is called _____ selection.

9 Genetic _____ allows pieces of DNA to be cut out of one organism and be _____ into another organism which then forms a substance new to that species.

10 _____ bacteria can be used as chemical _____ to produce useful products such as human _____.

11 Unlike selective _____, genetic engineering allows scientists to alter an organism's genotype _____ so that it will produce exactly what is required immediately.

Unit **3**

Animal Physiology

Vertebrate animals such as mammals are able to employ their physiology to make responses to changes in their internal and external environments that are of survival value.

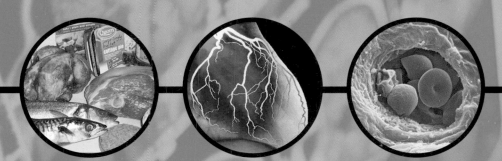

11   Mammalian nutrition

Requirement for food

To maintain good health, human beings need to consume a **balanced diet**. This must contain appropriate quantities of the main food groups: **carbohydrates**, **fats**, **proteins**, **vitamins** and **minerals**. Actual requirements vary from person to person. Compared with an inactive person, a person who leads a very active life, for example, needs more carbohydrate to provide energy (see also Chapter 4).

Chemical structure

Carbohydrates, fats and proteins all contain the chemical elements **carbon** (C), **hydrogen** (H) and **oxygen** (O). In addition, proteins contain **nitrogen** (N).

Carbohydrates

Carbohydrates are energy-rich compounds often referred to as 'fuel' foods. Examples include sugars, starch, glycogen and cellulose. Glucose is the most common simple sugar. Its chemical formula is $C_6H_{12}O_6$. A molecule of **glucose** is often represented in a simple way as a hexagonal shape (see Figure 11.1). A molecule of maltose sugar consists of two molecules of glucose joined together.

Molecules of more complex carbohydrates are built up from repeating units of simpler ones, as shown in Figure 11.1. In plants, excess glucose (which is soluble) is stored as insoluble **starch**. In animals (e.g. human beings), excess glucose is stored as insoluble **glycogen** in the liver.

Cellulose is a complex carbohydrate made of thousands of glucose molecules arranged in long chains which group together into cellulose fibres (the basic framework of plant cell walls). Although people are unable to digest cellulose, it is an essential part of their diet because it gives bulk to faeces. This stimulates the muscular action of the large intestine and prevents constipation. Plant foods rich in cellulose provide fibre (roughage). Figure 11.2 shows foods rich in carbohydrate.

Fat

Fats release about twice as much energy per gram as carbohydrates and are also referred to as 'fuel' foods. (The energy content of food is discussed in more detail on pages 64–6.)

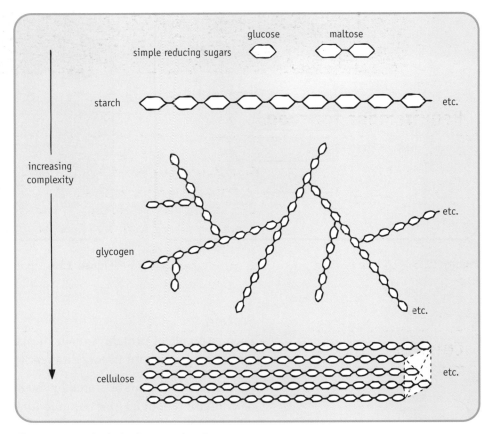

Figure 11.1 *Carbohydrate molecules*

Figure 11.2 *Foods rich in carbohydrate*

A molecule of fat consists of one molecule of **glycerol** combined with three **fatty acid** molecules, as shown in Figure 11.3. Fats are insoluble in water. In the human body, fat is stored around the kidneys and under the skin where it is especially effective in acting as a layer of insulation. Figure 11.4 shows foods rich in fat.

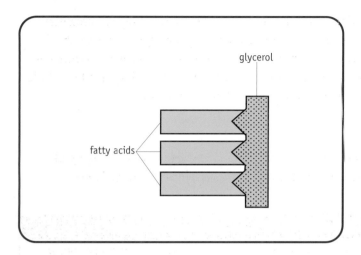

Figure 11.3 *Molecule of fat*

Figure 11.4 *Foods rich in fat*

Protein

Each molecule of **protein** is made of many sub-units called **amino acids**, of which there are about twenty different types (see Figure 11.5).

Excess protein cannot be stored by the human body. Therefore a minimum daily intake (about 80 g) is required for growth and tissue repair. Excess dietary protein is used to provide the body with energy. Structural proteins which make up the basic framework of the body's tissues and organs, only break down to release energy in a crisis (e.g. prolonged starvation). Figure 11.6 shows foods rich in protein.

Vitamins

Vitamins are complex chemical compounds required by the human body for good health. They are divided into two categories: fat-soluble vitamins (e.g. A and D) and water-soluble vitamins (e.g. B and C).

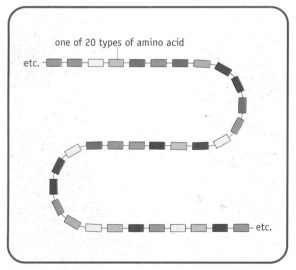

Figure 11.5 *Small part of a protein molecule*

Figure 11.6 *Foods rich in protein*

Although vitamins do not provide the body with energy, they are essential ingredients in the diet. They function primarily as **coenzymes** which are needed to promote various biochemical reactions. Like enzymes, vitamins are continuously reused and are therefore only required in small quantities. Table 11.1 gives some examples of vitamins and their importance.

Minerals

The term '**minerals**' refers to certain chemical elements essential in small quantities for a diverse variety of functions in the human body. These

vitamin	function	result of major deficiency	rich sources
A	component of visual pigment in retina; needed for maintenance of epithelial tissues	night-blindness; dry skin	carrots and other yellow vegetables
B$_1$	plays essential role in removal of CO$_2$ made during tissue respiration	weakening of heart; beri-beri (paralysis of limbs)	liver, yeast and unrefined cereal grains
B$_{12}$	essential for development of new red blood cells	anaemia	meat, eggs and milk
C	needed for growth and repair of skin and mucous membranes	scurvy (poor healing of wounds; soft, bleeding gums) (see Figure 11.7)	citrus fruits and green vegetables
D	required for absorption of calcium and phosphate from small intestine	rickets (soft bones that become deformed easily) (see Figure 11.8)	fish liver oil, butter, milk and egg yolk

Table 11.1 *Importance of vitamins*

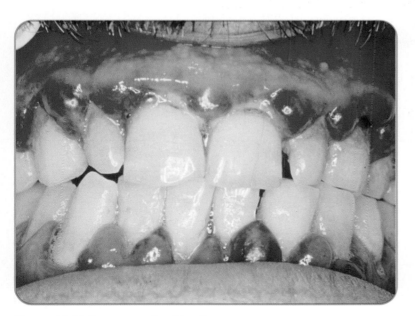

Figure 11.7 *Scurvy causing bleeding gums*

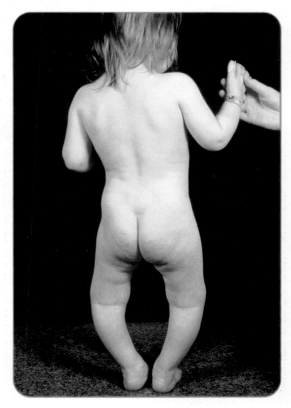

Figure 11.8 *This person is suffering from rickets*

elements are present as components of soluble compounds in foodstuffs consumed in a normal healthy diet. Six examples are given in Table 11.2

mineral element	function	rich sources
calcium	needed for hardening of bones and teeth, clotting of blood and contraction of muscles	milk, cheese and green vegetables
iron	acts as a structural component of haemoglobin and some enzymes	meat, eggs, cereals and green vegetables
phosphorus	needed for formation of bones and teeth, and synthesis of DNA and ATP	milk, fish, meat, nuts and cereals
iodine	required for formation of hormones produced by thyroid gland	sea foods and iodised salt
sodium	essential for contraction of muscles and transmission of nerve impulses	fish, meat, milk and salt
potassium	essential for contraction of muscles and transmission of nerve impulses	present in almost all foods

Table 11.2 *Importance of minerals*

Activity

Selecting and presenting information about foodstuffs

Information

Table 11.3 shows an analysis of 15 common foodstuffs with respect to their carbohydrate, protein and fat content.

foodstuff	number of grams present in 100 g portion of food		
	carbohydrate	protein	fat
almonds	4.3	19.5	53.2
beans (canned)	17.3	6.0	0.4
bread (brown)	49.0	9.2	1.8
chicken (roast)	0	29.6	7.3
coconut (desiccated)	6.5	6.5	62.0
cornflakes	88.0	7.5	0.5
fish fingers	21.0	13.0	7.0
jam (strawberry)	69.2	0.5	0
peanuts (roasted)	8.6	27.9	49.0

foodstuff (continued)	number of grams present in 100 g portion of food		
	carbohydrate	protein	fat
peas (frozen)	7.7	5.0	0
pork chops	0	19.5	53.2
potatoes (chipped)	38.0	4.0	9.0
rice (boiled)	86.8	6.2	1.0
spaghetti	84.0	10.0	1.0
yoghurt (fruit)	13.0	3.6	1.8

Table 11.3 *Analysis of foodstuffs*

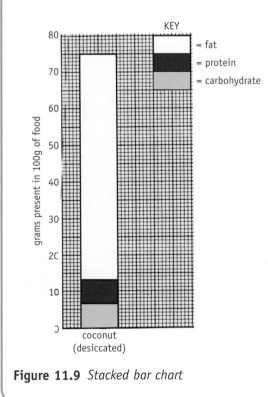

KEY
☐ = fat
■ = protein
▨ = carbohydrate

Figure 11.9 *Stacked bar chart*

What to do

1 Using only the information in the table, answer the following questions.

   a) Which food contains least protein?

   b) Which food contains most carbohydrate?

   c) How many foods contain more fat than brown bread?

   d) How many foods contain less carbohydrate than fruit yoghurt?

   e) Name TWO foods that contain the same mass of protein and fat as one another but differ in their carbohydrate content.

   f) By how many times is the protein content of roasted peanuts greater than that of rice?

   g) How many grams of fat would be present in a kilogram of roast chicken?

2 Figure 11.9 shows a 'stacked' bar chart for desiccated coconut. Using this style, draw bar charts to compare the carbohydrate, protein and fat content of fish fingers, chipped potatoes and spaghetti.

Food tests

Testing for starch

A small sample of the food to be tested is placed in a test tube. A few drops of **iodine solution** are added to the food (as shown in Figure 11.10). If a **blue-black** colour is produced, then starch is present in the food.

Testing for protein

A small sample of the food to be tested is placed in a test tube. A few drops of **Biuret reagent** are added to the food (as show in Figure 11.10). If a **lilac** colour is produced, then protein is present in the food.

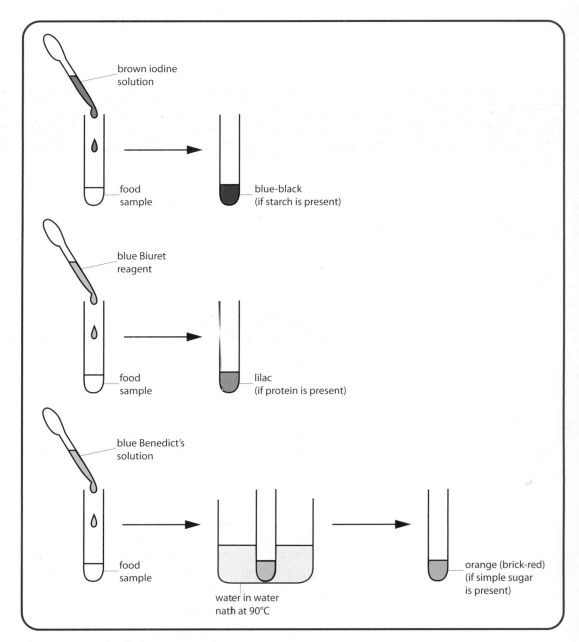

Figure 11.10 *Food tests*

Testing for simple sugar

A small sample of the food to be tested is placed in a test tube. A few drops of **Benedict's solution** are added to the food (as shown in Figure 11.10). Using a test tube holder, the test tube is placed in a water bath at 90°C for a few minutes. If an **orange** (**brick-red**) colour is produced, then simple sugar (e.g. glucose or maltose) is present in the food.

Testing for fat (or oil)

A small sample of the food to be tested is placed on the centre of a piece of **clean white paper**. If a **translucent stain** is produced, then fat (or oil) is present in the food.

(Note: A translucent stain is a greasy, semi-transparent mark that lets some light pass through it. It is useful to add a drop of water to a second sheet of the same white paper to compare a wet mark with a translucent stain.)

The four food tests are summarised in Table 11.4.

food being tested for	reagent or material used for test	original colour or state of testing reagent or material	heat or no heat	positive result
starch	iodine solution	brown	no heat	blue-black
protein	Biuret reagent	blue	no heat	lilac
simple sugar	Benedict's solution	blue	heat	orange (brick-red)
fat (or oil)	white paper	clean, non-greasy	no heat	greasy, translucent stain

Table 11.4 *Summary of food tests*

Practical Activity

Carrying out food tests to identify the contents of liquids X and Y

INFORMATION

Great care should be taken and goggles worn when carrying out this activity. Any spashes of chemical on the skin should be washed off immediately with cold water.

YOU NEED

pair of safety goggles
1 dropping bottle of starch (1% suspension)
1 dropping bottle of protein (1% suspension of egg albumen)
1 dropping bottle of simple sugar (1% solution of glucose)
1 dropping bottle of vegetable oil
1 dropping bottle of iodine solution
1 dropping bottle of Biuret reagent
1 dropping bottle of Benedict's solution

1 dropping bottle of unknown liquid X (containing some but not all of the 4 foods)

1 dropping bottle of unknown liquid Y (containing a different combination of the 4 foods from X)

10 or more squares of clean white paper (10cm × 10cm)

10 or more test tubes

1 test tube stand

1 test tube holder

access to water bath at 90°C

WHAT TO DO

1 Read all of the instructions in this section and prepare your table of results before carrying out the experiment.
2 Try out the food tests as summarised in Table 11.4 by testing each of the four foods (starch, protein, simple sugar and vegetable oil) with its appropriate testing agent to become familiar with the positive result that indicates the presence of each food type.
3 If time permits, try testing each of the four types of food with the other testing agents to establish that they do not give the positive result.
4 Test liquid X for each type of food and enter the results in your table using the symbols + to mean presence of the food and − to mean absence of the food.
5 Repeat step 4 using liquid Y.
6 If other students have carried out the same experiment, pool the results.

REPORTING

Write up your report by doing the following:

1 Copy the title at the start of this activity.
2 Put the subheading 'Aim' and state the aim of your experiment.
3 a) Put the subheading 'Method'
   b) Using the impersonal passive voice, briefly describe the experimental procedure that you followed and state how you obtained your results.
4 Put the subheading 'Results' and draw a final version of your results table.
5 Put the subheading 'Conclusion'. Draw a conclusion about the contents of liquids X and Y.

Need for digestion

Living cells need a constant supply of food to provide them with fuel for **energy** and building materials for **growth, tissue repair** and **antibody production**. This food is carried to the cells by the bloodstream. In order to be absorbed into the bloodstream in the first place, food must be **digested**.

Digestion involves the breakdown of large insoluble molecules of food into smaller soluble molecules in the gut. This allows them to be absorbed into the bloodstream through the wall of the small intestine (see Figure 11.11).

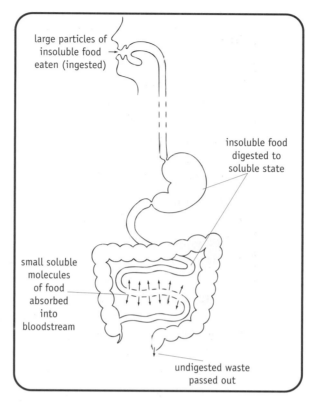

large particles of insoluble food eaten (ingested)

insoluble food digested to soluble state

small soluble molecules of food absorbed into bloodstream

undigested waste passed out

Figure 11.11 *Feeding and digestion*

Practical Activity and Report

Demonstration of the purpose of digestion and absorption using Visking tubing as a model gut

Information

- Food passing along the alimentary canal (gut) is a mixture of large particles of carbohydrate, protein and fat.
- Every cell in the human body needs a supply of food which it obtains from a nearby blood vessel. For this system to work food molecules must first gain access to the bloodstream from the gut.
- In the experiment that you are going to set up, Visking tubing acts as the wall of the gut, starch and glucose as food in the gut and water in the boiling tube as blood surrounding the gut.

You need

1 200 mm length of Visking tubing (soaking in warm water)

20 ml 1% starch suspension

20 ml 5% glucose solution

1 small lipped beaker (e.g. 100 ml)

1 boiling tube

1 teat pipette/dropper

1 Bunsen burner

1 tripod stand

1 250 ml glass beaker (to use as a simple hot water bath)

1 dropping bottle of Benedict's solution

1 dropping bottle of iodine solution

4 test tubes

1 test tube stand

1 stopclock

1 pair of safety spectacles

1 cavity tile

access to a thermostatically controlled water bath at 37°C

What to do

1 Read all the instructions in this section and prepare your table of results before carrying out the experiment.

2 Wearing safety spectacles, set up a simple water bath for the Benedict's solution test using the Bunsen burner, tripod stand and beaker of water.

3 Mix the starch suspension and glucose solution together in the small lipped beaker.

4 Tightly knot one end of the length of Visking tubing.

5 Tease open the untied end of the tubing and two-thirds fill it with the starch–glucose mixture.

6 Tightly knot the open end of the Visking tubing.

7 Rinse the Visking tubing 'sausage' under the cold tap.

8 Place the 'sausage' in a boiling tube and fill the remaining space in the boiling tube with cold tap water.

9 Switch on the clock and then use the teat pipette to immediately take a sample of the water in the boiling tube.

10 Test part of this water sample for simple sugar using Benedict's solution.

11 Test part of this sample for starch using iodine solution and a cavity tile.

12 Leave the boiling tube and its contents in the water bath at 37°C.

13 Repeat steps 10 and 11 at intervals of 10 minutes for a total of 30 minutes.

14 During this time, record your results in your table using the symbol + to mean presence of substance, (+) to mean faint presence of substance and − to mean absence of substance.

15 If other students have carried out the same experiment, pool the results.

Reporting

Write up your report by doing the following:

1 Copy the title given at the start of this activity.

2 Put the subheading 'Aim' and state the aim of your experiment.

3 a) Put the subheading 'Method'.

b) Draw a labelled diagram of the apparatus set up at the start of the experiment. State in brackets the roles being played by the Visking tubing, the starch–glucose mixture and the water in this model of the gut.

c) Using the impersonal passive voice, briefly describe the experimental procedure that you followed and state how you obtained your results.

4 Put the subheading 'Results' and draw a final version of your table of results.

5 Put a subheading 'Analysis of Results' and write a short paragraph to state what you have found out from a study of your results. This should include answers, in sentences, to the following questions:

a) Which food was unable to pass out of the Visking tubing 'sausage' through tiny pores in the membrane because its molecules are too large?

b) Which food was able to pass out of the Visking tubing because its molecules are very small?

c) If movement of food molecules from the gut into the bloodstream in a real human body works in the same way as in the model, which of the two foods would the body be unable to absorb?

6 Put a subheading 'Conclusion'. Draw a conclusion about the purpose of digestion based on the results of your experiment.

7 Put a final subheading 'Evaluation of Experimental Procedure' and then answer the following:

a) Explain why rinsing the Visking tubing 'sausage' is a valuable piece of experimental procedure.

b) State a possible source of error and give the steps that you would take to eliminate it.

c) Give a possible improvement that could be made to the experiment.

Testing Your Knowledge

1 a) Name THREE chemical elements always found in
   carbohydrates, proteins and fats. (3)
   b) Name a fourth chemical element found in proteins
   only. (1)
2 a) Describe the structure of a molecule of starch in
   terms of simple sugar molecules. (1)
   b) With the aid of a simple diagram, describe the
   structure of a molecule of protein. (2)
   c) Name the sub-units of which a molecule of fat is
   composed. (2)
3 a) Name TWO vitamins and explain why each is
   required by the body. (2)
   b) Explain why (i) sodium and (ii) calcium are needed
   for good health. (2)

c) Name a food rich in iodine. (1)
4 a) Name a chemical reagent used to test for each of
   the following foods: (i) protein; (ii) simple sugar;
   (iii) starch.
   b) Give the positive result obtained for each of the
   foods in a).
5 a) What must happen to large molecules of food
   before they can be absorbed into the
   bloodstream? (1)
   b) What name is given to this process? (1)
   c) Using the terms *soluble* and *insoluble* in your
   answer, write a sentence to describe what happens
   to food during this process. (2)

Alimentary canal and associated organs

The digestive system is made up of the **alimentary canal** (Figure 11.12), which
is a long tube running from mouth to anus. The salivary glands, liver and
pancreas (known as the associated organs) are connected to the alimentary
canal by tubes called ducts.

Mouth and salivary glands

Food in the mouth is broken down mechanically (physically) by the action of
the teeth which bite, chew and grind it into small fragments. These are mixed
up with saliva from the **salivary glands** (Figure 11.13). Saliva contains the
digestive enzyme **salivary amylase** which promotes the chemical breakdown
(digestion) of starch to maltose. Saliva also contains **mucus** which keeps the
mouth moist and lubricates the food, enabling it to be swallowed easily.

Oesophagus and peristalsis

The **oesophagus** (gullet) is a muscular tube that connects the back of the
mouth to the stomach. Once food has been swallowed, it is moved down the
oesophagus by a form of muscular activity called **peristalsis**.

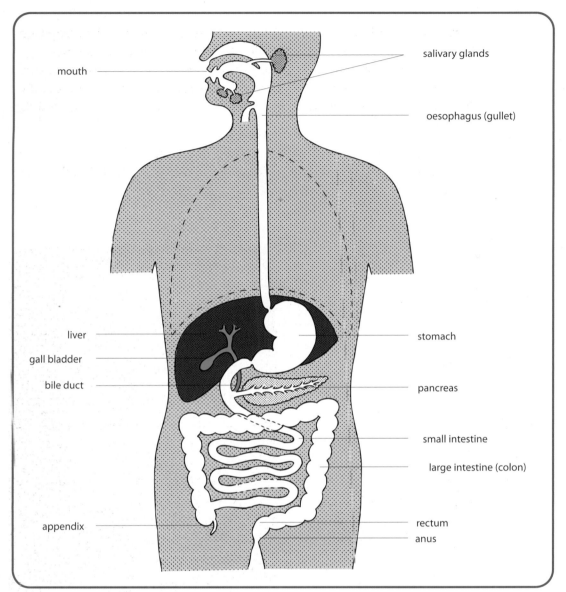

Figure 11.12 *Alimentary canal (gut)*

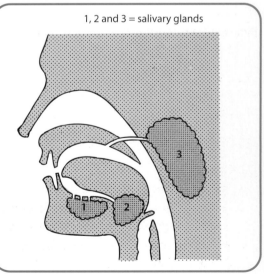

Figure 11.13 *Salivary glands*

Mechanism of peristalsis

Figures 11.14 and 11.15 show peristaltic activity in a region of the oesophagus. Part of the gut wall is composed of **circular muscle**. When this contracts behind a portion of food, the central hole of the tube becomes narrower and the food is pushed along. At the same time the circular muscle in front of the food becomes relaxed allowing the central hole to enlarge and let the food slip along easily.

Peristalsis is a wave-like motion since it results from the alternate contraction and relaxation of the gut wall muscle along the entire length of the alimentary canal.

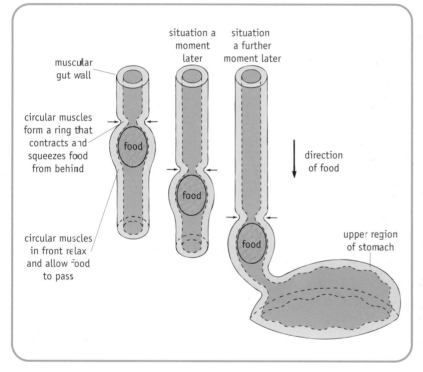

Figure 11.14 *Peristalsis in oesophagus*

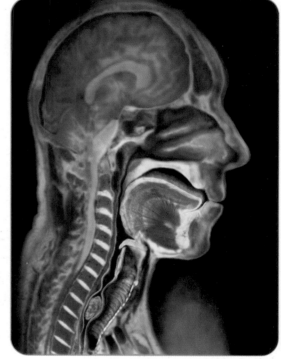

Figure 11.15 *Peristalsis*

Selecting and presenting information to illustrate peristalsis

Information

- The alimentary canal is a long muscular tube and peristaltic activity occurs all the way along it.
- Peristalsis ensures that food, once swallowed, is actively carried to the stomach. This movement of food does not depend on gravity and takes place even if the person is standing on his/her head.
- Peristalsis squeezes semi-solid food undergoing digestion along the small intestine and helps to mix it with digestive juices.
- Peristalsis keeps unwanted wastes on the move through the large intestine on their way to the rectum prior to elimination.

What to do

1 Study Figure 11.16. It shows three stages of a peristaltic wave in a small section of the gut. Part of stage (b) has been omitted.

2 Copy and complete stage (b) of the diagram.

3 Add the following labels to your diagram:
   muscular wall of alimentary canal
   region where circular muscle is contracted
   region where circular muscle is relaxed
   food undergoing digestion

4 Add a large arrow to your diagram to show the direction in which the food is travelling.

5 a) (i) Name the region of the alimentary canal represented by the diagram.
   (ii) Justify your choice.
   b) Name TWO other regions of the gut where peristalsis occurs.

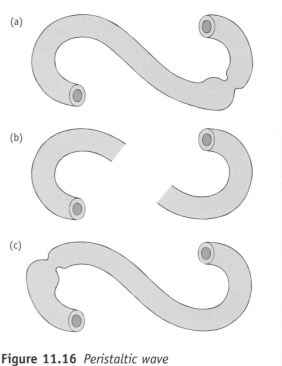

(a)

(b)

(c)

Figure 11.16 *Peristaltic wave*

Testing Your Knowledge

1 Arrange the following parts of the alimentary canal into the correct order: anus, large intestine, mouth, oesophagus, rectum, small intestine, stomach. (1)

2 a) Where is saliva produced? (1)

b) Name TWO substances present in saliva and state the function of each. (4)

3 Describe the mechanism of peristalsis with reference to a mouthful of food that has just been swallowed. (2)

Role of the stomach

The **stomach** is a muscular sac that is continuous with the base of the oesophagus at one end and the start of the small intestine at the other. The wall of the stomach contains layers of muscle (Figure 11.17).

The effect of contraction of the layer of **longitudinal** muscle on the shape of the stomach is shown in a simple way in Figure 11.18. The effect of localised contraction of **circular** muscle on the stomach is shown in a simple way in Figure 11.19. The combined effect of the different layers of muscle contracting and becoming relaxed in turn brings about **churning** of the stomach's contents (Figure 11.20).

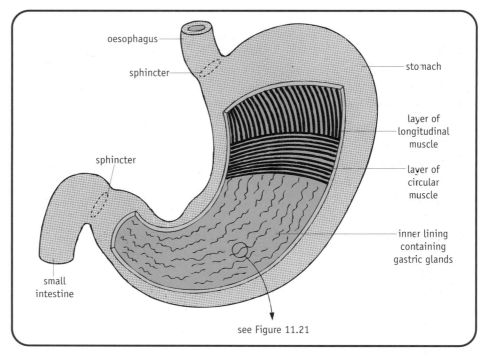

Figure 11.17 *Structure of the stomach*

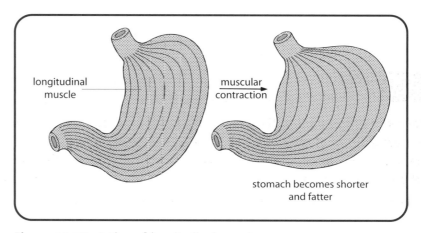

Figure 11.18 *Action of longitudinal muscle*

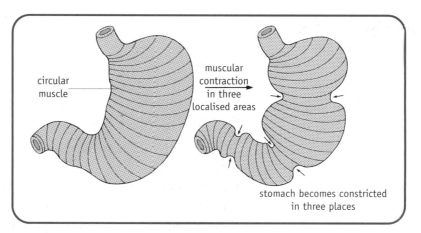

Figure 11.19 *Action of circular muscle*

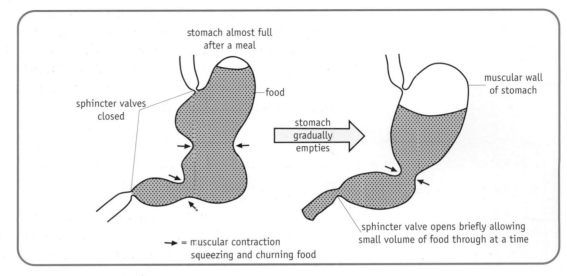

Figure 11.20 *Churning of food in the stomach*

During churning, food is held in the stomach by the closure of the muscular valves (sphincters) at either end of the stomach. Since the vigorous activity of the stomach wall mixes the food with the digestive (gastric) juices from the stomach wall, it helps the chemical breakdown of food.

Gastric glands
The inner lining of the stomach is folded and contains tubular **gastric glands** (Figure 11.21). These glands possess several types of secretory cell. The **mucus-secreting** cells are located in the 'neck' region of a gland. The **acid-secreting** cells and the **enzyme-secreting** cells are found towards the base of the gland.

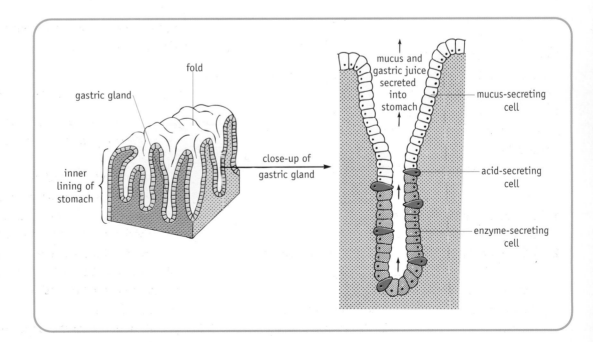

Figure 11.21 *Gastric gland*

On being secreted, slimy mucus adheres to the stomach lining and protects it from damage by digestive enzymes. The hydrochloric acid secreted by the gastric glands creates acidic conditions. The acid is needed to convert inactive pepsinogen (made by the enzyme-secreting cells) to active pepsin. This digestive enzyme then promotes the partial breakdown of protein molecules to simpler substances (peptides).

Practical Activity and Report

The effect of pH on the digestion of protein

Information

- Pepsin is an enzyme that digests insoluble protein to soluble end products.
- Cloudy egg white (albumen) suspension consists of molecules of insoluble protein suspended in water.
- When pepsin is active it promotes the following reaction:

cloudy suspension of insoluble protein molecules $\xrightarrow{\text{pepsin}}$ clear solution of soluble end products

- The experiment to be designed and carried out is to investigate the effect of pH on the digestion of protein by pepsin. Acidic, neutral and alkaline conditions are to be used.

You need

3 test tubes
3 labels or 1 marker pen
1 dropping bottle of 5% egg white suspension
1 dropping bottle of molar hydrochloric acid (HCl)
1 dropping bottle of water
1 dropping bottle of molar sodium hydroxide (NaOH) solution
1 dropping bottle of 1% pepsin solution
access to a thermostatically controlled water bath at 37°C
1 test tube rack
1 book of pH paper and a pH reference scale

What to do

1 Read all the instructions in this section before carrying out the experiment.
2 a) Draw a diagram of the three test tubes labelled A, B and C, with each containing egg white suspension and pepsin solution.
  b) Indicate on your design which tube will receive each of the following: acid (hydrochloric acid), water, alkali (sodium hydroxide).

3 Show the teacher your design.
4 Once the teacher has approved your design, set up your investigation by labelling the three test tubes A, B and C (and adding your initials) and then using 100-drop volumes of egg white suspension and 25-drop volumes of pepsin, acid, water and alkali as required.
5 Use pH paper to check that the contents of the tube that received acid are at acidic pH 2–3, that the contents of the tube that received water are at neutral pH 7 and that the contents of the tube that received alkali are at alkaline pH 10–11. If this is not the case, add further drops of acid, water or alkali as required.
6 Place the three test tubes in the water bath.
7 Draw up a table that refers to the three tubes and the appearance of their contents both before and after the time in the water bath.
8 After 30 minutes, examine the tubes and complete your table.
9 If other students have carried out the same experiment, pool the results.

Reporting

Write up your report by doing the following:

1 Copy the title given at the start of this activity.
2 Put the subheading '**Aim**' and state the aim of your experiment.
3 a) Put the subheading '**Method**'.
  b) Include a labelled diagram of your apparatus set up and ready to go into the water bath.
  c) Using the impersonal passive voice, briefly describe the experimental procedure that you followed and state how you obtained your results.
4 Put the subheading '**Results**' and draw a final version of your table of results.
5 Put a subheading '**Analysis of Results**' and write a sentence to state what you have found out from your

results. This should include answers to the following questions:

a) In which of the test tubes did pepsin promote the breakdown of protein?

b) How you were able to tell?

6 Put a subheading '**Conclusion**' and draw a conclusion based solely on the results of your experiment.

7 Put a final subheading '**Evaluation of Experimental Procedure**' and then answer the following:

a) State a possible source of error in this experiment.

b) Suggest how this source of error could be eliminated.

c) State how you could check the reliability of the results.

d) Feel free to comment on either of the following if you have an additional point you wish to make:

  (i) further improvements that you would include in a repeat of the experiment

  (ii) limitations of the equipment.

Practical Activity and Report

The effect of temperature on the digestion of protein

Information

- Pepsin is an enzyme that digests insoluble protein to soluble end products.
- Cloudy egg white (albumen) suspension consists of molecules of insoluble protein suspended in water.
- When pepsin is active it promotes the following reaction:

cloudy suspension of insoluble protein molecules $\xrightarrow{\text{pepsin}}$ clear solution of soluble end products

- The experiment to be designed and carried out is to investigate the effect of temperature on the digestion of protein by pepsin. The temperatures 2°C, 20°C, 37°C and 70°C are to be used.

You need

4 test tubes

4 labels or 1 marker pen

1 dropping bottle of 5% egg white suspension

1 dropping bottle of molar hydrochloric acid (HCl)

1 dropping bottle of 1% pepsin solution

1 beaker of icy water at 2°C

1 beaker of water at room temperature (e.g. 20°C)

access to thermostatically controlled water baths at 37°C and 70°C

1 test tube rack

1 book of pH paper and a pH reference scale

WHAT TO DO

1 Read all the instructions in this section before carrying out the experiment.

2 a) Draw a diagram of the four test tubes labelled A, B, C and D with each containing egg white suspension, pepsin solution and hydrochloric acid.

  b) Indicate on your design which test tube will be subjected to each of the four temperatures.

3 Show the teacher your design.

4 Once the teacher has approved your design, set up your investigation by labelling the four test tubes A, B, C and D (and adding your initials) and then using 100-drop volumes of egg white suspension and 25-drop volumes of pepsin and acid.

5 Use pH paper to check that the contents of the four tubes are all at acidic pH 2—3. If this is not the case, add further drops of acid as required.

6 Subject each test tube to its particular temperature by placing it in the appropriate container.

7 Draw up a table that refers to the four tubes and the appearance of their contents both before and after the time in the water bath.

8 After 30 minutes, examine the tubes and complete your table.

9 If other students have carried out the same experiment, pool the results.

Reporting

Write up your report by doing the following:

1 Copy the title given at the start of this activity.

2 Put the subheading '**Aim**' and state the aim of your experiment.

3 a) Put the subheading '**Method**'.

b) Include a labelled diagram of your apparatus set up with each test tube in its appropriate container.

c) Using the impersonal passive voice, briefly describe the experimental procedure that you followed and state how you obtained your results.

4 Put the subheading 'Results' and draw a final version of your table of results.

5 Put a subheading 'Analysis of Results' and write a sentence to state what you have found out from your results. This should include answers to the following questions:

a) In which of the test tubes did pepsin promote the breakdown of protein?

b) How could you tell?

6 Put a subheading 'Conclusion' and draw a conclusion based solely on the results of your experiment.

7 Put a final subheading 'Evaluation of Experimental Procedure' and then answer the following:

a) State a possible source of error in this experiment.

b) Suggest how this source of error could be eliminated.

c) State how you could check the reliability of the results.

d) Feel free to comment on either of the following if you have an additional point you wish to make:

(i) further improvements that you would include in a repeat of the experiment

(ii) limitations of the equipment.

Role of associated organs

Liver

The **liver** (see Figure 11.22) is not part of the alimentary canal but is closely associated with it and carries out many functions. One of these is the production of bile. (Further functions of the liver are discussed on page 250.)

Gall bladder

Bile leaves the liver by a tube and is stored in a sac called the **gall bladder** (see Figure 11.22). When required, bile flows down the bile duct and enters the small intestine along with digestive juices from the pancreas.

Role of bile

Bile in not an enzyme. On mixing with food, it converts large drops of fat (or oil) into tiny droplets. This process is called **emulsification**. It increases the surface area of the substrate upon which the enzyme lipase (see below) can act.

Pancreas

The pancreas (see Figure 11.22) is also an organ closely associated with the alimentary canal. It is the site of production of three further digestive enzymes – **pancreatic amylase**, **trypsin** and **lipase**. These enter the small intestine by the pancreatic duct.

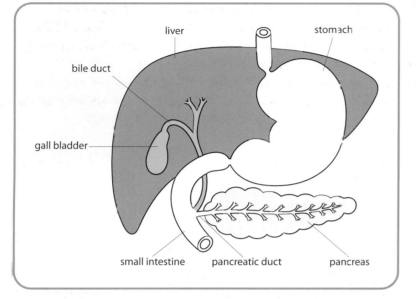

Figure 11.22 *Associated organs*

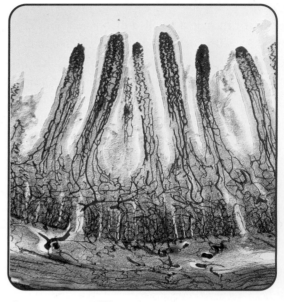

Figure 11.23 *Villi*

Role of the small intestine

Digestion

As partially digested food is moved along the small intestine by peristalsis, the three digestive enzymes from the pancreas continue to promote the breakdown of the remaining large insoluble molecules to smaller soluble molecules. This process, which is summarised in Table 11.5, prepares the food molecules for their absorption through the lining of the small intestine.

enzyme	substrate	end product
amylase	starch	maltose
trypsin	protein	peptides and amino acids
lipase	fat	fatty acids and glycerol

Table 11.5 *Digestive role of pancreatic enzymes*

Structure of small intestine

The small intestine is structurally suited to the function of food absorption in several ways:

- By being **long** and having a **folded** inner lining that bears thousands of fingerlike **villi** (see Figure 11.23), it presents a large absorbing surface area to digested food. This effect is further increased by the membranes of the epithelial cells that line the villi being folded into microvilli (Figure 11.24).

- The lining (epithelium) covering each villus is only **one cell thick**, allowing nutrient molecules to pass through it easily by diffusion.
- A **blood capillary** network and a **central lacteal** (tiny lymphatic vessel) are present in each villus to provide absorbed molecules with an efficient means of transport.

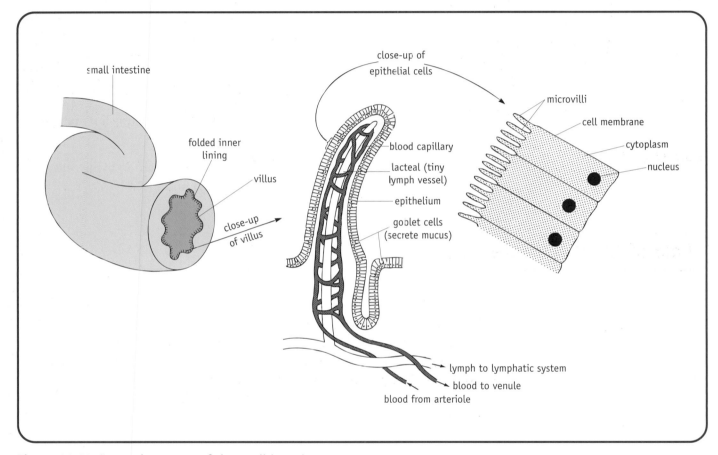

Figure 11.24 *Internal structure of the small intestine*

Absorption

Glucose and amino acids, the soluble end products of carbohydrate and protein digestion, are absorbed into the epithelial cells from where they pass directly into the **blood capillaries**.

The products of lipid digestion also pass into the epithelial cells but enter the central lacteal (not the blood capillary) and then pass into the **lymphatic system**. This is a transport system that eventually drains into the blood circulatory system through two ducts in the upper chest region.

Nutrients requiring no digestion by enzymes are also absorbed by the lining of the small intestine. These include calcium, iron and many vitamins.

Fate of absorbed materials

Blood rich in amino acids and glucose from the small intestine is transported in the hepatic portal vein (see page 282) to the **liver**. An appropriate concentration of glucose is released into general circulation for use as an energy source; any excess glucose is converted in the liver to **glycogen** and stored until required.

An appropriate concentration of amino acids is released into general circulation to be used as raw materials in protein synthesis during growth and tissue repair. Any excess amino acids are broken down in the liver to **urea**. This process is called **deamination**. The urea passes into general circulation and is removed later by the kidneys.

The products of fat digestion pass via the lymphatic system to the blood transport system. Some of these are used as an energy source. Excess fatty acids and glycerol are converted to **fat** and stored in the body's fatty tissues until required.

Vitamins and minerals are transported by the bloodstream to living cells where they play the roles described in Tables 11.1 and 11.2.

Role of large intestine

Material passing into the **large intestine** consists of undigested matter, bacteria and dead cells. The large intestine absorbs water from this unwanted material leaving **faeces**. Faeces are eliminated by being passed into the rectum from where they are later expelled through the anus.

Activity

Selecting and presenting information on colonic cancer

INFORMATION

The **colon** is another name for the main part of the large intestine. It is a muscular tube about 1.5 metres in length and composed of four regions, as shown in Figure 11.25 (on page 251). **Cancer** is an abnormal, uncontrolled growth of cells. In colonic cancer it leads to the development, somewhere in the colon, of a lump of cells called a tumour.

Table 11.6 shows the results of a recent survey into the incidence of colonic cancer. After lung cancer, it is the second most common cause of cancer death in the western world. Each year in Scotland over 2000 people are diagnosed with cancer of the colon and over 1000 of them die despite the fact that the disease can be cured if detected early enough.

Figure 11.26 (on page 251) shows the relationship between age and incidence of colonic cancer in the UK. Table 11.7 lists some aspects of lifestyle that may be related, directly or indirectly, to increased or decreased risk of colonic cancer.

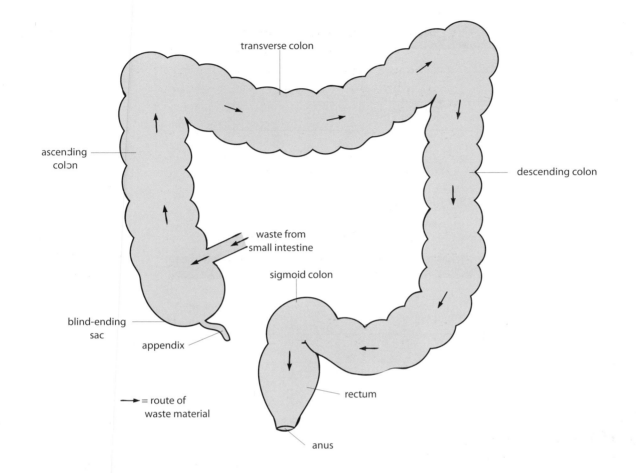

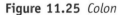

Figure 11.25 *Colon*

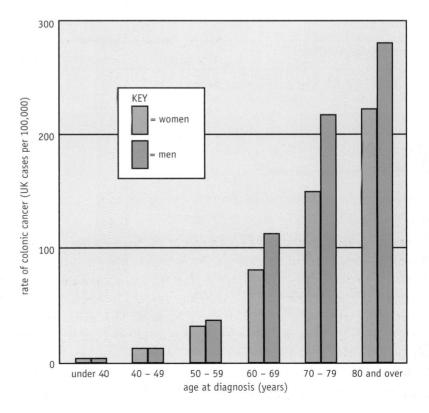

Figure 11.26 *Incidence of colonic cancer with increasing age*

country	rate of colonic cancer (cases per 100000 population)	
	female	male
Scotland	28.3	38.6
England	23.7	32.7
Wales	24.9	36.3
N. Ireland	26.7	34.0

Table 11.6 *Incidence of colonic cancer in UK*

aspect of lifestyle	risk score
being grossly overweight	-
consuming a diet rich in fruit and vitamins	++
consuming a diet rich in saturated fats	-
consuming a diet rich in vegetables of the cabbage family	++
consuming a diet with a low fibre content	--
drinking alcohol to excess	-
drinking green tea	+
living a very stressful lifestyle	-
smoking tobacco	-
taking regular exercise	+

Key to risk score

– = possible link with increased risk of colonic cancer

– – = very probable link with increased risk of colonic cancer

+ = possible link with decreased risk of colonic cancer

++ = very probable link with decreased risk of colonic cancer

Table 11.7 *Lifestyle and possible risk factors*

WHAT TO DO

Answer the following questions:

1 a) What is the *colon* and whereabouts in the human body is it situated?
   b) What is the normal function of the colon? (3)
2 a) In general, what is meant by the term *cancer*?
   b) What form does colonic cancer take? (2)
3 a) Present the data in Table 11.6 as a bar graph.
   b) Draw TWO conclusions from this data. (5)
4 Draw TWO conclusions from the graph shown in Figure 11.26. (2)
5 Using the information in Table 11.7, write a paragraph to describe the lifestyle that should be adopted to reduce the risk of developing colonic cancer later in life. (5)

Testing Your Knowledge

1 Name TWO types of muscle whose activities bring about churning of food in the stomach. (2)

2 a) Where exactly are gastric glands located? (1)

  b) (i) Name THREE chemicals secreted by gastric glands.

    (ii) Describe the roles played by TWO of these in the chemical breakdown of food.

    (iii) State the function of the third chemical. (6)

3 a) State THREE ways in which the structure of the small intestine is suited to its function. (3)

  b) Explain why these features enable digested food to be easily absorbed. (1)

4 What feature of the lining epithelium of the small intestine enables digested food to be easily absorbed? (1)

5 a) Name the TWO types of structure present inside a villus which transport absorbed materials. (2)

  b) For each of these give an example of an end product of digestion that it transports away from the small intestine. (2)

6 Briefly describe the role played by the large intestine. (1)

Applying Your Knowledge

1 Match the terms in list X with their descriptions in list Y.

List X
1) amylase
2) Benedict's solution
3) Biuret reagent
4) capillary
5) carbohydrate
6) digestion
7) fat

8) gastric gland
9) iodine solution
10) lacteal
11) lipase
12) mineral
13) peristalsis

14) protein
15) salivary gland
16) trypsin
17) villus
18) vitamin

List Y
a) chemical element such as iron needed for good health
b) food composed of amino acids and needed for tissue repair
c) form of muscular activity that moves food along the alimentary canal
d) enzyme present in pancreatic juice that promotes the digestion of fat
e) enzyme present in pancreatic juice that promotes the digestion of protein
f) enzyme present in saliva and pancreatic juice that promotes the digestion of starch
g) complex chemical compound required in tiny quantities to promote biochemical reaction
h) energy-rich food such as sugar or starch
i) process by which insoluble particles are broken down to soluble ones
j) structure that secretes mucus and starch-digesting enzyme
k) energy-rich food composed of fatty acids and glycerol
l) one of the many finger-like projections on the surface of the small intestine
m) tiny lymphatic vessel that absorbes the end products of fat digestion from the small intestine
n) tiny blood vessel that absorbs glucose and amino acids from the small intestine
o) structure that secretes mucus, acid and protein-digesting enzyme
p) chemical reagent used to test for the presence of simple sugar
q) chemical reagent used to test for the presence of starch
r) chemical reagent used to test for the presence of protein

2 The two pie charts in Figure 11.27 show the actual and the recommended consumption of the three main classes of food for the average person in Britain.

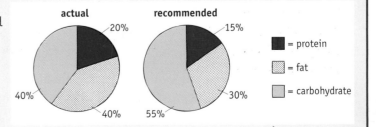

Figure 11.27

a) Present all of this information in one table. (3)

b) State THREE ways in which the average British citizen should change his/her eating habits according to these recommendations. (3)

3 Some amino acids can be synthesised by the body from simple compounds; others cannot be synthesised and must be supplied in the diet. The latter type are called the essential amino acids. The graph in Figure 11.28 shows the results of an experiment using rats where group 1 was fed zein (maize protein), group 2 was fed casein (milk protein) and group 3 was fed a varied diet.

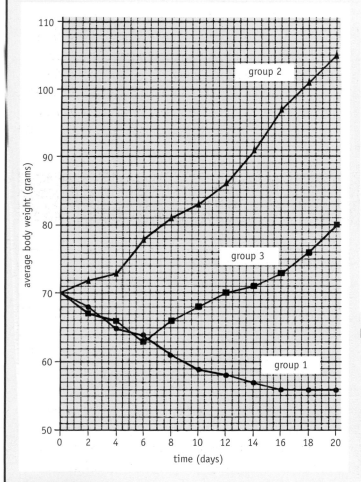

Figure 11.28

a) One of the proteins contains all of the essential amino acids whereas the other lacks two of them. Identify each protein and explain how you arrived at your answer. (3)

b) (i) State which protein was given to the rats in group 3 during the first six days of the experiment.

(ii) Suggest TWO different ways in which the diet of these rats could have been altered from day 6 onwards to account for the results shown in the graph. (3)

c) Calculate the percentage decrease in average body weight shown by the rats in group 1 over the 20-day period. (1)

4 A rat's normal heart rate is 500 beats per minute. Figure 11.29 shows the effect of withdrawing vitamin B_1 from the diet of a group of rats.

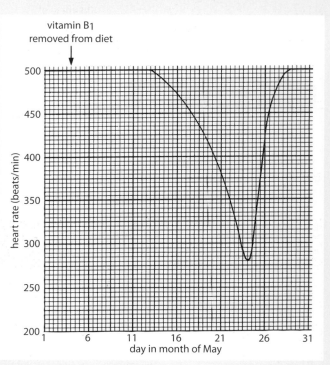

Figure 11.29

a) On which day in the month of May was vitamin B_1 removed from the rats' diet? (1)

b) For how many days did the rats' heart rate remain normal after the withdrawal of vitamin B_1? (1)

c) From where were the rats obtaining vitamin B_1 during this time? (1)

d) What effect does lack of vitamin B_1 have on heart rate? (1)

e) (i) What was the lowest heart rate recorded in the experiment?

(ii) Express this drop in heart rate as a percentage decrease from the normal heart rate. (2)

f) (i) Suggest what was done from 24th May onwards.

(ii) Justify your answer. (2)

5 Table 11.8 refers to the results of an experiment where three groups of 50 rats were fed the same basic diet except for the type of orange juice supplement given.

group	details of preparation of orange juice supplement	long-term effect on rats
1	freshly squeezed	absence of scurvy
2	squeezed and left to stand for a period of time	development of scurvy
3	squeezed, boiled and cooled	development of scurvy

Table 11.8

a) Which vitamin is required to prevent scurvy? (1)
b) Draw TWO conclusions from the table. (2)

6 Choose suitable forms of calculation and then copy and complete Table 11.9. (5)

7 Table 11.10 shows the factors that create a continuous demand for iron by the human body.
a) Give ONE reason why the human body needs iron. (1)
b) With reference to the table, explain why a fully grown healthy man still needs to consume iron in his diet. (1)
c) (i) Why does a pregnant woman have the highest demand for iron?
(ii) The iron content of porridge is 4 mg/100 g but only 10% of this iron is absorbed from the gut into the bloodstream. How much porridge would a pregnant woman need to eat to meet the demand given in the table?
(iii) Name TWO other foods rich in iron that the pregnant woman could also eat. (5)

vitamin	recommended daily allowance (mg)	mass of vitamin present in 100 g of named food (mg)		minimum mass of this food needed to supply recommended daily allowance (g)
A	1.0	carrots	2.0	
B_1	1.5	porridge		300.0
B_{12}	0.003	cheese	0.0015	
C		apple	5.0	900.0
D	0.01	kipper		50.0

Table 11.9

factor creating demand for iron	average mass of iron required to satisfy demand (mg/day)
basic loss in skin cells, bile and urine	0.5 – 1.0
growth: 0 – 1 year	0.7 – 0.8
1 – 11 years	0.3
adolescence	0.5
normal menstruation	0.5 – 1.6
pregnancy	6.0

Table 11.10

8 The undigested waste passed each day from the small intestine to the large intestine contains, on average, 2 litres of water. Faeces passed each day contain, on average, a total of 200 ml of water. Calculate the percentage of water reabsorbed from waste material by the large intestine on an average day. (2)

9 Figure 11.30 (on page 256) shows the daily turnover of calcium in a human adult who has consumed 900 mg of calcium. Bone is continually being remodelled by the processes of bone formation and reabsorption which are equal in an adult whose bones have reached their final size.

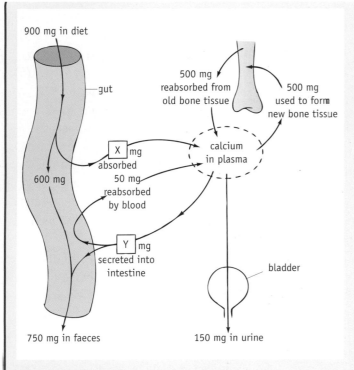

Figure 11.30

a) State the figures that should have been inserted in boxes X and Y. (2)

b) Assume that a 13-year-old girl also consumes 900 mg of calcium each day. Suggest a way in which her overall calcium turnover would differ from that of an adult. (1)

10 a) Present the data in Table 11.11 as a bar chart. (2)

b) By how many times is the length of time spent in the stomach by roast chicken longer than that spent by scrambled egg? (1)

foodstuff	average length of time spent in human stomach (h)
rice	1.5
scrambled eggs	2.5
steak	3.5
peas	4.5
roast chicken	6.0
tinned sardines	8.5

Table 11.11

c) Explain fully why an equal mass of each food must be used in this investigation. (1)

d) Suggest why people often feel hungry again an hour or two after consuming a Chinese meal. (1)

11 Figure 11.31 shows a diagram of the human stomach in action after a heavy meal.

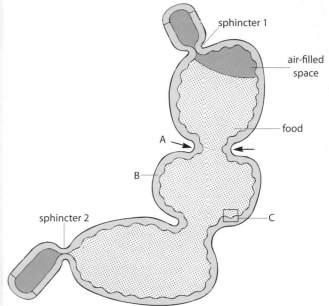

Figure 11.31

a) Describe the state of the circular muscle at sites A and B. (2)

b) State TWO ways in which food leaving the stomach at sphincter 2 will differ from food entering at sphincter 1. (2)

c) Make a simple diagram of the type of gland that would be found at position C and label the three types of secretory cell that it would contain. (4)

d) Suggest what happens when a person burps. (1)

12 In the investigation shown in Figure 11.32, tubes A and C received one large cube of solid egg white (protein) and tubes B and D received an equal mass of egg white chopped into small pieces. Tubes A and B were kept at room temperature; tubes C and D were kept at body temperature.

a) Give a simple equation to summarise the chemical reaction promoted by pepsin. (1)

b) (i) Name TWO tubes that could be compared to draw a conclusion about the effect of temperature on pepsin activity.

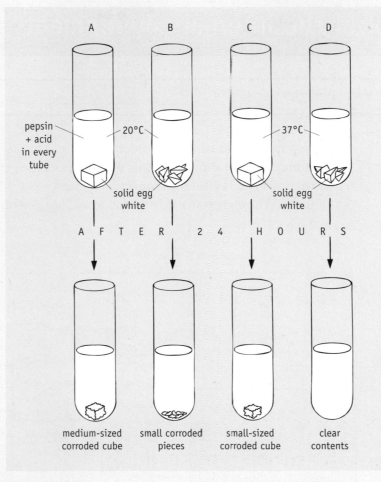

pepsin + acid in every tube

20°C   37°C

solid egg white   solid egg white

A F T E R   2 4   H O U R S

medium-sized corroded cube   small corroded pieces   small-sized corroded cube   clear contents

Figure 11.32

(ii) What conclusion can be drawn from the two tubes that you chose, after 24 hours?

(iii) In what way would the investigation need to be extended to allow you to make a generalisation about the effect of temperature on pepsin activity?　(3)

c) (i) Name the second variable factor being investigated in this set-up.

(ii) Name TWO tubes that could be compared to draw a conclusion about its effect on pepsin activity.

(iii) What conclusion can be drawn from the two tubes that you chose, after 24 hours?　(3)

d) (i) Which of the four tubes contains the conditions closest to the optimum for the digestion of egg white by pepsin?

(ii) Justify your answer.　(2)

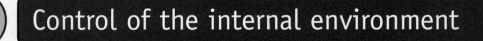

12 Control of the internal environment

Water balance

In order to maintain internal water balance, a human being must gain sufficient water through **eating**, **drinking** and the generation of **metabolic** water. This is needed to compensate for water lost in **sweat**, **breath**, **urine** and **faeces**. The daily water balance for an average person is summarised in Figure 12.1.

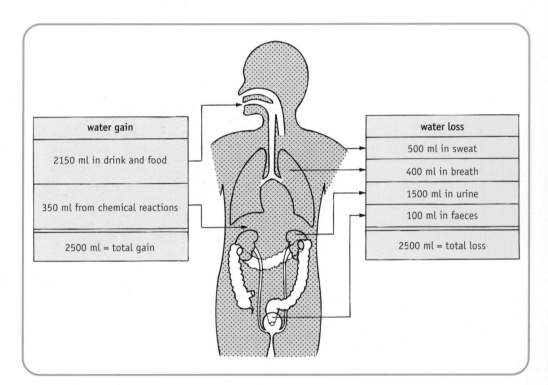

water gain		water loss
2150 ml in drink and food		500 ml in sweat
		400 ml in breath
350 ml from chemical reactions		1500 ml in urine
		100 ml in faeces
2500 ml = total gain		2500 ml = total loss

Figure 12.1 *Daily water balance*

Osmoregulation

About 70% of human body weight consists of water. This remains fairly constant from day to day because the **kidneys** regulate the water content of the body. **Osmoregulation** is the name given to this process. If a person consumes hardly any water then the kidneys produce a small volume of urine; if the person drinks a lot of water then the kidneys produce a large volume of urine. By this means the body's internal water content is kept constant.

The volume of water lost in sweat, faeces and exhaled air cannot be altered in order to maintain water balance. Only the kidneys are able to regulate the body's water content. This process of **osmoregulation** involves **antidiuretic hormone** (**ADH**) and is discussed more fully later in this chapter.

Human urinary system

The urinary system consists of the **kidneys**, **ureters**, **bladder** and **urethra** as shown in Figure 12.2. A **renal artery** delivers unpurified blood to each kidney; a **renal vein** transports purified blood away from each kidney.

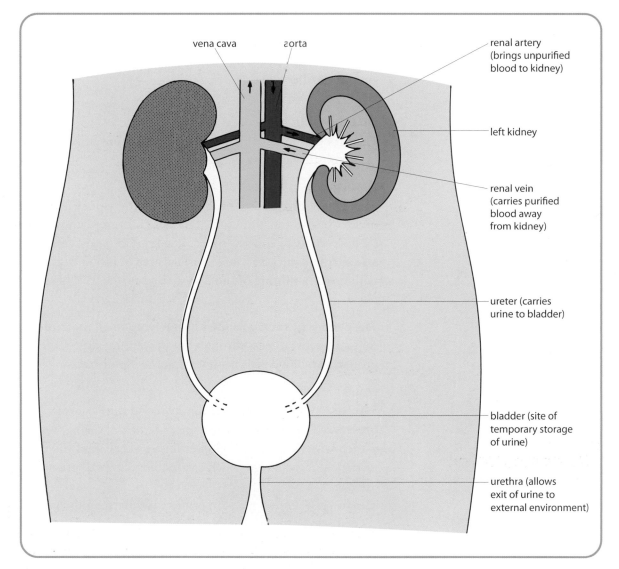

Figure 12.2 *Human urinary system*

Excretion

Excretion is the extraction and elimination from the body of the **waste** products of metabolism. **Urea** is a poisonous nitrogenous waste made in the liver by the breakdown (deamination) of surplus amino acids (formed during protein digestion). Urea passes from the liver into the bloodstream. It is then transported round the body and removed by the kidneys in **urine**. In order to eliminate urea, it is essential for the body to produce a minimum daily volume of urine of about 500 ml. This excretory role of the **kidneys** (in addition to their osmoregulatory role) is essential for survival.

Structure of kidney

Each kidney (see Figure 12.3) contains about a million microscopic functional units called **nephrons**. Each nephron (see Figures 12.4 and 12.5) is composed of several parts. A **glomerulus** (knot of blood vessels) is enclosed in a cup-shaped **Bowman's capsule**, which leads into a long kidney **tubule** surrounded by a dense network of blood capillaries. Several kidney tubules share a common **collecting duct**.

Urine production

The production of urine involves the **filtration** of blood and the **reabsorption** of useful substances only, from the filtrate.

Filtration

Blood containing waste products enters the kidney by the **renal artery** which divides into about a million branches each supplying a glomerulus. Since each glomerulus consists of a coiled knot of blood capillaries, this arrangement enables a large surface area of blood vessels to be in contact with the inner lining of the Bowman's capsule. It is here that **filtration** takes place.

The glomerulus acts as a filter by allowing small molecules (e.g. glucose, water, salts and urea) to pass through but preventing large molecules of plasma protein and blood cells from leaving the bloodstream.

Blood pressure

Successful filtration depends on the blood in the glomeruli being at **high pressure**. This is achieved as a result of the following two factors:
- each vessel supplying a glomerulus is a branch of the **renal artery**, itself a branch of the aorta, which carries blood at high pressure directly from the heart
- each vessel entering a glomerulus is **wider** than the vessel leaving it; this creates a 'bottle-neck' effect and causes the blood in a glomerulus to be squeezed.

Reabsorption

As glomerular filtrate passes through a kidney tubule, useful substances (all glucose, some salt and much water) are reabsorbed into the branching

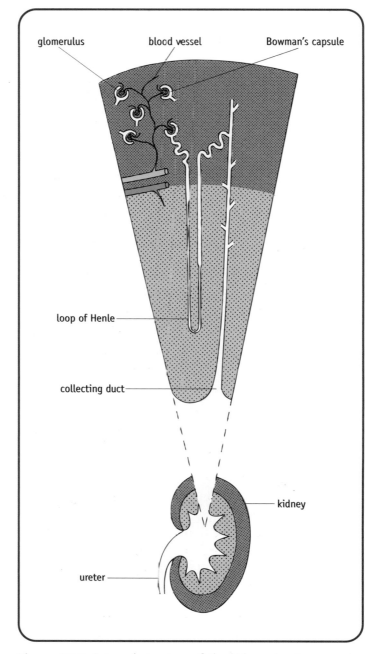

Figure 12.3 *Internal structure of the kidney showing a nephron*

network of capillaries surrounding the tubule. Further water is reabsorbed from the liquid flowing down the collecting duct.

The whole process is so effective that about 99% of the water originally present in the glomerular filtrate is reabsorbed. Therefore of the 120 ml of glomerular filtrate produced every minute, only 1 ml leaves in urine. In addition to excess unwanted water, final urine contains urea and excess salts.

Table 12.1 compares the composition of glomerular filtrate with urine. To bring about such **selective reabsorption**, kidney cells need energy. This is generated during tissue respiration in kidney cells using oxygen.

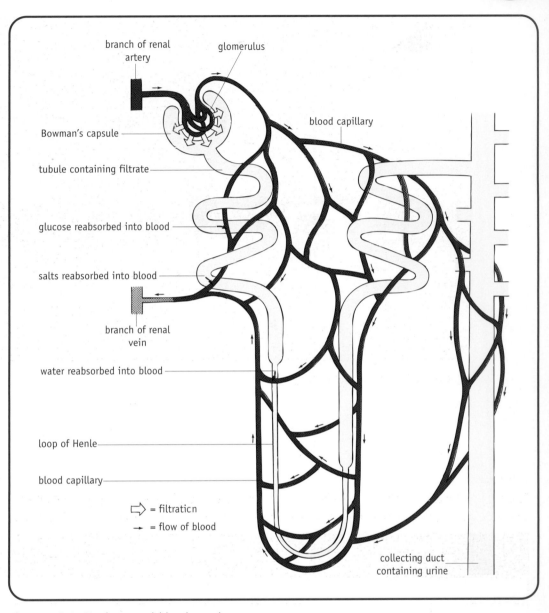

Figure 12.4 *Nephron and blood supply*

The blood capillaries surrounding the tubules unite to leave the kidney as the **renal vein**, which contains purified blood.

substance	glomerular filtrate (%)	final urine (%)
glucose	0.1	0
salts (sodium, calcium, etc.)	1.0	1.8
urea	0.02	2.0
water	98.5	96.0

Table 12.1 *Comparison of glomerular filtrate and urine*

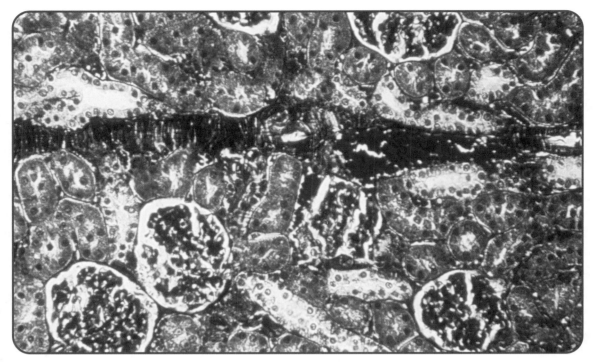

Figure 12.5 *Microscopic view of kidney*

Selecting and presenting data relating to volume and concentration of urine

Information

The data in Table 12.2 shows the effect of consuming a litre of water on the volume and concentration of urine produced by a student volunteer. Urine was collected from the student every 30 minutes over a 5-hour period.

time on 24-hour clock	volume of urine produced (ml)	concentration of urea (g/l)	colour of urine (see scale below)
12.00	55	24.1	3
12.30	60	24.5	3
13.00	50	23.8	3
13.30	55	24.6	3
14.00	355	4.2	1
14.30	490	2.9	0
15.00	215	6.8	2
15.30	60	24.1	3
16.00	45	24.9	3
16.30	55	24.5	3
17.00	60	24.1	3
0 = colourless, 1 = faint yellow, 2 = yellow, 3 = amber			

Table 12.2 *Volume and concentration of urine*

WHAT TO DO

1 Plot a line graph of volume of urine produced against time.

2 Using the same x axis but a new y axis on the right-hand side of the graph paper, plot a second line graph of concentration of urea against time.

3 Answer the following questions:

a) (i) At what time did the student consume the litre of water?

(ii) Indicate this event on your graph using a large labelled arrow.

b) What was the student's average output of urine per 30-minute period before the consumption of the litre of water?

c) How long did it take for the rate of urine production to return to normal after the litre of water had been consumed?

d) What was the student's average output of urine per 30-minute period after production had returned to normal?

e) (i) What relationship exists between volume of urine produced and concentration of urea?

(ii) Explain why.

Testing Your Knowledge

1 Name THREE ways by which the human body gains water. (3)

2 State FOUR ways in which the body loses water. (4)

3 Which organs are the main regulators of the body's water content? (1)

4 a) Which organ in the body makes urea? (1)

b) Which type of chemical molecules are broken down to form urea? (1)

c) Which organs remove urea from the bloodstream? (1)

d) In which liquid is urea eliminated from the body? (1)

5 a) What name is given to each knot of blood vessels surrounded by a Bowman's capsule in a kidney? (1)

b) (i) Name FOUR substances that pass from the blood into a Bowman's capsule.

(ii) Name ONE substance that fails to pass through. (5)

6 Which substance is completely reabsorbed back into the bloodstream from the kidney tubules? (1)

7 Name THREE substances normally present in urine. (3)

Water balance in aquatic animals

Animals such as the jellyfish whose cell contents are **isotonic** (equal in water concentration) to the surrounding environment do not have the problem of maintaining water balance since their cells neither gain nor lose water by osmosis.

A water balance problem does exist, however, in aquatic animals whose cell contents are **hypertonic** (lower in water concentration) compared with the surrounding environment since they constantly **gain** water by osmosis.

A water balance problem also exists in aquatic animals whose cell contents are **hypotonic** (higher in water concentration) compared with the surroundings since they constantly **lose** water to their environment by osmosis.

In each case, these animals possess certain adaptations that enable them to maintain their water concentration at the correct level.

Osmoregulation in freshwater bony fish

The scaly skin of a fish is impermeable to water. However, in a freshwater fish (see Figure 12.6), the delicate membranes of the mouth and gills are selectively permeable and the fish has to deal with a continuous influx (inflow) of water by osmosis as it breathes. To avoid the body tissues swelling up and becoming damaged, this excess water must be removed. Freshwater bony fish are adapted in several ways to carry out this function:

- Their kidneys possess **many large glomeruli** (see Figure 12.6), which allow a **high filtration rate** of the blood. This results in the production of very dilute glomerular filtrate.
- The kidney tubules are very efficient at **reabsorbing mineral salts** from this glomerular filtrate back into the bloodstream.

The fish therefore excretes a **large volume** of very dilute urine, which contains only a trace of salts and some nitrogenous waste.

The membranes of a freshwater bony fish's gills contain specialised cells that **absorb** salts from the water passing over them to replace the salts lost in urine.

Osmoregulation in saltwater bony fish

Although its scaly skin is impermeable to water, this type of fish (see Figure 12.7) suffers constant dehydration of its hypotonic tissues. This is because water is continuously being lost to the surrounding hypertonic sea water by

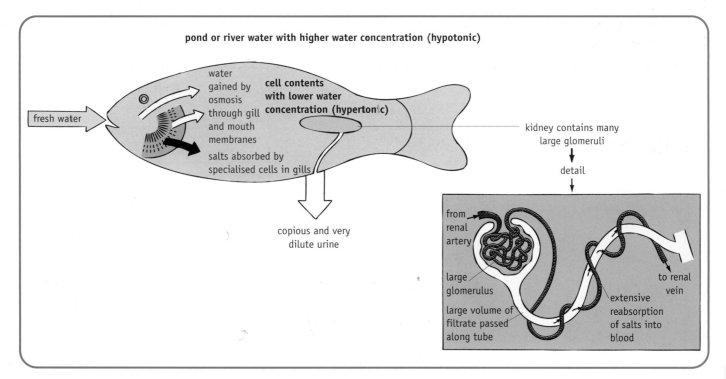

Figure 12.6 *Osmoregulation in freshwater bony fish*

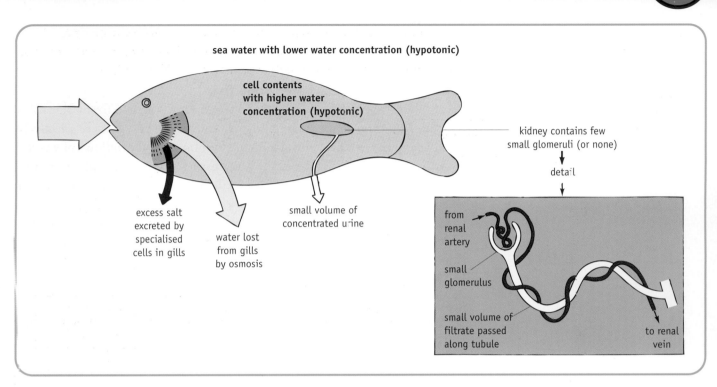

Figure 12.7 *Osmoregulation in saltwater bony fish*

osmosis through the gill and gut membranes, which are selectively permeable. The problem is overcome in several ways:

- **Sea water is drunk** to replace losses.
- The kidney contains only a **few small glomeruli** (see Figure 12.7) or none at all.

This results in a **low filtration** rate of blood. The fish therefore excretes a small volume of very concentrated urine containing the **minimum volume** of water necessary to get rid of its nitrogenous waste.

Specialised cells in the gill membranes work in the reverse direction to those of the freshwater fish. In a marine bony fish they **excrete** the excess salt gained from drinking sea water back out into the sea.

Negative feedback control

The millions of cells that make up the human body and the tissue fluid that bathes them are collectively known as the **internal environment**. For the human body to function efficiently, the state of the internal environment must be maintained within tolerable limits.

When some factor affecting the body's internal environment deviates from its normal optimum level (called the **norm** or **set point**), the change in the factor is detected by **receptors**. These send out nerve or hormonal messages that are received by **effectors**.

The effectors bring about certain responses that counteract the original deviation from the norm and return the system to its set point. This corrective mechanism is called **negative feedback control** (see Figure 12.8).

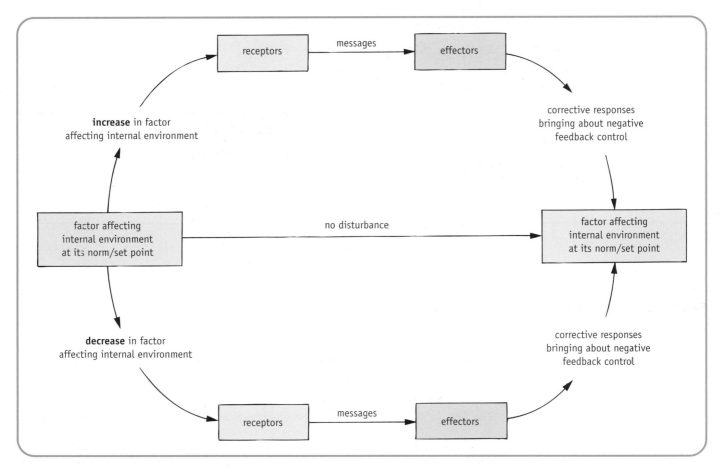

Figure 12.8 *Principle of negative feedback control*

Negative feedback control by ADH in humans

Water is absorbed from the tubules and collecting ducts in the kidneys. The volume of water returning to the bloodstream varies depending on the water concentration of the blood plasma.

If the water concentration of the blood decreases (due to increased sweating, lack of drinking water or consumption of salty food) then **osmoreceptors** in the **hypothalamus** (see Figure 12.9) are stimulated. These trigger the release of an increased concentration of **ADH** (antidiuretic hormone) by the **pituitary gland** into the bloodstream. On arriving in the blood capillary network in a kidney, ADH increases the permeability to water of the tubules and collecting ducts. As a result more water is reabsorbed into the bloodstream. This increases the blood's water concentration until it is returned to normal. Under these conditions a small volume of concentrated urine is produced.

If the water concentration of the blood becomes too high (due to consumption of excessive quantities of hypotonic liquids), less ADH is released and the kidney tubules and collecting ducts become less permeable to

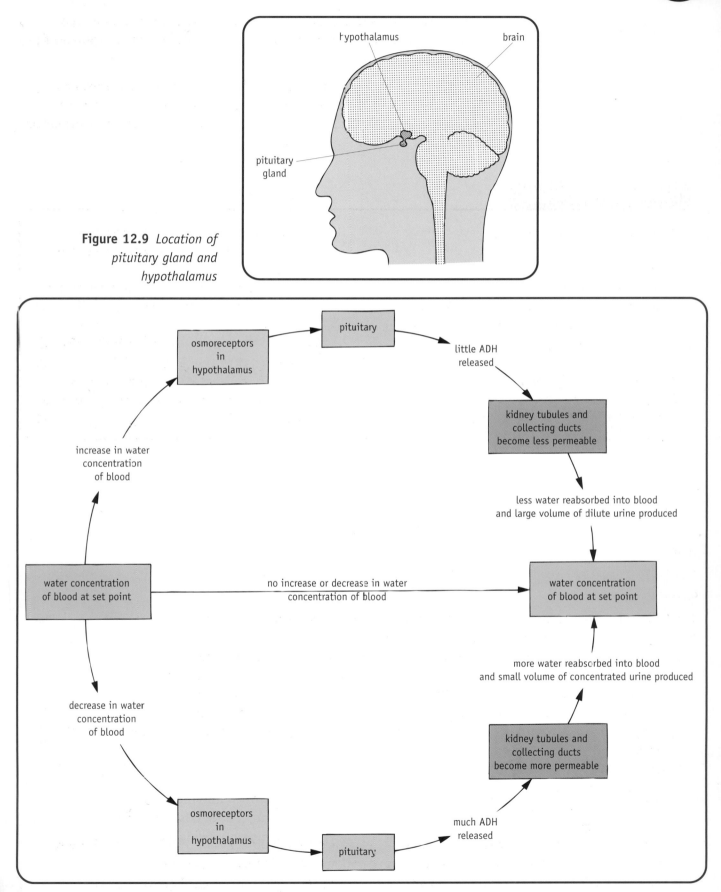

Figure 12.9 *Location of pituitary gland and hypothalamus*

Figure 12.10 *Negative feedback control by ADH*

water. **Less water** is reabsorbed and the blood's water concentration soon drops to around its normal level. Under these conditions a large volume of dilute urine is formed.

This process is summarised in Figure 12.10. The kidneys act as **effectors**, which respond to a hormonal message (ADH) and bring about negative feedback control. By this means the water balance of the internal environment is maintained.

Testing Your Knowledge

1 Rewrite the following sentences choosing only the correct answer from each choice in brackets.
The tissues of freshwater fish are (hypertonic/hypotonic) to the surrounding aquatic environment and so the fish constantly (loses/gains) water by osmosis. The problem is solved by the fish producing a (small/large) volume of (dilute/concentrated) urine. (4)

2 Describe the means by which osmoregulation is brought about in a saltwater fish. (3)

3 a) (i) Where in the human body are the osmoreceptors found that respond to a decrease in the water concentration of the blood?
  (ii) Which part of the body releases an increased concentration of ADH under these circumstances? (2)
 b) By what means does ADH bring about its effect so that water is conserved by the body? (2)
 c) Suggest why a corrective mechanism such as osmoregulation is described as a form of *negative feedback control*. (1)

Activity

Selecting and presenting information on the role of ADH

INFORMATION
Figure 12.11 represents the mechanism of osmoregulation in the human body.

WHAT TO DO
1 Copy or trace the diagram and complete the blank boxes with appropriate statements.
2 Use your diagram to answer the following questions:
 a) (i) Under what circumstances is a high concentration of ADH released by the pituitary gland?
  (ii) By what means does this result in the blood's water concentration returning to normal?
 b) (i) When is a low concentration of ADH released into the bloodstream by the pituitary gland?
  (ii) How does this result in the blood's water concentration returning to set point?
 c) Imagine an athlete running a marathon on a hot day. She sweats profusely during the race but drinks little fluid.
What effect will this have on her:
  (i) ADH production?
  (ii) water reabsorption?
  (iii) urine output?
 d) Some people suffer a malfunction of their pituitary gland and make insufficient ADH. This condition is called *diabetes insipidus* and it can be treated by taking nasal snuff containing ADH.
  (i) Suggest ONE of the symptoms of untreated *diabetes insipidus*.
  (ii) Predict what would happen if a sufferer exceeded the recommended dose of the treatment.

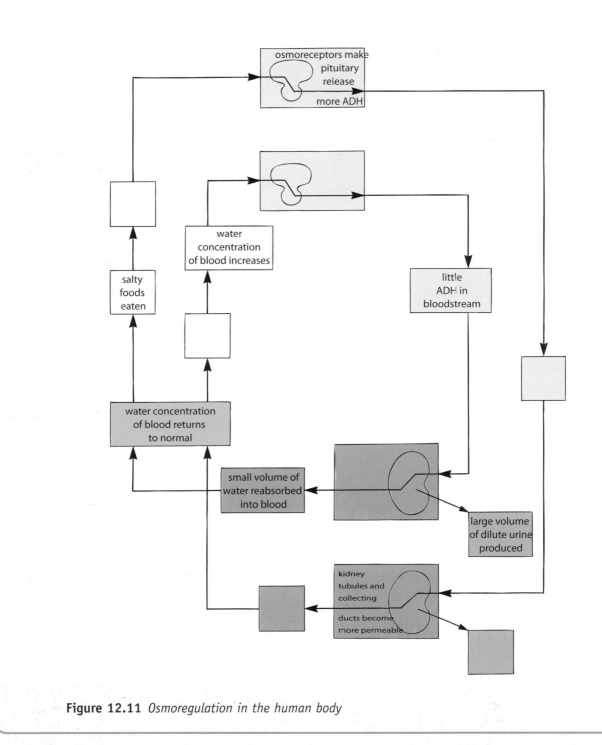

Figure 12.11 *Osmoregulation in the human body*

Applying Your Knowledge

1 Match the terms in list X with their descriptions in list Y.

List X
1) antidiuretic hormone
2) bladder
3) filtration
4) glomerulus
5) hypertonic
6) hypothalamus
7) hypotonic
8) kidney
9) kidney tubule
10) negative feedback
11) nephron
12) osmoregulation
13) pituitary gland
14) reabsorption
15) urea
16) ureter
17) urethra
18) urine

List Y
a) regulation of the body's water content
b) one of two bean-shaped organs that excrete urea and maintain water balance
c) general name for the system of control where effectors counteract a deviation by a factor from the norm
d) one of a million functional units in a kidney
e) structure attached to the mid-brain that releases ADH
f) region of the brain containing osmoreceptors
g) process by which useful molecules are returned from kidney tubules to the bloodstream
h) tube that allows the exit of urine from the bladder to the external environment
i) tube that carries urine from a kidney to the bladder
j) sac where urine is stored temporarily
k) chemical messenger that brings about osmoregulation by negative feedback control
l) poisonous waste material made in the liver from excess amino acids
m) process by which small molecules pass out of the blood into Bowman's capsules
n) solution of urea and salts excreted by kidneys
o) region of a nephron from which glucose is reabsorbed
p) knot of blood vessels surrounded by a Bowman's capsule
q) term describing a solution with a relatively low water concentration
r) term describing a solution with a relatively high water concentration

2 a) Copy and complete Tables 12.3 and 12.4, which give the water balance figures for a human volunteer over a period of 24 hours. (4)
b) Present the percentage gains and percentage losses as two separate pie charts. (4)

c) If the person had consumed a large volume of drinking water during this period, which form of water loss would have increased to maintain water balance? (1)

means by which water was gained	volume of water gained (ml)	percentage of total gain
in food	800	
in drink	1350	
metabolism		
total	2500	100

Table 12.3

means by which water was lost	volume of water lost (ml)	percentage of total loss
by breathing	400	
in sweat		
in urine	1450	
in faeces	100	
total	2500	100

Table 12.4

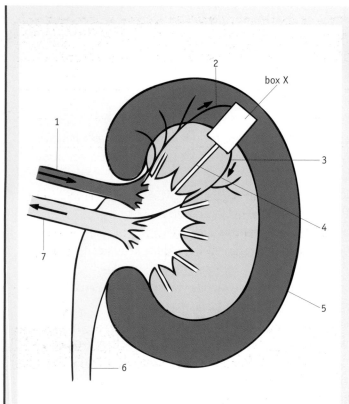

Figure 12.12

time on 24-hour clock	volume of urine collected (ml)	
	day 1	day 2
12.00	54	50
12.30	51	53
13.00	54	56
1 litre of liquid consumed immediately after 13.00 hours		
13.30	320	53
14.00	440	56
14.30	150	51
15.00	53	54

Table 12.5

3 a) Identify parts 1 to 7 in Figure 12.12 (7)
   b) Name THREE structures that would be found in box X. (3)
   c) What liquid would be present in structure
   (i) 1? (ii) 6? (2)
   d) Give ONE difference in chemical composition between the liquids present in tubes 1 and 7. (1)

4 Which of the following are BOTH functions of the kidney?
   (Choose ONE answer only.)
   A excretion and defecation
   B osmoregulation and excretion
   C defecation and urea production
   D urea production and osmoregulation (1)

5 The data in Table 12.5 refer to the responses of a student to consuming, on day 1, a litre of water and, on day 2, a litre of salt solution isotonic to blood plasma. On each day, urine was collected every 30 minutes over a period of 3 hours.
   a) Why was the same person used in both parts of the experiment? (1)
   b) What was the one variable factor under investigation? (1)

   c) What is the normal rate of urine production per 30-minute interval before consumption of the litre of liquid? (1)
   d) (i) What effect did consumption of a litre of water have on urine production?
      (ii) How long did it take for negative feedback control to take effect and return the system to set point? (2)
   e) (i) What effect did consumption of the litre of salt solution have on urine production?
      (ii) Suggest why this effect occurred with reference to ADH. (2)

6 Figure 12.13 shows a nephron from a human kidney.

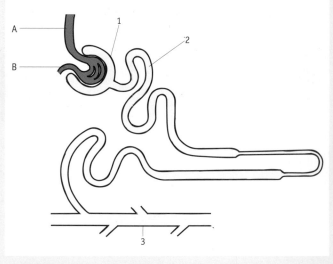

Figure 12.13

type of renal fluid	concentration of substance in renal fluid (g/100 ml)				
	urea	glucose	protein	salts	amino acids
plasma	0.03	0.1	8.0	0.72	0.05
glomerular filtrate	0.03	0.1	0	0.72	0.05
urine	2.1	0	0	1.44	0

Table 12.6

a) State the TWO factors that make the blood pressure at A higher than that at B. (2)

b) Name structures 1 to 3 and state ONE function of each. (3)

c) (i) In what way is structure 3 affected by an increase in concentration of ADH in the bloodstream?

(ii) Relate this change in ADH concentration to the volume and concentration of urine produced. (3)

7 Table 12.6 gives the results of an investigation into the composition of three types of renal fluid in an adult male.

a) Name TWO substances that were completely reabsorbed back into the bloodstream. (2)

b) (i) Which TWO substances became concentrated in urine?

(ii) For each of these, state the concentration factor. (4)

c) The man produced a total of 180 litres of glomerular filtrate in one day. Calculate his GFR (volume of glomerular filtrate in millilitres produced per minute by both kidneys). (1)

d) If the man produced 1.8 litres of urine per day:

(i) what percentage of glomerular filtrate was passed as urine?

(ii) what happened to the rest of the glomerular filtrate? (2)

8 Copy and complete Table 12.7, which compares freshwater bony fish and saltwater bony fish with respect to certain features relating to osmoregulation. (5)

feature	freshwater bony fish	saltwater bony fish
relative number of glomeruli in kidney		
relative size of glomeruli		
rate of filtration of blood		
state of fish's tissues compared with environment (hypo/hypertonic)		
relative volume of urine produced		

Table 12.7

Word bank

absorbed	negative
acid	nitrogen
ADH	osmoreceptors
carbon	osmoregulation
churn	peristalsis
circular	permeable
digestion	pituitary
ducts	proteins
filtration	reabsorbed
fresh	salt
glucose	sweat
glycerol	urea
lubricated	urine
lymphatic	villi
metabolic	vitamins

Table 12.8 *Word bank for Chapters 11 and 12*

What You Should Know

(Chapters 11 and 12) (See Table 12.8 for word bank)

1 The main groups of foods are carbohydrates (composed of simple sugars), _____ (composed of amino acids) and fats (made of fatty acids and _____).

2 The three main groups of foods all contain the chemical elements _____, hydrogen and oxygen. In addition, proteins contain _____.

3 A healthy diet must also contain small quantities of _____ and minerals.

4 The breakdown of insoluble molecules of food to soluble molecules by enzymes is called _____. This process is necessary to enable food to be _____ into the bloodstream.

5 Food in the mouth is _____ by saliva before being moved down the oesophagus by muscular activity called _____.

6 The wall of the stomach contains longitudinal and _____ muscle whose activities _____ food and mix it with mucus, _____ and enzymes secreted by the gastric glands.

7 The small intestine's inner surface area is enormously increased by the presence of finger-like projections called _____. These absorb food once it has undergone further digestion by pancreatic enzymes.

8 _____ and amino acids pass into the blood transport system and the products of fat digestion pass into the _____ system.

9 The human body gains water in food, drink and _____ water. It loses water in _____, breath, urine and faeces.

10 The kidneys are the organs of _____ and excretion.

11 _____ is a waste product made from the breakdown of excess amino acids in the mammalian liver. Some _____ must be made every day by the kidneys to remove urea from the bloodstream.

12 Unpurified blood enters a kidney and undergoes _____ in glomeruli. Useful materials are _____ from kidney tubules. Unwanted substances are eliminated in urine.

13 _____water bony fish gain water continuously from their surroundings. They solve the problem by producing a large volume of dilute urine.

14 _____water bony fish lose water continuously to their surroundings. They solve the problem by drinking sea water and excreting the excess salt.

15 When the water concentration of blood in the human body decreases, _____ in the hypothalamus respond and trigger an increased release of antidiuretic hormone (ADH) from the _____.

16 On arriving in the kidneys, _____ makes the tubules and collecting _____ become more _____. More water is reabsorbed and the blood's water concentration is returned to normal. The reverse situation occurs when the blood's water concentration increases above its set point. This mechanism is an example of _____ feedback control.

13 Circulation and gas exchange

Function of the heart

The **circulatory system** is composed of a continuous system of tubes called **blood vessels** which carry blood to all parts of the body. Blood must be continuously transported round this system to allow exchanges of materials between the bloodstream and:

- the **external environment** (e.g. absorption of oxygen and release of carbon dioxide at alveoli)
- the **internal environment** (e.g. release of oxygen and removal of carbon dioxide at living cells).

It is the function of the **heart** to maintain this uninterrupted flow of blood by pumping it round the body's circulatory system.

Heart

The **heart** is a muscular organ that is divided into four chambers: two **atria** and two **ventricles** (see Figure 13.1). The right atrium receives deoxygenated

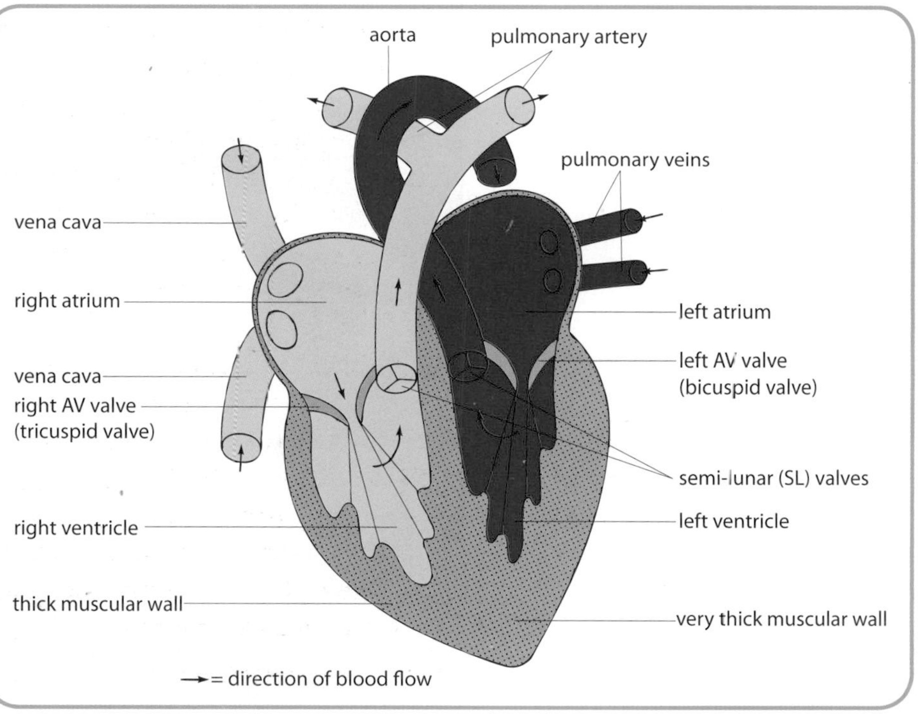

Figure 13.1 *Human heart*

blood from all parts of the body via two main veins called the **venae cavae**. This deoxygenated blood passes into the right ventricle and then leaves the heart by the **pulmonary artery** which divides into two branches each leading to a lung.

Following oxygenation in the lungs, blood returns to the heart by the **pulmonary veins** and enters the left atrium. It flows from the left atrium into the left ventricle and leaves the heart by the **aorta**, the largest artery in the body.

Thickness of ventricle walls

The wall of the left ventricle is particularly thick and muscular since it is required to pump blood all round the body. The wall of the right ventricle is less thick since it pumps blood only the short distance to the lungs.

Valves

Figure 13.1 shows the four heart **valves**. Two of these, situated between the atria and the ventricles, are called the **atrio-ventricular** (AV) valves. The AV valves allow blood to flow from atria to ventricles but prevent backflow from ventricles to atria.

The other two heart valves, situated at the origins of the pulmonary artery and aorta, are called the **semi-lunar** (SL) valves. These valves open during ventricular contraction allowing blood to flow into the arteries. When blood pressure in the arteries is greater than blood pressure in the ventricles, they close, preventing backflow. The presence of the valves ensures that blood is only able to flow in **one direction** through the heart. Figure 13.2 shows an artificial heart valve.

Figure 13.2 *Artificial heart valve*

Coronary arteries

Since the heart wall is composed of actively respiring muscle cells, it needs a continuous supply of oxygenated blood. The first two branches of the aorta are the left and right **coronary arteries** (see Figures 13.3 and 13.4). These vessels spread out over the surface of the heart and divide into an enormous number of tiny branches leading to a dense network of capillary beds in the muscular wall of the heart itself. This arrangement allows rapid diffusion of oxygen and food into the actively respiring cardiac muscle tissue.

Coronary veins return deoxygenated blood from the heart wall to the vena cava.

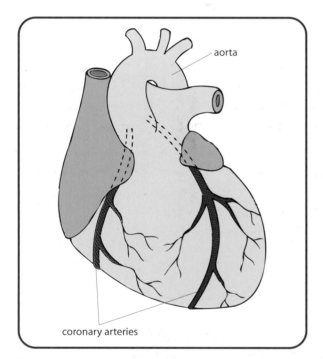

Figure 13.3 *Coronary arteries*

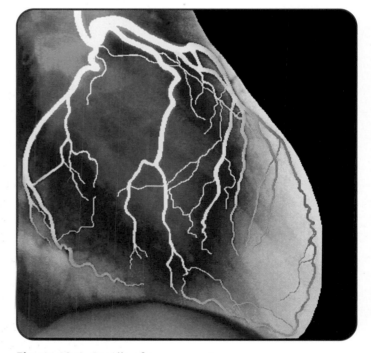

Figure 13.4 *Details of coronary artery*

Coronary heart disease

This general term refers to any disease that results in restriction or blockage of the coronary blood supply to part of the heart's muscular wall. The blockage deprives the affected part of oxygen and may result in death of the muscle cells.

If a small region of the heart is affected, the person may make a satisfactory recovery; if a large part is affected, the person may die instantly. Coronary heart disease is the most common form of premature death in many developed countries.

Practical Activity

Examining a mammalian heart

INFORMATION

- This activity should be carried out by a small group of students acting as a discussion group while they take turns in carrying out the instructions given below.
- Figure 13.5 shows a sheep's heart with about 10–20 mm of each main blood vessel left intact. Ideally your specimen should resemble this.

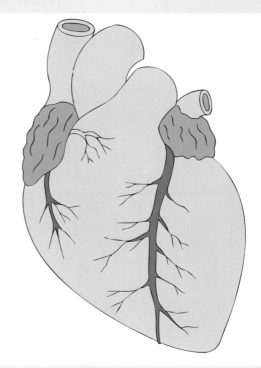

Figure 13.5 *Sheep's heart*

YOU NEED

1 sheep's heart
1 dissecting board
1 pair of dissecting scissors
1 pair of forceps
several pairs of disposable plastic gloves

WHAT TO DO

1 Wearing plastic gloves, examine the heart and, with the aid of Figure 13.6, distinguish between the atria – (1) and (2) in the diagram – and the region occupied by the two ventricles (3).

2 Distinguish between the two atria and note the wide but thin-walled veins (4) and (5) entering them. Try to identify these veins and state their functions.

3 Feel the thick muscular wall of the ventricles (3).

4 Try to distinguish the locations of the left and right ventricles and note the thick-walled arteries (6) and (7) leaving them. Try to identify these arteries and state their functions.

5 Note the presence of the coronary arteries (8) running over the outer surface of the heart. Discuss their role with the members of your group.

6 Cut the heart open by making the first incision down through the pulmonary artery and right ventricle to the apex (9) of the heart, as shown in Figure 13.6.

7 Examine the semi-lunar (SL) valve at the base of the pulmonary artery. Discuss its function.

8 Note the thick muscular ventricle wall. What role does it play?

9 Make the second line of incision to open the right atrium and note how thin its wall is compared with that of the ventricles.

10 Look for the right AV valve. What role does it play?

11 Cut open the left side of the heart and compare the thickness of the left and right ventricle walls. What is the reason for this obvious difference?

12 Now attempt the *Testing Your Knowledge* exercise that follows this activity.

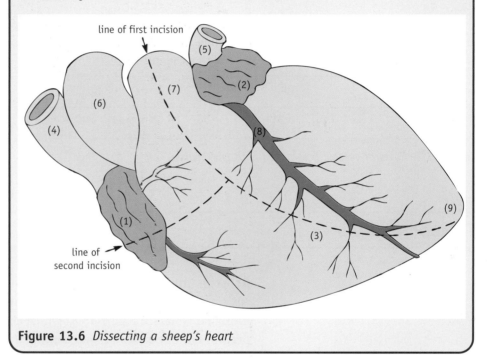

Figure 13.6 *Dissecting a sheep's heart*

Testing Your Knowledge

1 Give the names of the FOUR chambers of the mammalian heart. (4)

2 a) Name the veins that enter (i) the right atrium and (ii) the left atrium. (2)
   b) State whether the type of blood carried by each of these veins is oxygenated or deoxygenated. (1)
   c) From where does each of these veins bring its blood? (2)

3 a) Give the name of the artery that leaves (i) the right ventricle and (ii) the left ventricle. (2)
   b) State whether the type of blood carried by each is oxygenated or deoxygenated. (1)

c) State where each of these arteries is transporting blood to. (2)

4 a) From which artery do the coronary arteries arise? (1)
   b) Which tissue do the coronary arteries supply with oxygenated blood? (1)
   c) Describe the circumstances involving the coronary arteries that lead to a heart attack. (2)

5 a) Describe the positions of the TWO types of heart valve. (2)
   b) State the function of each type of heart valve. (2)

6 Explain why one ventricle's muscular wall is thicker than the other ventricle's muscular wall. (2)

Heart and associated blood vessels

The path taken by blood as it flows through the heart and its associated blood vessels is shown in a simplified way in Figure 13.7 and summarised in Table 13.1.

number in Figure 13.7	blood vessel	transports blood from	transports blood to
1	vena cava	all parts of the body	right atrium
2	pulmonary artery	right ventricle	lung
3	pulmonary vein	lung	left atrium
4	aorta	left ventricle	all parts of the body

Table 13.1 *Roles of vessels associated with the heart*

Blood vessels

Arteries and veins

An **artery** is a vessel that carries blood away from the heart. It has a thick muscular wall (see Figure 13.8) which enables it to withstand the **high pressure** of oxygenated blood coming from the heart. (The pulmonary artery is exceptional in carrying deoxygenated blood.)

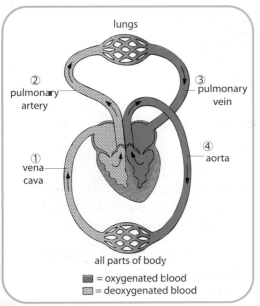

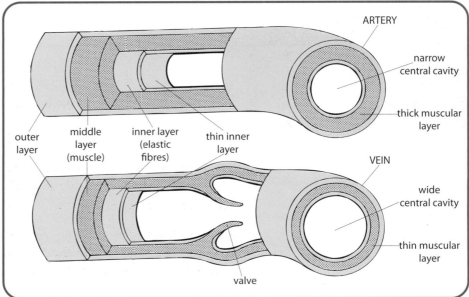

Figure 13.7 *Heart and associated blood vessels*

Figure 13.8 *Comparison of an artery and a vein*

Each time the heart beats, contraction of the muscular ventricle walls forces blood along the arteries at high pressure. This can be felt as a **pulse beat** in an artery. The presence of a pulse indicates that blood is being pumped through arteries.

A **vein** is a vessel that carries blood back to the heart. Although muscular, its wall (see Figure 13.8) is thinner than that of an artery since deoxygenated blood flows along a vein at **low pressure**. (The pulmonary vein is exceptional in carrying oxygenated blood.) Compared with an artery the central cavity of a vein is wider. This helps to reduce resistance to the flow of blood along a vein. **Valves** are present in veins to prevent backflow.

Capillaries

Blood is transported from arteries to veins through a series of smaller blood vessels. Each artery branches into smaller and smaller vessels called arterioles. Each vein, on the other hand, is formed from the convergence of many tiny venules. Blood passes from arterioles to venules via a dense network of tiny microscopic vessels called **capillaries** (see Figure 13.9).

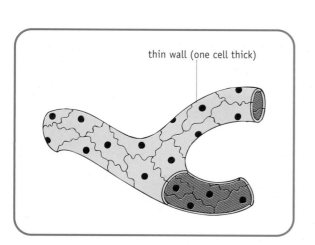

Figure 13.9 *Capillary*

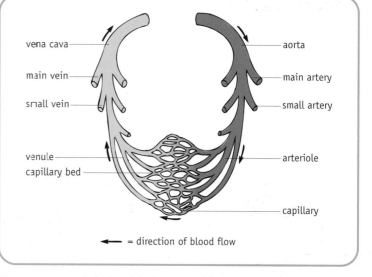

Figure 13.10 *Relationship between blood vessels*

Capillaries are the most numerous type of blood vessel in the body. They are referred to as **exchange vessels** since all exchanges of materials between blood and living tissues take place through their thin walls (only one cell thick).

The relationship between blood vessels and the direction of blood flow in the circulatory system is summarised in Figure 13.10. Figure 13.11 shows a simplified version of the human circulatory system.

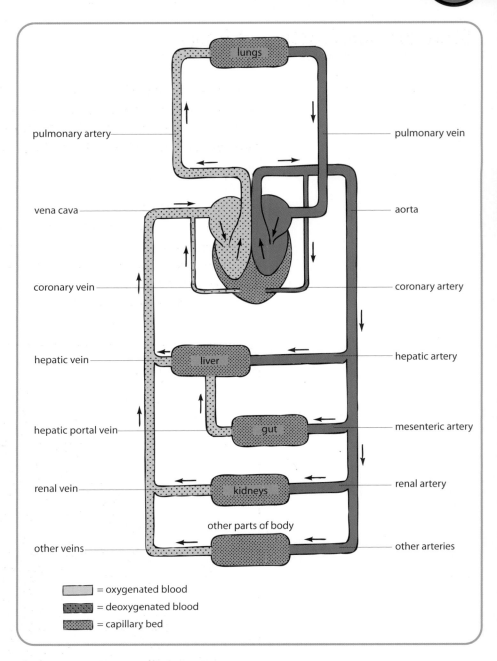

Figure 13.11 *Human circulatory system*

Testing Your Knowledge

1 Describe the route taken by blood as it flows from the vena cava to the aorta. (6)

2 Rewrite the following sentence choosing the correct answer from each set of brackets. Blood leaves the heart in (an artery/a capillary/a vein), flows through a dense network of thin-walled (arteries/capillaries/

veins) and returns to the heart in (an artery/a capillary/a vein). (3)

3 Name the vessel that supplies oxygenated blood to (i) the heart muscle (ii) the gut (iii) the kidney. (3)

4 Name the vessel that carries deoxygenated blood away from the (i) the liver (i) the gut (iii) the kidney. (3)

Structure and function of lungs

The **lungs** (see Figure 13.12) are a mammal's organs of gaseous exchange. Air enters the nose or mouth and passes down through a system of tubes of decreasing diameter – the **trachea**, **bronchi** (singular bronchus) and **bronchioles**. The trachea and bronchi are supported and held open by rings of cartilage.

Internal structure

Air passing along the narrow bronchioles ends up in tiny air sacs deep in the lungs (see Figure 13.13). These air sacs are called **alveoli** (singular alveolus). The alveoli are so numerous that they provide a very **large surface area** for gas exchange. The total internal surface area of the two lungs is approximately 90m^2 (the area of a tennis court).

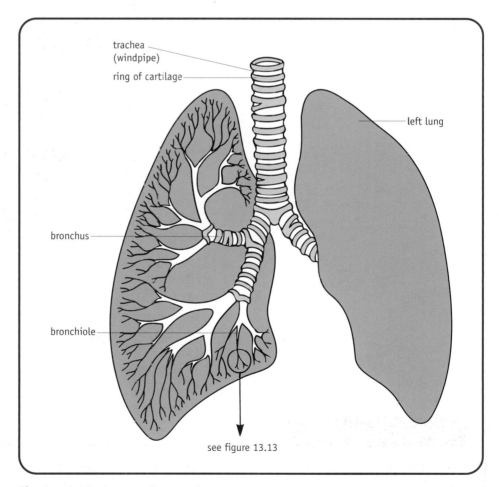

trachea (windpipe)

ring of cartilage

left lung

bronchus

bronchiole

see figure 13.13

Figure 13.12 *Organs of gas exchange*

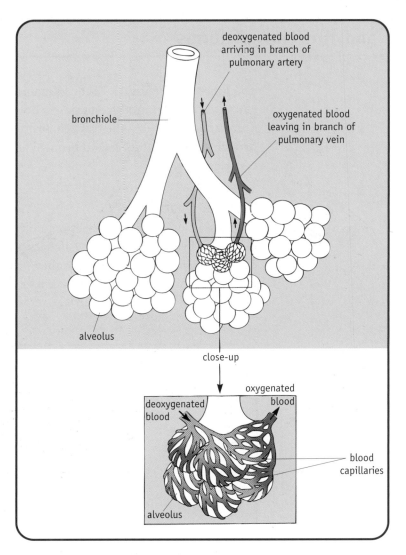

Figure 13.13 *Alveoli and capillary network*

Capillary network

Each alveolus is surrounded by a dense network of blood capillaries. The lining of an alveolus is very **thin** and in **close proximity** to the walls of the blood capillaries which are themselves only one cell thick. This combination of large surface area, short distance and thin walls presents the ideal conditions for gas exchange to occur between alveolar air (the external environment) and blood (the internal environment).

Gas exchange

Blood arriving in a lung is said to be **deoxygenated** because it contains a low concentration of oxygen. Since air breathed into an alveolus contains a higher concentration of oxygen, diffusion occurs.

Oxygen first dissolves in the moisture on the surface of the thin lining of an alveolus (see Figures 13.14 and 13.15) and then diffuses into the blood in the surrounding blood vessels (capillaries). The blood therefore becomes **oxygenated** (rich in oxygen) before leaving the lung and passing to all parts of the body.

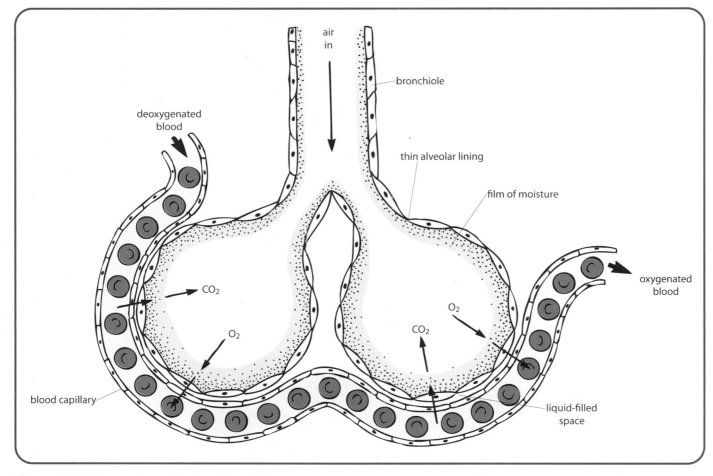

Figure 13.14 *Gas exchange in alveoli*

Since deoxygenated blood contains a higher concentration of carbon dioxide than the air breathed into an alveolus, carbon dioxide diffuses from the blood into the alveolus, from where it is exhaled.

Important features of a respiratory surface

The alveoli and associated blood vessels make the lungs efficient gas exchange structures for the reasons given in Table 13.2.

feature	function
alveoli present a large surface area	to absorb oxygen
alveolar surface is moist	to allow oxygen to dissolve
alveolar lining is thin	to allow oxygen to diffuse through into blood easily
network of tiny blood vessels surrounds alveoli	to pick up and transport oxygen

Table 13.2 *Features of human respiratory surface*

Features of the capillary network

Capillaries are very narrow tubes that branch repeatedly, forming such a **dense network** that every cell in a body tissue is close to a capillary. The combined **surface area** of the capillary network is vast and the capillary walls are only **one cell thick**. These properties of the capillary network allow efficient gas exchange to occur between the bloodstream and the tissue cells.

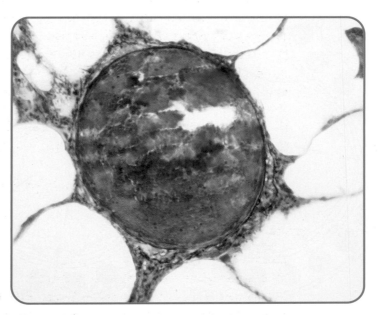

Figure 13.15 *Blood capillary and alveoli*

Testing Your Knowledge

1 Rewrite the following sentence including only the correct words from the brackets.
   During breathing (oxygen/carbon dioxide) is absorbed into the body and (oxygen/carbon dioxide) is released. (2)
2 Arrange the following structures into the order in which inhaled air would pass through them: bronchus, air sac, trachea, bronchiole. (1)
3 At the ends of which tubes are alveoli found? (1)
4 Briefly describe the process of gas exchange between an air sac and a surrounding blood vessel. (2)
5 What features make the lungs efficient gas exchange structures? (3)
6 Describe TWO features of a capillary network that allow efficient gas exchange. (2)

Applying Your Knowledge

1 Match the terms in list X with their descriptions in list Y.

List X	List Y
1) alveolus	**a)** tiny thin-walled blood vessel in contact with tissue cells
2) bronchiole	**b)** term used to describe blood that contains a high concentration of oxygen
3) bronchus	**c)** term used to describe blood that contains a low concentration of oxygen
4) capillary	**d)** tiny branch of a bronchus that ends in a group of air sacs in a lung
5) cartilage	**e)** organ of gaseous exchange in mammals
6) deoxygenated	**f)** process by which oxygen passes into and CO_2 out of the bloodstream from a high to a low concentration
7) diffusion	**g)** tube that connects the voice box with the bronchi and allows entry and exit of air
8) lung	**h)** tough material laid down in rings that support the trachea
9) oxygenated	**i)** tiny air sac with a thin lining that allows efficient gas exchange with the bloodstream
10) trachea	**j)** branch of the trachea that allows air to enter a lung

2 Match the terms in list X with their descriptions in list Y.

List X	List Y
1) artery	**a)** chamber of the heart that receives blood from the body
2) AV valve	**b)** chamber of the heart that pumps blood to the body
3) capillary	**c)** structure that prevents the backflow of blood from artery to ventricle
4) coronary artery	**d)** general name for a large thick-walled vessel carrying blood away from the heart
5) left atrium	**e)** chamber of the heart that pumps blood to the lungs
6) left ventricle	**f)** vessel that supplies oxygenated blood to heart muscle
7) right atrium	**g)** one of a dense network of microscopic blood vessels
8) right ventricle	**h)** chamber of the heart that receives blood from the lungs
9) SL valve	**i)** general name for a thin-walled vessel carrying blood towards the heart
10) vein	**j)** structure that prevents the backflow of blood from ventricle to atrium

3 Match the blood vessels in list X with their functions in list Y.

List X	List Y
1) aorta	**a)** carries deoxygenated blood to the lungs
2) hepatic artery	**b)** supplies the kidney with oxygenated blood
3) hepatic portal vein	**c)** carries deoxygenated blood to the right atrium
4) hepatic vein	**d)** removes deoxygenated blood from the kidney
5) mesenteric artery	**e)** carries oxygenated blood to the left atrium
6) pulmonary artery	**f)** removes deoxygenated blood from the liver
7) pulmonary vein	**g)** carries oxygenated blood to the body
8) renal artery	**h)** supplies the gut with oxygenated blood
9) renal vein	**i)** carries deoxygenated blood from the gut to the liver
10) vena cava	**j)** supplies the liver with oxygenated blood

4 Figure 13.16 shows the changes in pressure that occur in the left ventricle and the aorta during one heart beat.

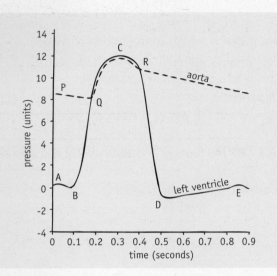

Figure 13.16

a) What is the highest pressure recorded for the left ventricle? (1)

b) Contraction of the left ventricle is represented by the part of the graph lettered: A–B, B–C, C–D, D–E. (Choose the ONE correct answer.) (1)

c) (i) At which lettered point on the graph does ventricular pressure first become equal to the pressure in the aorta?

(ii) After this point on the graph, for a short time ventricular pressure slightly exceeds aortic pressure. Which valve(s) will be open during this period?

(iii) At which lettered point on the graph will these valves start to close? (3)

5 Figure 13.17 shows the result of an international survey carried out during the 1990s on the annual death rate from coronary heart disease.

a) Which country had: (i) the highest, (ii) the lowest death rate among men due to coronary heart disease? (2)

b) Which country had: (i) the highest, (ii) the lowest death rate among women? (2)

c) How many Scottish men out of 100 000 died of coronary heart disease annually? (1)

d) What percentage of Australian women died of coronary heart disease per year? (1)

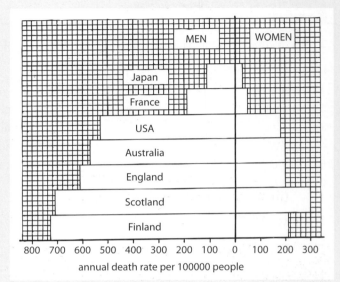

Figure 13.17

e) By how many men did the annual death rate in England exceed that in the USA? (1)

f) By how many times was a Scottish woman more likely to die of coronary heart disease than a French woman? (1)

6 Figure 13.18 shows a transverse section of part of a vein and the outline for the equivalent part of an artery.

a) Copy or trace the diagram and name part X. (1)

b) Complete the diagram to show the structure of an artery and then label it. (2)

c) (i) State a further structural difference between the two types of vessel that is not shown in this diagram.

(ii) Describe the role played by these structures. (2)

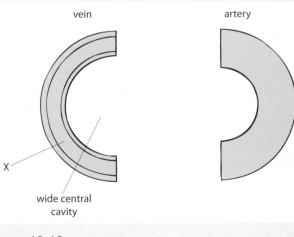

Figure 13.18

7 The data in Table 13.3 refer to the pulse rates of two students P and Q.
   a) Draw a line graph of the data. (2)
   b) What causes a pulse beat? (1)
   c) State the resting pulse rate of each student. (1)
   d) (i) Between which TWO times did exercise begin?
      (ii) Between which TWO times did exercise come to a halt? (2)
   e) (i) How long did it take each student to return to resting pulse after the exercise stopped?
      (ii) Which student was fitter? Explain your answer. (4)

time (min)	pulse rate (beats/min)	
	P	Q
0	60	70
1	60	71
2	60	69
3	60	70
4	130	140
5	140	150
6	143	155
7	145	157
8	147	160
9	90	130
10	80	110
11	75	100
12	67	90
13	60	80
14	60	75
15	60	70
16	60	70

Table 13.3

8 Table 13.4 shows the rate of blood flow in various parts of a person's body under differing conditions of exercise.

	rate of blood flow (ml/min)		
	at rest	light exercise	strenuous exercise
skeletal muscle	1200	4500	12500
gut	1400	1100	600
skin	500	1500	1900
kidneys	1100	900	600
brain	750	750	750
heart muscle	250	350	750

Table 13.4

   a) (i) What effect does increasingly strenuous exercise have on blood flow in skeletal muscle?
      (ii) Suggest the reason for this. (2)
   b) Which other body part(s) show the same trend in response to increase in exercise as
      (i) skeletal muscle? (ii) gut? (2)
   c) (i) Which body part's rate of blood flow remains unaffected by exercise?
      (ii) Suggest why. (2)
   d) (i) In what way would the appearance of facial skin change as a result of strenuous exercise?
      (ii) Explain your answer. (2)
9 Describe the route that would be taken through the human circulatory system by a molecule of urea on its way from the liver to the kidneys to be excreted. (Include the names of SIX blood vessels and four heart chambers in your answer.) (5)
10 The data in Table 13.5 refer to an athlete before and after a race.

	breathing rate (breaths per minute)	average volume of each breath (ml)	percentage of carbon dioxide in exhaled air
before race	15	500	5
after race	25	1000	5

Table 13.5

a) What happened to the athlete's rate and depth of breathing as a result of running the race? (1)

b) Calculate the volume of air passing in and out of the athlete's lungs per minute before the race. (1)

c) Assuming that the athlete's rate of breathing remained unchanged for the first 5 minutes after the race, calculate the volume of carbon dioxide that would be breathed out during that time. (1)

11 The relative concentrations (in units) of carbon dioxide in the air in an alveolus and in the blood in a pulmonary capillary are shown in Figure 13.19.

a) At which site is the blood (i) deoxygenated? (ii) oxygenated? (iii) Explain your choice. (3)

b) Match the following relative concentrations (in units) of oxygen with sites A, B and C: (i) 40 (ii) 95 (iii) 100. (3)

c) Match X and Y with the two main vessels that connect the lungs with the heart. (2)

d) Briefly explain why the lungs are regarded as organs of excretion. (1)

e) The following list gives the substances and layers through which a molecule of oxygen passes on its way from being part of alveolar air to entering a red blood cell. Put them into the correct order.

1) blood plasma
2) capillary wall
3) film of moisture
4) liquid-filled space
5) thin alveolar lining (1)

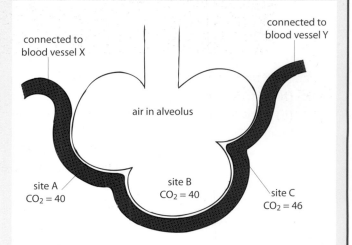

Figure 13.19

14 Blood

Composition

Blood contains **red** blood cells and **white** blood cells suspended in a watery, yellowish fluid called **plasma**.

Role of plasma

Plasma contains various types of protein, such as antibodies (see page 299), and many other water-soluble substances, including the end products of digestion. Soluble foods such as **glucose** and **amino acids** are carried round the body dissolved in plasma. By this means they are transported to within **diffusion distance** of all living cells. Glucose and amino acids are then able to pass from a region of high concentration in a blood capillary to a region of low concentration in nearby cells, providing the cells with an energy source and a supply of building materials.

Carbon dioxide is produced by living cells during respiration. This waste product is then transported in a mammal's bloodstream to the lungs to be excreted. Some carbon dioxide is carried dissolved in the plasma. However, the concentration that can be transported by this means is limited by the fact that carbon dioxide combines with water to form an acid. The presence of large quantities of acid in the bloodstream would lead to problems since blood functions best within a pH range of 7.36 to 7.44. Most of the carbon dioxide is transported in blood plasma as bicarbonate (hydrogen carbonate) ions. Some carbon dioxide is also carried inside red blood cells attached to other molecules.

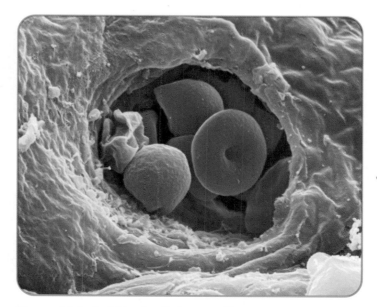

Figure 14.1 *Red blood cells in a capillary*

Red blood cells

Red blood cells (see Figure 14.1) are very small and extremely numerous (approximately 5.5 million per mm³ of blood). Each red blood cell is shaped like a **biconcave disc**. It contains cytoplasm rich in **haemoglobin** (the oxygen-carrying pigment) but lacks a nucleus (see Figure 14.2).

A red blood cell's biconcave shape gives it a **larger surface area** in relation to its volume than it would have if it were spherical (see Figure 14.3). This increases its ability to absorb oxygen since it exposes a larger surface area of haemoglobin molecules to the surrounding environment. (The total surface area of all the red blood cells in an adult's body provides an area larger than a football pitch for the uptake of oxygen!)

Since red blood cells are **tiny** in size and **flexible** in shape they are able to squeeze through the narrowest of blood capillaries and deliver oxygen to nearby body cells.

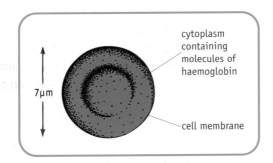

Figure 14.2 *Red blood cell*

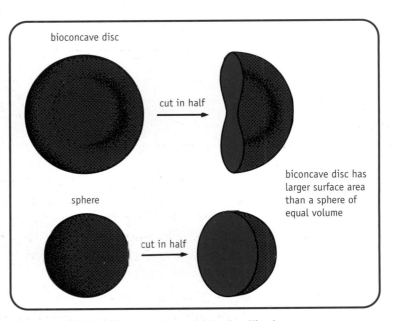

Figure 14.3 *Significance of a red blood cell's shape*

Examining a prepared slide of blood

You need

1 prepared slide of stained human blood
1 microscope with a source of illumination

What to do

1 View the prepared slide under the high power of the microscope.
2 Look for red blood cells which will be circular in shape and will probably
   appear faint grey/purple due to staining.
   a) Decide whether or not they have nuclei.
   b) Discuss with your partner how it is possible to tell without seeing them
      side on or in cross-section that red blood cells are biconcave discs and
      not spheres.
3 Look for white blood cells which will be slightly irregular in shape and
   stained purple with dark purple areas inside the cells.
   a) Decide whether or not they contain nuclei.
   b) Look for two to three different types of white blood cell.
   c) Discuss with your partner the ways in which these differ from one
      another (and then check with Figure 14.5).
4 Decide whether red or white cells are more numerous.
5 a) Count how many white cells you can see in one field of view at high
      power.
   b) Estimate the number of red blood cells in the same field of view by
      counting the number of red cells from the centre of the field along an
      imaginary radius line to the edge of the field of view. This gives you a
      measurement of r. Apply r to the formula πr^2 to give you the number of
      red blood cells.
   c) Express your result as a ratio of red to white cells. (Actual ratio =
      approximately 600 red:1 white.)

Haemoglobin

Since oxygen is only slightly soluble in water, the quantity that could be
carried dissolved in plasma would be inadequate to satisfy the needs of
respiring cells. This problem is overcome by the presence of **haemoglobin** (a
respiratory pigment), which combines with oxygen and significantly increases
the oxygen-carrying capacity of blood.

Association and dissociation

To be effective a respiratory pigment must be able to combine readily
(**associate**) with oxygen on those occasions when the oxygen concentration in

the surroundings is high, and rapidly release (**dissociate** from) oxygen when the surrounding oxygen concentration is low.

Haemoglobin meets these requirements exactly by having a **high affinity** for oxygen when the oxygen concentration in the surrounding environment is high (e.g. at the respiratory surface of the lungs) and a **low affinity** for oxygen when the oxygen concentration is low (e.g. in actively respiring cells). (Affinity means tendency to combine with oxygen.)

When haemoglobin combines with oxygen it forms **oxyhaemoglobin**. This reversible chemical reaction is summarised by the equation:

$$\text{haemoglobin} + \text{oxygen} \xrightleftharpoons[\text{dissociation (in tissues)}]{\text{association (in lungs)}} \text{oxyhaemoglobin}$$

Oxygen tension

The partial pressure (tension) of oxygen is a measure of its concentration and is expressed in kilopascals (kPa). The oxygen tension of inhaled alveolar air, for example, is about 13 kPa.

Oxygen dissociation curve

Percentage saturation of haemoglobin with oxygen decreases with decreasing oxygen tension of the surroundings. However, the relationship between the two is not a linear one; when graphed it gives an S-shaped curve called the **oxygen dissociation curve** (Figure 14.4).

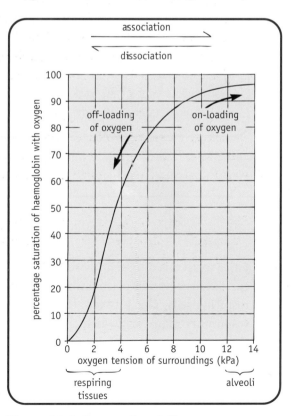

Figure 14.4 *Oxygen dissociation curve*

At the extreme right-hand side of the curve, the oxygen tension in the environment (e.g. alveoli) is high and haemoglobin's level of saturation with oxygen is close to 100%.

Moving gradually to the left along the graph, the oxygen concentration of the surroundings is found to decrease, yet haemoglobin still remains loaded up with oxygen to levels of over 85% saturation even when the oxygen tension of the surroundings has dropped to 8 kPa. This is due to the fact that haemoglobin has a high affinity for oxygen.

Moving further to the left, as the oxygen tension of the surroundings drops to below 6 kPa, the percentage saturation of haemoglobin with oxygen drops rapidly. This is because haemoglobin's affinity for oxygen decreases rapidly in surroundings of low oxygen concentration. As a result it unloads its oxygen. This process is represented by the steep part of the S-shaped dissociation curve.

Respiring cells
Actively respiring cells (e.g. working muscles and liver) consume much oxygen and their oxygen tension is found to be low, around 2.7 kPa or less. At the other extreme the oxygen tension of alveolar air is high, about 13 kPa.

When haemoglobin from respiring cells returns to the lungs, it becomes **loaded up** with oxygen, which moves along the diffusion gradient from alveoli to blood. This process of association continues as before until haemoglobin is almost 100% saturated.

When this haemoglobin is transported to actively respiring cells with an oxygen tension of 2.7 kPa, haemoglobin's percentage saturation with oxygen drops to a low level (about 35%). This is because haemoglobin rapidly dissociates from its oxygen and **unloads** it. As a result oxygen becomes available to satisfy the demands of the actively respiring cells.

Effectiveness of haemoglobin
The oxygen dissociation curve is especially steep between oxygen tensions of 6 and 2 kPa. This means that any slight drop in oxygen tension of body cells within this range results in a **rapid release** of oxygen by haemoglobin to these cells.

So effective is haemoglobin at this loading up (association) and unloading (dissociation) of oxygen, that it is responsible for the transport of 97% of the oxygen carried in the bloodstream.

White blood cells

Compared with red blood cells, white blood cells are less numerous (4000–13 000 per mm^3 of blood). They also differ from red blood cells in that they contain **nuclei** and they can change shape and squeeze through tiny pores in capillary walls. This ability makes them perfectly suited to their function of **defending** the body since it enables them to reach sites of microbial infection outwith the circulatory system.

Several types of white blood cell exist. Two of these are **monocytes** and **lymphocytes**, as shown in Figure 14.5.

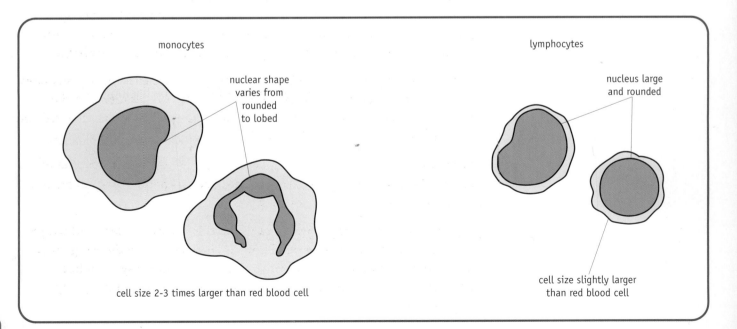

monocytes

nuclear shape varies from rounded to lobed

cell size 2-3 times larger than red blood cell

lymphocytes

nucleus large and rounded

cell size slightly larger than red blood cell

Figure 14.5 *Two types of white blood cell*

Phagocytosis

Phagocytosis (see Figure 14.6) is the process by which bacteria (and other foreign bodies) are **engulfed** and destroyed by phagocytic cells such as **monocytes** and **macrophages**. Macrophages are cells derived from monocytes. They are found throughout the body in the connective tissue (cellular and fibrous 'background' material in which specialised structures such as nerves and blood vessels are embedded).

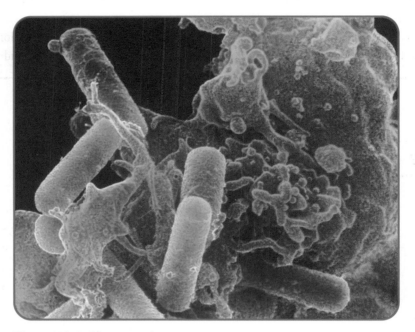

Figure 14.6 *Phagocytosis*

The process of phagocytosis is illustrated in Figure 14.7 (on page 298). A phagocytic cell detects chemicals released by a bacterium and moves up a concentration gradient towards it. The phagocyte adheres to the bacterium and **engulfs** it in a vacuole formed by an infolding of the cell membrane.

A phagocyte's cytoplasm contains a rich supply of organelles called **lysosomes** which contain **digestive enzymes**. Some of these lysosomes fuse with the vacuole and release their enzymes into it. The bacterium becomes digested and the breakdown products are absorbed by the phagocyte.

During infection hundreds of phagocytes migrate to the infected area and engulf many bacteria by phagocytosis. Dead bacteria and phagocytes often accumulate at a site of injury forming **pus**.

Large numbers of macrophages are also found in the liver, spleen and lymph nodes. They are fixed to the inside lining of channels within these organs and are immobile. However, they operate in exactly the same way as mobile phagocytes by removing foreign bodies from passing blood or lymph by phagocytosis.

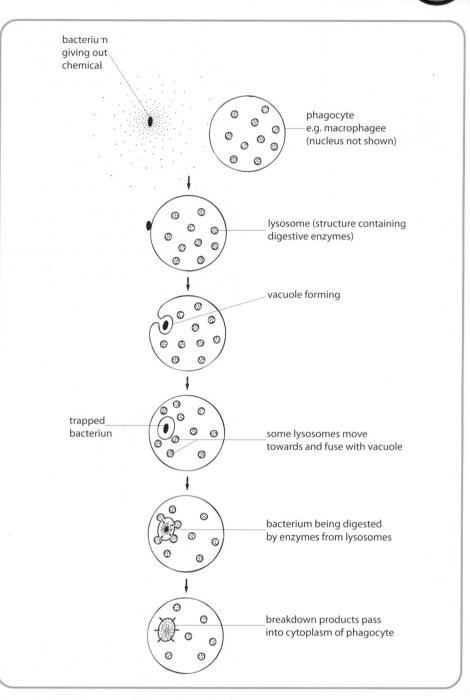

Figure 14.7 *Phagocytosis in a macrophage*

Immune response

Immunity is an organism's ability to **resist** infectious disease.

Non-specific immune response
Phagocytosis is an example of a **non-specific** immune response since it provides general protection against a wide range of micro-organisms.

Specific immune response (antibody formation)

An **antigen** is a complex molecule, such as a protein, that is recognised as an alien by the body's lymphocytes. The antigen's presence in the bloodstream stimulates lymphocytes to produce special protein molecules called **antibodies** that are specific to that antigen. The body possesses thousands of different types of lymphocytes each capable of recognising and responding to one antigen.

A virus particle is surrounded by a coat of protein. This protein possesses several sites that act as antigens on the surface of the virus.

An **antibody** is a Y-shaped molecule, as shown in Figure 14.8. Each of its arms bears a **receptor** (binding) site whose shape is **specific** to a particular antigen. When an antibody meets its complementary antigen, the two combine at their specific sites like a lock and key and the antigen is rendered harmless (see Figure 14.9).

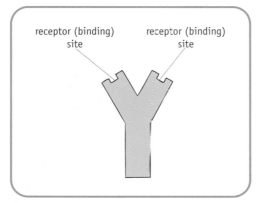

Figure 14.8 *Antibody*

Immunological memory

Primary and secondary responses

When a person is infected by a disease-causing organism, the body responds by producing antibodies. This is called the **primary response** (see Figure 14.10 on page 301). Because there is a latent period before the appearance of the antibodies, the primary response is often unable to prevent the person from suffering the disease.

If the person survives, exposure to the same antigen at a later date results in the **secondary response**. This time the disease is usually prevented because:

- antibody production is much **more rapid**
- the concentration of antibodies produced reaches a **higher level**
- the higher concentration of antibodies is maintained for a **longer time**.

Memory cells

The secondary response is made possible by the presence in the body of **memory cells**. These are lymphocytes specific to the antigen that result from

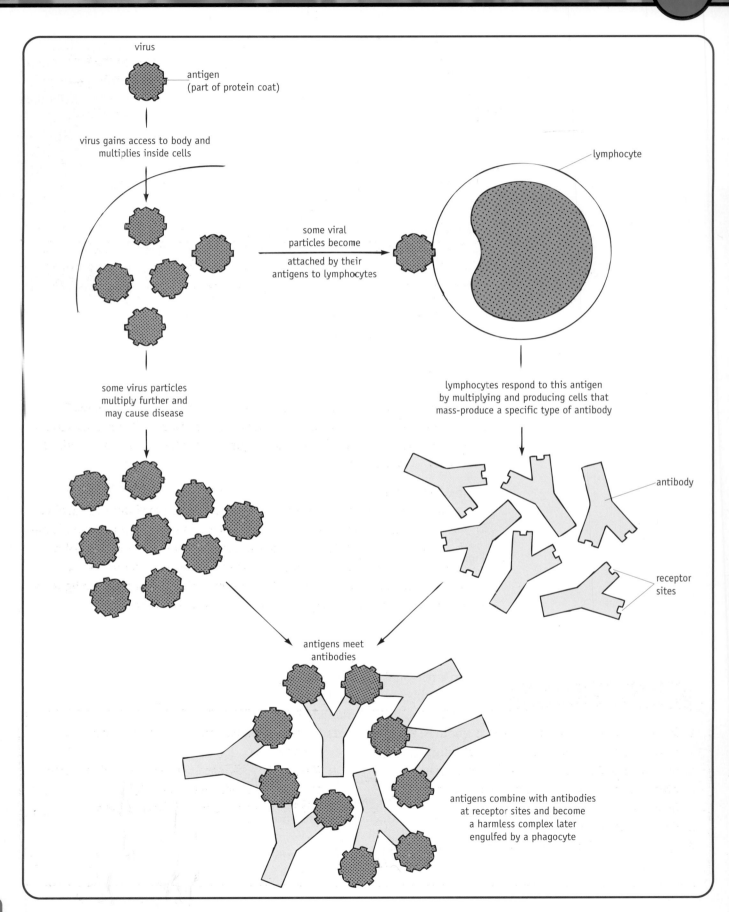

virus

antigen
(part of protein coat)

virus gains access to body and
multiplies inside cells

some viral
particles become
attached by their
antigens to lymphocytes

lymphocyte

some virus particles
multiply further and
may cause disease

lymphocytes respond to this antigen
by multiplying and producing cells that
mass-produce a specific type of antibody

antibody

receptor
sites

antigens meet
antibodies

antigens combine with antibodies
at receptor sites and become
a harmless complex later
engulfed by a phagocyte

Figure 14.9 *Action of antibodies*

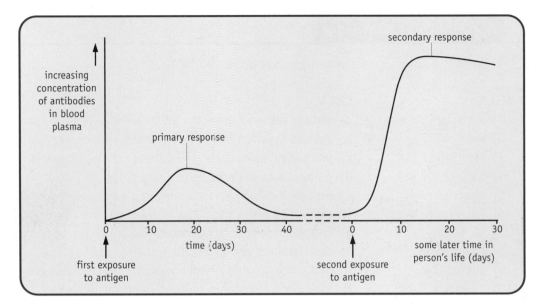

Figure 14.10 *Primary and secondary response*

the first exposure to it. When the body becomes exposed to the disease-causing micro-organism for a second time, the memory cells produce clones of antibody-forming lymphocytes and fight it off. The person has acquired immunity by **natural** means.

Artificially acquired immunity (immunisation)

For artificially acquired immunity a person receives a small dose of vaccine containing antigens that have been treated in a certain way so that they induce antibody formation but do not cause the disease. Examples include polio vaccine and BCG immunisation. If the person is exposed to the disease-causing antigen at some later time, their body is already prepared to fight the infection. The person has acquired immunity by **artificial** means.

Testing Your Knowledge

1 State TWO ways in which the structure of a white blood cell differs from that of a red blood cell. (2)
2 Draw up a table to compare TWO named types of white blood cell with respect to their structure and function. (3)
3 Write a brief account of phagocytosis. (5)
4 a) Explain the difference between an antigen and an antibody. (2)

b) Is antibody formation a specific or a non-specific response? (1)
c) What is the difference between naturally acquired and artificially acquired immunity? (2)
d) Briefly describe the role of memory cells. (2)

Applying Your Knowledge

1 Match the terms in list X with their descriptions in list Y.

List X
1) antibody
2) antigen
3) association
4) biconcave disc
5) dissociation
6) haemoglobin
7) immunity
8) lymphocyte
9) macrophage
10) monocyte
11) phagocytosis
12) plasma

List Y
a) watery, yellowish liquid in which blood cells are suspended
b) process by which haemoglobin unloads its oxygen in the presence of respiring cells
c) phagocytic white blood cell found in the bloodstream
d) protein molecule specific to one antigen
e) process by which bacteria are engulfed and destroyed by certain white blood cells
f) process by which haemoglobin loads up with oxygen in the lungs
g) complex molecule recognised as foreign to the body by lymphocytes
h) description of the shape of a red blood cell
i) iron-containing respiratory pigment that combines reversibly with oxygen
j) ability of an organism to resist infectious disease
k) phagocytic cell derived from a monocyte and found in connective tissues
l) type of white blood cell that makes antibodies

2 a) Name the type of blood vessel shown in Figure 14.11. (1)
   b) (i) Identify the type of blood cell present in the vessel.
      (ii) State its function.
      (iii) Give THREE ways in which this type of cell's structure can be related to its function. (5)
   c) (i) Name another type of blood cell that could change shape and squeeze out through the tiny pore.
      (ii) Describe the function of this type of blood cell. (2)

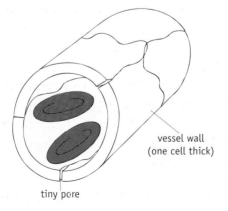

tiny pore

Figure 14.11

3 Figure 14.12 shows haemoglobin's affinity for oxygen at two different temperatures.

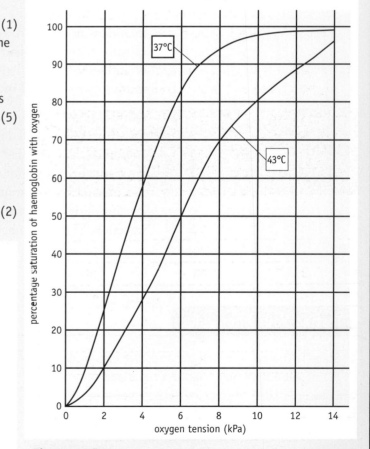

Figure 14.12

a) State the percentage saturation of haemoglobin with oxygen at 37°C and at oxygen tension
   (i) 14 kPa
   (ii) 6 kPa

(iii) 2 kPa. (3)

b) (i) Calculate the decrease in percentage saturation of haemoglobin with oxygen that occurred at 37°C as oxygen tension decreased from 6 to 2 kPa.

(ii) Does this decrease involve the process of association or dissociation?

(iii) Would it be more likely to occur in respiring tissues or in the lungs? Explain your answer. (4)

c) State the percentage saturation of haemoglobin with oxygen at an oxygen tension of 8 kPa and temperatures of

(i) 37°C

(ii) 43°C.

(iii) State the effect of increase in temperature on haemoglobin's affinity for oxygen. (3)

4 One cubic millimetre of a person's blood was found to contain 5 350 904 red blood cells and 8948 white blood cells. Express these figures as a whole number ratio. (1)

5 The data in Table 14.1 refer to the affinity for oxygen of fetal and adult haemoglobin.

partial pressure of oxygen (kPa)	percentage saturation of haemoglobin with oxygen	
	fetal haemoglobin	adult haemoglobin
0	0	0
1	10	8
2	30	20
3	60	40
4	77	60
5	87	72
6	91	79
7	94	85
8	95	90
9	96	93
10	97	95
11	97.5	96
12	98	97
13	98	98
14	98	98

Table 14.1

a) Draw a line graph of the data and join up the points to form two curves similar in shape to the one in Figure 14.4 on page 294. (4)

b) Describe the effect that increasing partial pressure of oxygen has on percentage saturation of haemoglobin, as shown by the two curves. (1)

c) State the partial pressure of oxygen at which

(i) adult haemoglobin shows 60% saturation

(ii) fetal haemoglobin shows 60% saturation. (2)

d) State the percentage saturation with oxygen of

(i) adult haemoglobin when the partial pressure is 6 kPa

(ii) fetal haemoglobin when the partial pressure is 6 kPa. (2)

e) (i) What important difference between fetal and adult haemoglobin is highlighted by the fact that the curve for fetal haemoglobin is to the left of that for adult haemoglobin?

(ii) Suggest why this difference is of survival value to the fetus. (2)

6 Arrange the following steps into the correct sequence in which they occur during phagocytosis:

1) bacterium digested by enzymes

2) bacterium engulfed

3) lysosomes fuse with vacuole

4) phagocyte meets bacterium. (1)

7 a) Which type of blood cell produces antibodies? (1)

b) Using a diagram to illustrate your answer, explain what is meant by the specificity of an antibody for an antigen. (2)

8 Table 14.2 refers to antibody proteins called immunoglobulins found in human blood. The graph in Figure 14.13 refers to the sequence of events that occurs in response to two separate injections of a type of antigen into a small mammal.

a) Which immunoglobulin in the table would be found in the blood of an unborn baby? (1)

b) Of the five types of immunoglobulin molecule, which is

(i) the largest?

(ii) the rarest? (2)

c) With reference to IgG in Figure 14.13, state THREE differences between the primary and the secondary response. (3)

d) Antibodies such as IgG are now known to be produced by the activity of long-lived

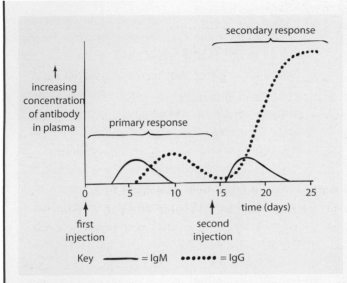

Figure 14.13

9 Figure 14.14 shows a disease-causing virus (V) and an attenuated (weakened) version of it (W).
  a) Suggest why receiving a vaccine containing W gives immunity against attacks in the future by V. (2)
  b) Would such immunity be naturally or artificially acquired? (1)

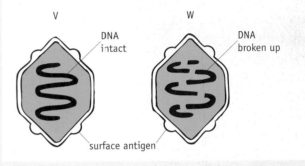

lymphocytes. With reference to the graph, suggest why the latter are called 'memory cells'. (2)

Figure 14.14

immunoglobulin (Ig)				
	IgA	IgD	IgE	IgG	IgM
molecular weight	170 000	184 0C0	188 100	150 000	960 000
normal serum concentration	1.4 – 4.0 g/l	0.1 – 0.4 g/l	0.1 – 1.3 mg/l	8.0 – 16.0 g/l	0.5 – 2.0 g/l
ability to cross placenta	no	no	no	yes	no

Table 14.2

Word bank

antigens	macrophages
aorta	mesenteric
arteries	phagocytosis
arterioles	plasma
atrium	portal
backflow	pulmonary
body	pump
capillaries	red
carbon dioxide	renal
coronary	respiring
deoxygenated	specific
dissociates	thicker
dissolved	valve
haemoglobin	vein
heart	vena cava
hepatic	ventricles
lungs	venules
lymphocytes	white

Table 14.3 *Word bank for Chapters 13 and 14*

What You Should Know

(Chapters 13 and 14) (See Table 14.3 for word bank)

1 The heart is a muscular _____ consisting of two upper chambers called atria and two lower chambers called _____.

2 _____ blood returns to the heart via the _____; it is pumped by the heart to the lungs via the _____ artery. Oxygenated blood returns to the heart from the lungs by the pulmonary _____; it is pumped by the heart to the body via the _____.

3 An AV valve prevents _____ of blood from a ventricle to an _____; an SL _____ prevents backflow from an associated artery to a ventricle.

4 The wall of the left ventricle is _____ than that of the right ventricle because the left pumps blood all round the _____.

5 The aorta divides into arteries that supply oxygenated blood to all parts of the body. The _____ artery supplies the heart wall, the _____ artery supplies the liver, the _____ artery takes blood to the gut, the _____ artery transports blood to the kidneys. Veins from all of these organs, except the gut, return deoxygenated blood directly to the vena cava.

6 The gut is exceptional in that its deoxygenated blood is first carried to the liver by the hepatic _____ vein.

7 _____ are thick-walled vessels that carry blood away from the heart; veins are thinner-walled vessels with valves that carry blood back to the _____.

8 Arteries divide into smaller vessels called _____; veins are formed by the convergence of smaller vessels called _____. Blood passes from arterioles to venules through tiny microscopic vessels called _____.

9 Blood contains red and _____ blood cells suspended in liquid called _____. _____ food is carried in plasma. Some _____ is transported in plasma and some is carried in _____ blood cells.

10 Oxygen is carried in red blood cells combined with _____. Haemoglobin associates with oxygen at high oxygen levels in the _____ and _____ from it at low oxygen levels in _____ tissues.

11 White blood cells called monocytes (and _____ derived from them) engulf and digest disease-causing bacteria by _____, which is a non-specific response.

12 White blood cells called _____ produce antibodies in response to alien molecules called _____. Antibody formation is a _____ response.

15 Brain and nervous system

Brain

Structure

The **brain** (see Figure 15.1) is a complex organ consisting of several different regions. The **cerebrum** is the largest part of the human brain. It is divided into two connecting halves called **cerebral hemispheres**.

Figure 15.2 shows an external view of the brain and the 'internal' view that would be revealed by removing the left cerebral hemisphere and continuing to cut a longitudinal section down through the lower regions where the cerebellum, medulla and hypothalamus are located.

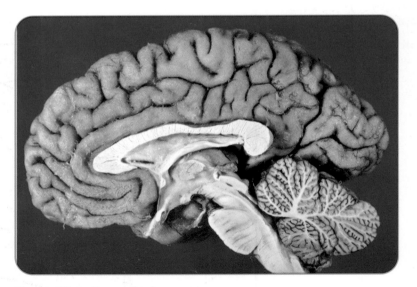

Figure 15.1 *Human brain*

Functions

The **cerebellum** controls balance and muscular co-ordination. The **medulla** contains vital centres that control the rate of breathing and heartbeat. The **hypothalamus** contains the centres that regulate water balance (see pages 267–8) and body temperature (see pages 318–9).

The **cerebrum** is the site of conscious thought. It is responsible for higher mental faculties such as reasoning, imagination, creativity, conscience and memory.

Discrete areas of cerebrum

A **cerebral hemisphere** consists of several different regions. Each of these areas is discrete. This means that it performs its own function distinct from the

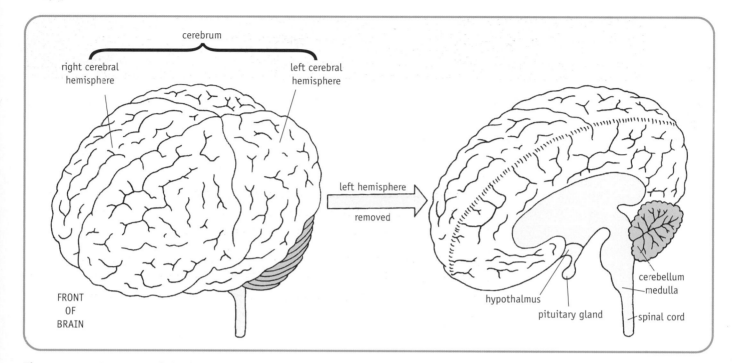

Figure 15.2 *Structure of the brain*

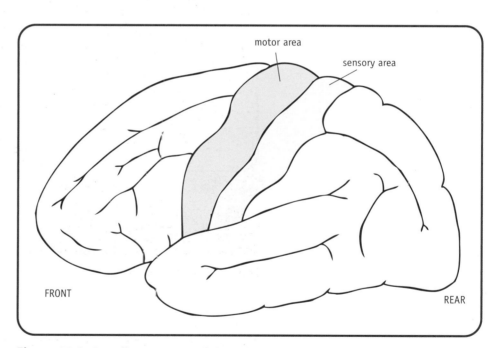

Figure 15.3 *Two discrete areas of the left cerebral hemisphere*

others. Two discrete areas of the left cerebral hemisphere are shown in Figure 15.3. These are sometimes referred to as the **sensory** and **motor strips**. They are also present as mirror images in the right cerebral hemisphere.

Sensory area

Each **sensory area** (strip) receives information as sensory impulses from the body's receptors (e.g. sense organs). It then passes this information on to other parts of the brain to be analysed, interpreted and perhaps acted on.

Each part of the body that possesses receptors capable of sending nerve impulses to the sensory strip is represented by an area on it. However, the size of each of these regions is not found to be in proportion to the actual size of the body part. Instead, the extent of the sensory area allocated is in proportion to the **relative number of receptor cells** present in the body part.

Figure 15.4 shows the extent of the sensory area given over to each body part containing receptors. The more sensitive the body part, the larger the region of sensory strip.

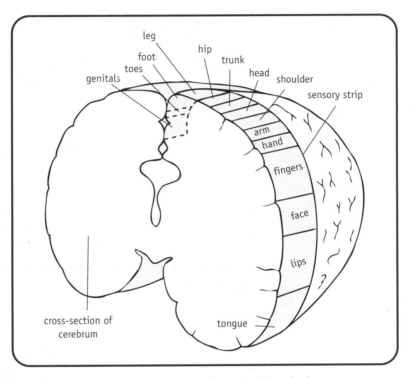

Figure 15.4 *Sensory area of the left cerebral hemisphere*

Sensory homunculus
Figure 15.5 shows an imaginary figure ('sensory homunculus') whose body parts have been drawn in proportion to their **sensitivity** as opposed to their actual size.

Motor area
Each **motor area** (strip) consists of motor neurones which, in the light of nerve impulses received from other parts of the brain, send out impulses to bring about appropriate voluntary movements of skeletal muscles.

However, the size of the part of the motor area devoted to any one part of the body operated by skeletal muscles is not in proportion to the actual size of the body part. Instead, the extent of motor area allocated to a particular body part is found to be in proportion to its mobility which results from the **relative number of motor endings present** in it.

Figure 15.6 shows the extent of motor area given over to each body part capable of movement. The more mobile the part, the larger the region of motor area.

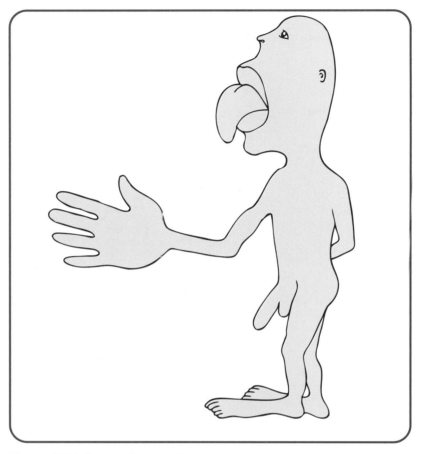

Figure 15.5 *Sensory homunculus*

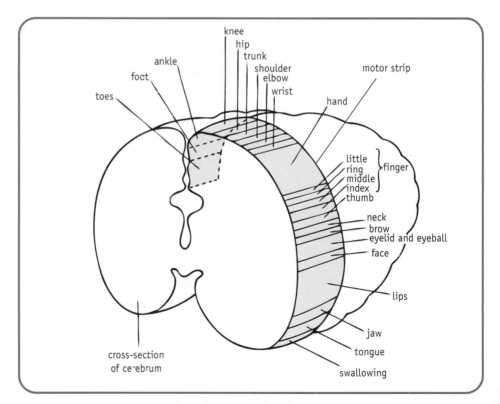

Figure 15.6 *Motor area of the left cerebral hemisphere*

Motor homunculus

Figure 15.7 shows an imaginary human figure ('motor homunculus') whose body parts have been drawn in proportion to their **mobility** and fine motor control as opposed to their actual size.

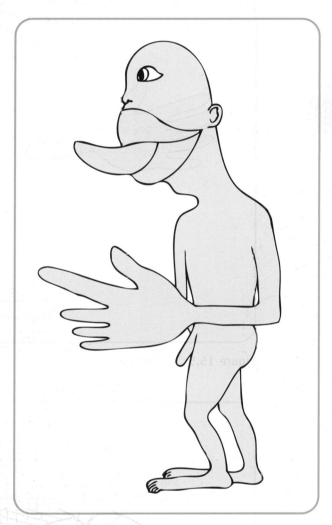

Figure 15.7 *Motor homunculus*

Testing Your Knowledge

1 a) State THREE functions of the cerebrum. (3)
   b) Name THREE other regions of the human brain and give ONE function of each. (6)

2 a) What is meant by a discrete area of the cerebrum? (1)
   b) Name TWO types of discrete area. (2)

3 a) What relationship exists between the extent of the sensory strip employed to receive impulses from a body part and the sensitivity of that body part? (1)

   b) What relationship exists between the extent of the motor strip employed to transmit motor impulses to skeletal muscle and the mobility of the body part affected? (1)

   c) Why are sensory and motor homunculi out of proportion compared with real humans? (2)

Organisation of human nervous system

The human nervous system (see Figures 15.8 and 15.9) is composed of three parts: the **brain**, the **spinal cord** and associated **nerves**. The brain and spinal cord make up the **central nervous system (CNS)**. The CNS is connected to all

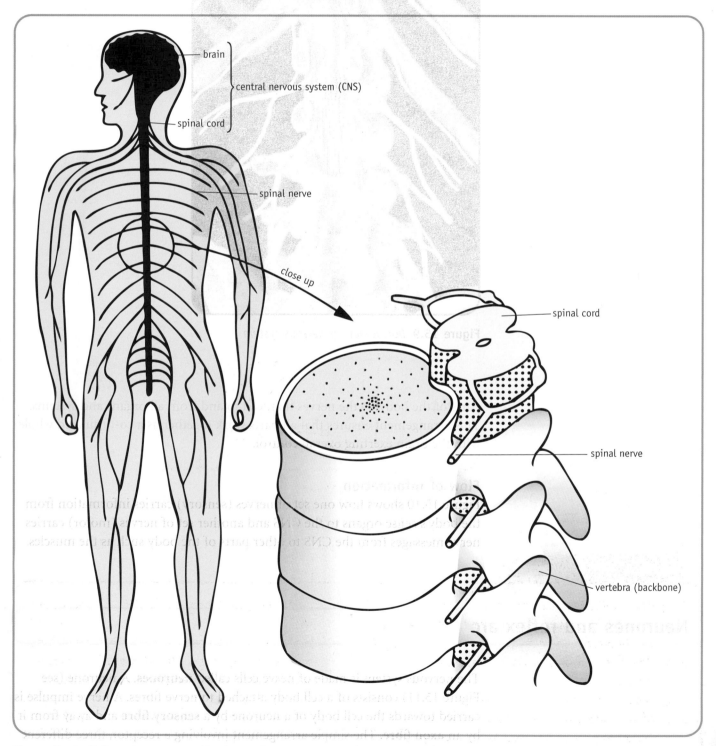

Figure 15.8 *Human nervous system*

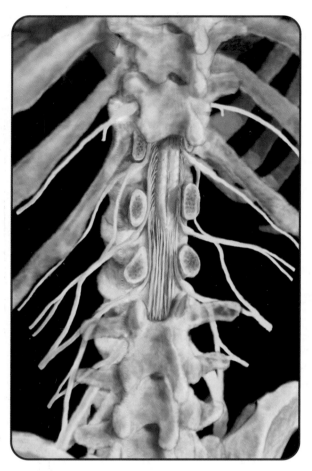

Figure 15.9 *Dorsal view of nervous system*

parts of the body by the nerves that lead to and from all organs and systems. This arrangement ensures that all parts work together as a co-ordinated whole with the brain exerting overall control.

Flow of information

Figure 15.10 shows how one set of nerves (sensory) carries information from the body's sense organs to the CNS and another set of nerves (motor) carries nerve messages from the CNS to other parts of the body such as the muscles.

Neurones and reflex arc

The nervous system is made of nerve cells called **neurones**. A neurone (see Figure 15.11) consists of a cell body attached to nerve fibres. A nerve impulse is carried towards the cell body of a neurone by a **sensory fibre** and away from it by an **axon fibre**. The simple arrangement involving a receptor, three different types of neurone and an effector as shown in Figure 15.12 is called a **reflex arc**.

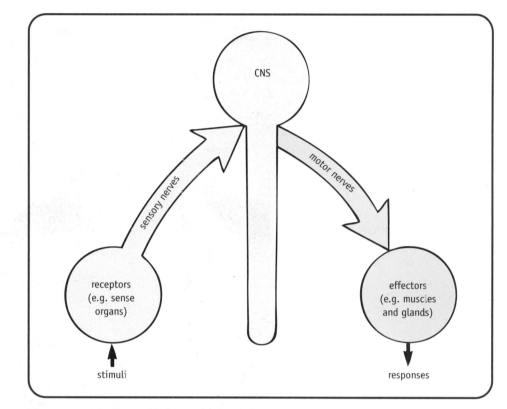

Figure 15.10 *Flow of information*

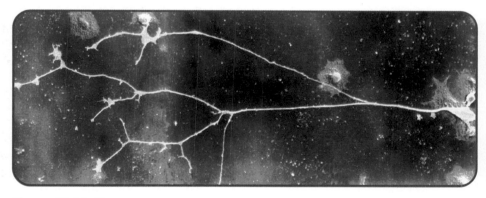

Figure 15.11 *Neurone*

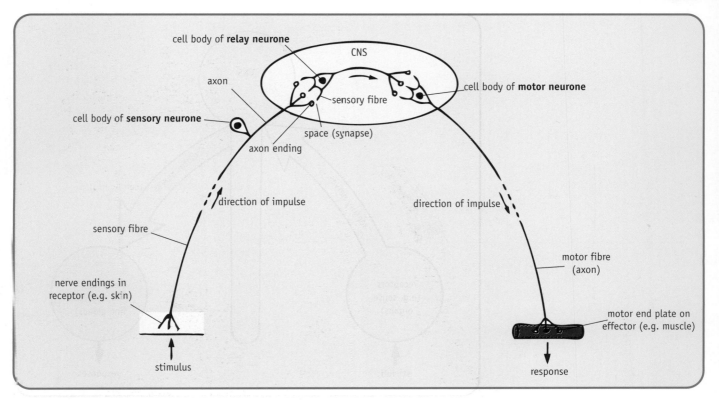

Figure 15.12 *Reflex arc*

A tiny space (**synapse**) is present between the axon ending of one neurone and the sensory fibre of the next. When a nerve impulse arrives, the tiny knob at the end of the axon branch releases a **chemical** that diffuses across the space and triggers off an impulse in the sensory fibre of the next neurone in the arc.

Reflex action

The transmission of a nerve impulse through a reflex arc results in a **reflex action**. A reflex action is a *rapid, automatic, involuntary response to a stimulus.*

Figure 15.13 shows an example of a reflex action called limb withdrawal. When the back of the hand accidentally comes in contact with intense heat, this stimulus is picked up by the pain receptors in the skin (1) and an impulse is immediately sent up the fibre of the sensory neurone (2). In the grey matter of the spinal cord, the impulse crosses the first synapse (3) and passes through the relay neurone (4). Once across the second synapse, the impulse is picked up by the motor neurone (5) and quickly conducted to the axon endings (6) (motor end plates), which are in close contact with the flexor muscle of the arm. Here a chemical is released that brings about muscular contraction (the response), making the arm bend and move out of harm's way.

Protective role

Reflex actions **protect** the body from damage (see Table 15.1). Since they do not need conscious thought by the brain, many reflex actions may still be performed for a short period by an animal whose brain has been destroyed.

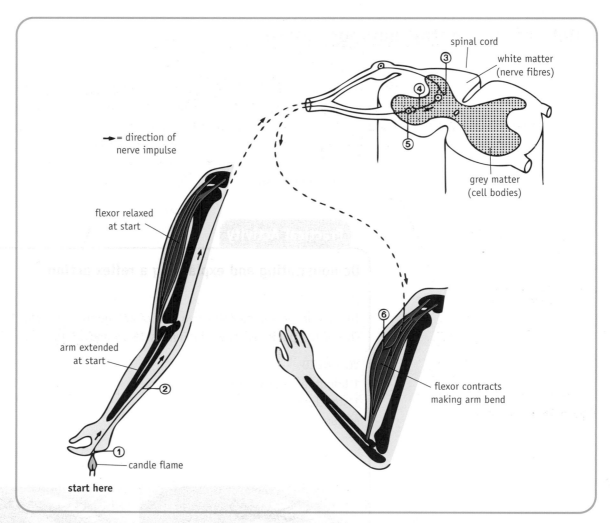

Figure 15.13 *Reflex action*

reflex action	stimulus	response	protective function
limb withdrawal	heat from naked flame	contraction of flexor muscle	removal of limb to safety
blinking	harmful object approaching eye surface	contraction of eyelid muscle	prevention of damage to eye
sneezing	foreign particles in nasal tract	sudden contraction of chest muscles	expulsion of unwanted particles from nose

Table 15.1 *Reflex actions*

Role of the central nervous system

The messages (in the form of nerve impulses) arriving from the senses keep the CNS informed about all aspects of the body and its surroundings. The CNS sorts out all this information and stores some of it. When the body is required to perform a particular physical activity, the CNS sends out messages (in the form of nerve impulses) to certain muscles which then make the appropriate response.

Practical Activity

Demonstrating and explaining a reflex action

INFORMATION

The iris of the eye contains two sets of antagonistic muscles that control the size of the pupil, as shown in Figures 15.14 and 15.15.

YOU NEED

1 bright lamp or torch
1 stopclock
mirror

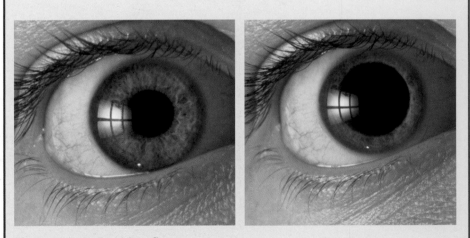

Figure 15.14 *Eye iris reflex*

WHAT TO DO

1 Work with a partner and decide who is to be the 'subject'and who is to be the 'operator'.
2 Arrange the mirror in a position that allows easy viewing by the subject of his/her own eyes.
3 If you are the operator, observe the size of the subject's pupils in their 'normal' state while the subject uses the mirror to do the same.
4 Ask the subject to keep his/her eyes closed and covered until told by you to uncover and open them.

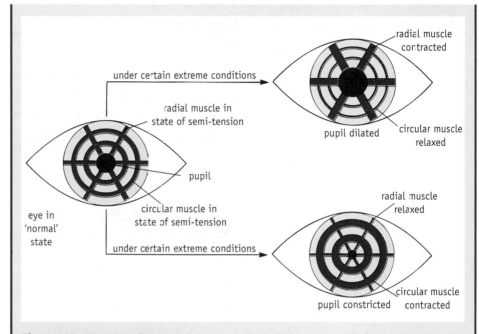

Figure 15.15 *Iris reflex muscles*

5 Start the clock.

6 After three minutes ask the subject to uncover and open his/her eyes. Observe the size of his/her pupils while he/she does the same in the mirror.

7 Repeat steps 3 to 6 to increase the reliability of the results.

8 If time permits, exchange roles and repeat the experiment.

9 Answer the following questions.

   a) Was the diameter of each of the subject's pupils large, medium or small at the start?

   b) Was the diameter of each of the subject's pupils larger, unchanged or smaller *immediately* after three minutes in darkness?

   c) Explain this reflex action in terms of stimulus, response (see Figure 15.15 for the names of muscles) and protective function.

10 Repeat step 3 after one minute in normal lighting.

11 Carefully shine the bright lamp close to the subject's eyes for five seconds and observe any changes in the diameter of his/her pupils while he/she does the same in the mirror.

12 After one minute of normal lighting, repeat step 11 to increase the reliability of the results.

13 Exchange roles and repeat the experiment if time permits.

14 Answer the following questions.

   a) Was the diameter of each of the subject's pupils large, medium or small at the start?

   b) Was the diameter of each of the subject's pupils larger, unchanged or smaller *immediately* after five seconds of bright light?

   c) Explain this reflex action in terms of stimulus, response (see Figure 15.15 for the names of muscles) and protective function.

Testing Your Knowledge

1 a) Name THREE parts of the human nervous system. (3)

   b) Which of these make up the CNS (central nervous system)? (2)

2 a) Name the type of nerves that
   (i) lead into the CNS
   (ii) emerge from the CNS. (2)

   b) Compare these two types of nerve with respect to their function. (1)

3 Make a simple labelled diagram of a reflex arc. (5)

4 a) Define the term *reflex action*. (2)

   b) Sneezing in response to foreign particles entering the nasal tract is a reflex action. Describe how it works and explain why it is protective. (4)

5 Briefly describe the overall role of the CNS. (2)

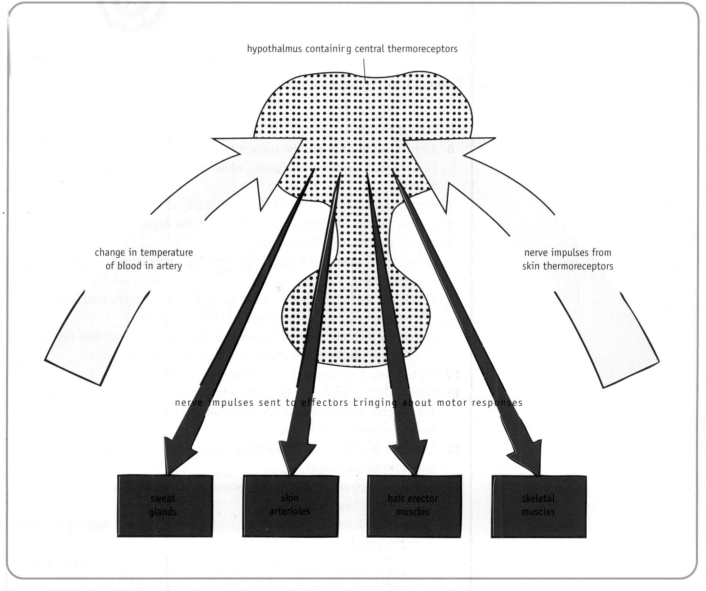

hypothalmus containing central thermoreceptors

change in temperature of blood in artery

nerve impulses from skin thermoreceptors

nerve impulses sent to effectors bringing about motor responses

sweat glands

skin arterioles

hair erector muscles

skeletal muscles

Figure 15.16 *Hypothalamus as a temperature-monitoring and control centre*

Regulation of body temperature

Figure 15.2 (on page 307) shows the location of the **hypothalamus**, which contains the body's temperature-monitoring centre. It receives nerve impulses from heat and cold **thermoreceptors** in the skin, which convey information about the surface temperature of the body.

In addition, the hypothalamus itself possesses **central thermoreceptors** (see Figure 15.16). These are sensitive to a change in the temperature of blood, which indicates a change in temperature of the **body core** (see Figure 15.17). The hypothalamus responds to this information by sending appropriate motor nerve impulses to **effectors**. These trigger motor responses that return the body temperature to its normal level.

Role of the skin

The **skin** plays a leading role in temperature regulation. In response to nerve impulses from the hypothalamus, the skin acts as an effector.

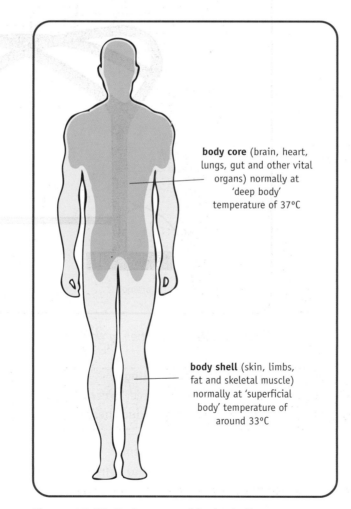

body core (brain, heart, lungs, gut and other vital organs) normally at 'deep body' temperature of 37°C

body shell (skin, limbs, fat and skeletal muscle) normally at 'superficial body' temperature of around 33°C

Figure 15.17 *Body core and body shell*

Correction of overheating

The skin helps to correct overheating of the body by employing the following mechanisms which promote heat loss.

Increase in rate of sweating

Heat energy from the body is used to convert the water in sweat to **water vapour**. This brings about a lowering of body temperature.

Vasodilation

Arterioles leading to skin become **dilated**, as shown in Figures 15.18 and 15.19. This allows a large volume of blood to flow through the capillaries near the skin surface from where it is able to lose heat by **radiation**.

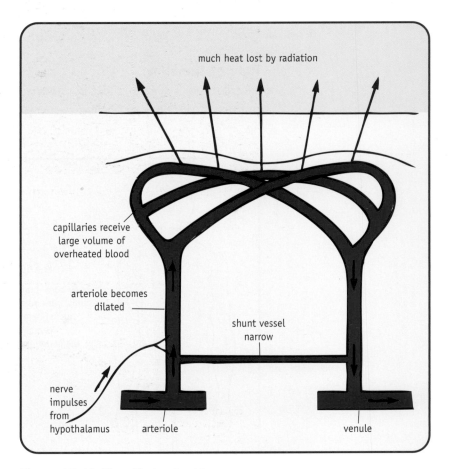

Figure 15.18 *Vasodilation in skin*

Correction of overcooling

The skin helps to correct overcooling of the body by employing the following mechanisms which reduce heat loss.

Decreased rate of sweating

Very little heat is used to convert water in sweat to water vapour.

Vasoconstriction

Arterioles leading to the skin become **constricted**, as shown in Figure 15.20.

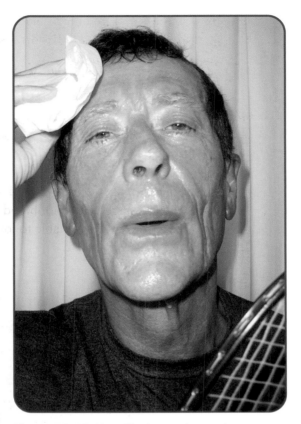

Figure 15.19 *Vasodilation and sweating*

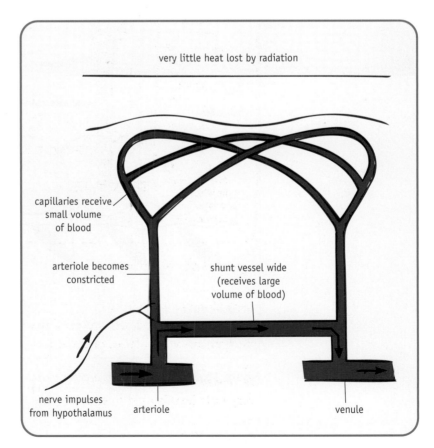

Figure 15.20 *Vasoconstriction in skin*

This allows only a small volume of blood to flow to the surface capillaries. Little heat is therefore lost by radiation.

Contraction of erector muscles

This process (see Figure 15.21) results in hairs being raised and a **wider layer of air** (which is a poor conductor of heat) being trapped between the mammal's body and the external environment. This layer of insulation reduces heat loss from the animal's body.

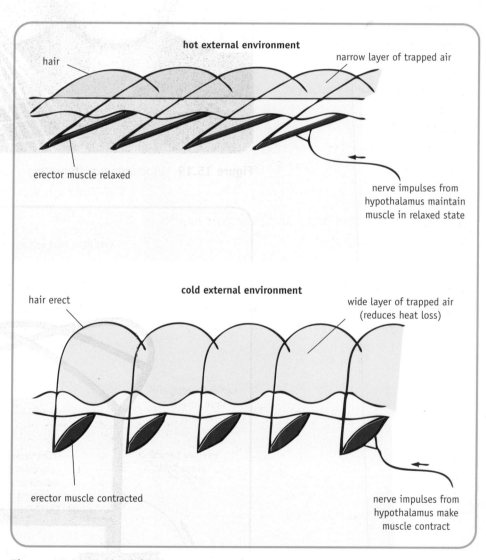

Figure 15.21 *Action of hair erector muscles*

Practical Activity and Report

Changes in body surface temperature in changing environmental conditions

INFORMATION

- A **thermistor** is a device that responds to tiny changes in temperature.
- In this investigation you are going to measure any changes that occur in the surface temperature of one part of the body (the left hand) when another part of the body (the right hand) undergoes sudden heat loss.
- At the same time you are going to measure any changes that occur in the temperature of the armpit.
- The left hand represents the body shell; the armpit represents the body core.

YOU NEED

1 bucket of crushed ice in water
2 thermistors
2 digital multimeters
2 sets of leads and wheatstone bridges containing appropriate resistors (see Figure 15.22)
2 6-volt power packs
1 stopclock
(Alternatively the thermistors can be connected up to interface with a computer.)

WHAT TO DO

1 Read all the instructions in this section and prepare your results table before carrying out the experiment.
2 With the aid of your partner, secure thermistor 1 between two fingers of your left hand, as shown in Figure 15.22.
3 Position thermistor 2 in your left armpit.
4 Switch on the multimeters and allow 2 minutes for the thermistors to equilibrate with their surroundings.
5 Note the starting temperature of thermistor 1 (left hand, body shell) and that of thermistor 2 (armpit, body core).
6 Start the clock and then plunge your right hand into the bucket of icy water.
7 Record the temperatures of the thermistors every 30 seconds for as long as you can bear it, up to a maximum of 5 minutes.
8 If other students have carried out the same investigation, pool your results.

REPORTING

Write up your report by doing the following:

1 Copy the title given at the start of this activity.
2 Put the subheading '**Aim**' and state the aim of your experiment.
3 a) Put the subheading '**Method**'.
   b) Draw a simple labelled diagram of a thermistor in use.
   c) Using the impersonal passive voice, briefly describe the experimental procedure that you followed and state how you obtained your results.

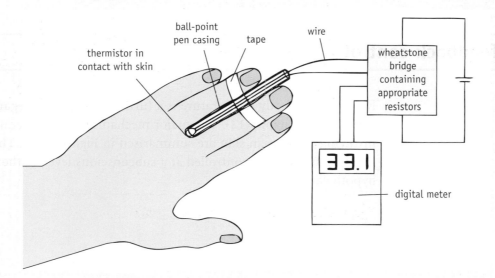

Figure 15.22 *Use of thermistor*

4 Put the subheading '**Results**' and draw a final version of your table of results.

5 Put a subheading '**Presentation of Results**' and present your results as two line graphs with shared axes on the same sheet of graph paper.

6 Put a subheading '**Conclusions**' and write a short paragraph to state what you have found out from a study of your results. This should include answers, in sentences, to the following questions:
   a) What overall trend is shown by each set of results?
   b) Which part of the body (core or shell) remains unaffected by a sudden drop in temperature of the right hand?
   c) Which part of the body (core or shell) responds to a sudden drop in temperature of the right hand?
   d) Why is such a compensatory reduction in temperature (which is brought about by constriction of blood vessels) of survival value?

7 Put a final subheading '**Evaluation of Experimental Procedure**' and then answer the following:
   a) State a possible source of error in this investigation.
   b) By what means was the experiment controlled so that there was only one variable factor being investigated at a time?
   c) State how you could check the reliability of the results.
   d) Feel free to comment on either of the following if you have a further point that you wish to make:
      (i) possible improvements that you would include in a repeat of the experiment
      (ii) limitations of the equipment.

Role of other effectors

In addition to skin, the body possesses other effectors that play an important part in temperature regulation by generating heat energy when necessary.

Shivering by skeletal muscles

When the hypothalamus detects a drop in body temperature, nerve impulses to skeletal muscles cause them to undergo brief repeated contractions. This process, called **shivering**, generates heat energy and helps to return body temperature to its normal level.

Liver

The high **metabolic rate** that occurs in active organs such as the liver produces heat and helps to maintain body temperature at its set point.

Negative feedback control

Regulation of body temperature is a further example of **negative feedback control** (see Chapter 12). The major mechanisms of temperature regulation involving the human skin are summarised in Figure 15.23. They are **involuntary** and are controlled at a subconscious level by the hypothalamus.

Voluntary responses

Nerve impulses also transmit information about the state of the internal environment to the cerebrum, which allows the person to be aware of the situation. Depending on the circumstances, this makes the person aware of 'feeling cold' or 'feeling hot'. He or she can then make a **voluntary** response appropriate to the situation such as putting on extra clothing or consuming a cold drink.

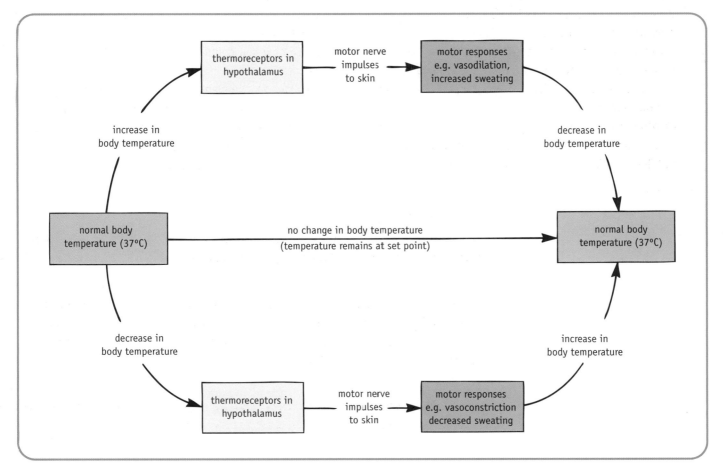

Figure 15.23 *Regulation of body temperature by negative feedback control*

Testing Your Knowledge

1 a) Which part of the brain contains the centre responsible for regulation of body temperature? (1)
   b) By what means does this structure obtain information about the body's surface temperature? (1)
2 Give TWO differences between the body core and the body shell. (2)
3 a) Name TWO effectors to which the hypothalamus sends motor impulses when the body temperature increases to above the normal level. (2)
   b) Describe the motor responses made by these effectors and explain how they return the body temperature to normal. (4)
4 What general name is given to a means of regulation where effectors make responses that correct a deviation by a factor from its normal level? (1)

Applying Your Knowledge

1 Match the terms in list X with their descriptions in list Y.

List X	List Y
1) cerebellum	a) region of cerebrum that sends impulses to effectors
2) cerebrum	b) simple arrangement of receptor, three neurones and an affector through which a nerve impulse passes
3) effector	c) process by which the bore of skin arterioles becomes wider
4) hypothalamus	d) secretion of liquid onto the skin surface from where it evaporates using the body's excess heat energy
5) medulla	e) largest part of the brain, divided into two hemispheres
6) motor strip	f) region of cerebrum that receives information from the body's receptors
7) neurone	g) region of the brain responsible for balance and muscular co-ordination
8) receptor	h) process by which the bore of skin arterioles becomes narrower
9) reflex action	i) region of the brain responsible for control of heart rate
10) reflex arc	j) region of the brain containing the centre that regulates body temperature
11) sensory strip	k) rapid automatic involuntary response to a stimulus
12) sweating	l) structures that detect changes in body temperature
13) thermoreceptors	m) sense organ that converts an environmental stimulus into a nerve impulse
14) vasoconstriction	n) muscle or gland that responds to a nerve impulse by making a response
15) vasodilation	o) nerve cell composed of a cell body and nerve fibres

2 Figure 15.24 shows a section of the human brain.

a) Would the person's face be at side 1 or side 2? (1)

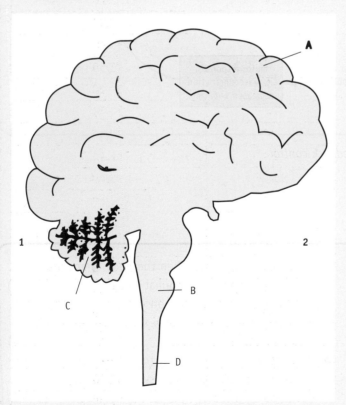

Figure 15.24

b) Name parts A, B, C and D. (4)

c) Decide whether each of the following statements about the brain is true or false and then use T or F to indicate your choice. Where a statement is false, give the word that should have been used in place of the one in **bold print**.

  (i) The **medulla** brings about an increase in the rate of breathing after vigorous exercise.

  (ii) The **cerebrum** is responsible for creativity and personality.

  (iii) The **hypothalamus** makes the heart beat more rapidly during exercise.

  (iv) The **cerebellum** contains the long-term memory.

  (v) The **medulla** receives information from the semi-circular canals in the inner ears.

  (vi) The **cerebrum** contains a centre that regulates water balance.

  (vii) The **hypothalamus** sends out nerve impulses to effectors to control body temperature.

  (viii) The **cerebellum** is needed for effective muscular coordination. (8)

3 This question refers to Figures 15.4 and 15.5.

a) Account for the fact that a leg is a large part of a

normal human body yet it is represented by a fairly small part of the cerebrum's sensory region. (2)

b) Which of the following body parts contains the largest number of sensory receptors relative to its actual size? (Choose one answer only.)

A leg **B** shoulder **C** hip **D** tongue (1)

c) Which of the following structures have fewest sensory nerve endings in relation to their actual size? (Choose one answer only.)

A lips **B** arms **C** fingers **D** genitals (1)

4 This question refers to the 'motor homunculus' shown in Figure 15.7.

a) Identify TWO especially mobile parts of the body. (2)

b) Which third part of the body is sufficiently mobile to be used, with much practice, to operate a pencil or paint brush? (1)

c) (i) With reference only to the pinna (ear flap), predict how the 'motor homunculus' would differ if a rabbit had been drawn.

(ii) Explain your answer. (2)

5 Figure 15.25 shows the reflex action that happens when the foot comes into contact with a sharp tack.

a) Match numbers 1 to 6 with the following statements:

(i) leg muscles contract, removing foot from danger

(ii) impulse enters spinal cord

(iii) impulse travels through motor neurone

(iv) pain receptors in sole of foot are stimulated

(v) impulse passes through spinal cord

(vi) nerve impulse travels up sensory neurone (3)

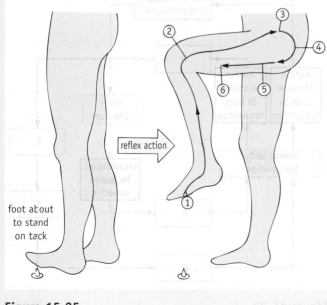

foot about to stand on tack

reflex action

Figure 15.25

b) Explain why this response is protective. (1)

6 a) When a piece of dirt lands in the eye, it makes the eye water. Explain this reflex action by identifying the stimulus, receptor, effector, response and protective value. (6)

b) Some reflex actions are completely involuntary; others can be partly altered by voluntary means. Classify the following into these two categories: blinking, churning of food in stomach, constriction of eye pupil, coughing, dilation of eye pupil, flushing of skin, laughing, muscular contraction making knee jerk, peristaltic contraction, sneezing. (5)

time (min)	body temperature (°C)
0	37.00
2	37.00
4	37.00
6	37.05
8	37.10
10	37.10
12	37.15
14	37.20
16	37.30
18	37.40
20	37.50
22	37.60
24	37.60
26	37.50
28	37.45
30	37.40
32	37.40
34	37.35
36	37.35
38	36.90
40	36.65
42	36.70
44	36.80
46	36.85
48	37.00
50	37.00

Table 15.2

7 The data in Table 15.2 refer to the body temperature of a student who exercised vigorously and then, after a rest, plunged into a cold bath. Body temperature was measured every two minutes by inserting a sterilised thermistor under the student's tongue.

a) Plot a line graph of the data. (3)

b) Using four arrows, indicate on your graph that:
   (i) exercise was begun at minute 2
   (ii) exercise was stopped at minute 22
   (iii) immersion in the cold bath occurred at minute 34
   (iv) exit from the cold bath took place at minute 40. (4)

c) In general what trend in body temperature occurs during
   (i) the period of vigorous exercise?
   (ii) the time in the cold bath?
   (iii) Why is there a slight delay before each of these trends begins? (3)

d) The student's skin was flushed from minute 18 onwards.
   (i) Suggest why.
   (ii) What is the benefit to the body of flushed skin? (2)

e) (i) At which ONE of the following times in minutes was the student found to be shivering?
   A 2 **B** 22 **C** 32 **D** 42
   (ii) What is the survival value of shivering? (2)

8 Figure 15.26 represents a section through human skin.

a) Name the part of the brain to which heat and cold receptors relay information about the external environment. (1)

b) By what means does the thermoregulatory centre of the brain communicate information to structures X and Y in order to effect control of body temperature? (1)

c) (i) In what way would structure X respond following a drop in body temperature?
   (ii) Explain how this response would help to conserve heat. (2)

d) (i) In what way would structure Y respond to an increase in body temperature?
   (ii) Explain how this response would help to promote heat loss. (2)

9 a) (i) Copy Figure 15.27 which shows involuntary regulation of body temperature.
   (ii) Complete boxes A, B, C, D and E.

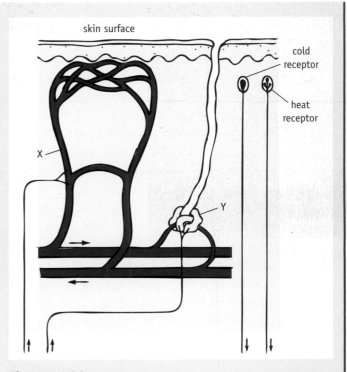

Figure 15.26

(iii) Identify the effector involved in this system of control. (6)

b) Describe a further involuntary response the body could make to bring about an increase in body temperature which involves a different effector from the one you gave as your answer to a)(iii). (1)

c) Name TWO voluntary responses that could be made to complement the effect of circuit A–D in the diagram. (2)

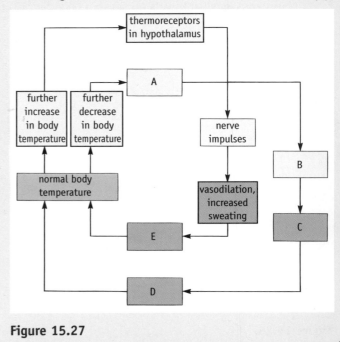

Figure 15.27

Word bank

action	motor
arc	negative
automatic	neurone
brain	receptor
cerebellum	responses
cerebrum	sensory
effectors	set point
heart	spinal
hypothalamus	stimulus
impulses	temperature
medulla	

Table 15.3 *Word bank for Chapter 15*

What You Should Know

(Chapter 15) (See Table 15.3 for word bank)

1 The largest part of the human brain is called the _____. It contains centres responsible for higher mental faculties. It has many discrete areas, two of which are the _____ strip and the motor strip, each with its own function.

2 The brain possesses many other parts in addition to the cerebrum. The _____ is the centre responsible for controlling balance and movement; the _____ contains centres that control breathing and _____ rate; the _____ has centres that regulate water balance and body _____.

3 The central nervous system consists of the _____ and the _____ cord. The CNS receives information as nerve _____ via sensory nerves from the sense organs. It sorts out this information and sends messages via _____ nerves to muscles, which make appropriate _____.

4 A reflex _____ is a simple arrangement of a _____, three nerve cells (a sensory _____, a relay neurone and a motor neurone) and an effector.

5 A reflex _____ is a rapid, _____, involuntary response to a _____. It depends on the transmission of a nerve impulse through a reflex arc.

6 Regulation of body temperature is an example of _____ feedback control. The hypothalamus responds to an increase in body temperature by sending nerve impulses to _____ such as the skin. Responses are made that correct the deviation and bring the body temperature back to its _____.

Index

Answers

Answers to Section 1

1 Structure and function of cells

Activity (Selecting information on the commercial use of a micro-organism)

a) Yeast.

b) (i) Favourable temperature, lack of competition.

  (ii) Temperature maintained by thermostatic control, other microbes killed by boiling wort in kettle.

c) To give flavour.

d) Cattle food (from barley grains), fertiliser (from spent hops), yeast extract in health foods (from dead yeast).

e) The yeast is killed by this concentration of alcohol.

f) Filtration.

Applying your knowledge

1 1) d
   2) q
   3) i
   4) a
   5) h
   6) c
   7) r
   8) m
   9) j
   10) p
   11) n
   12) f
   13) e
   14) g
   15) l
   16) k
   17) o
   18) b

2 a) W = rhubarb stalk epidermis
     X = *Elodea*
     Y = onion leaf epidermis
     Z = yeast.

   b) 1 = vacuole
      2 = cell wall
      3 = chloroplast
      4 = cytoplasm

5 = cell wall
6 = nucleus
7 = cell membrane
8 = cytoplasm
9 = cytoplasm
10 = cell wall
11 = nucleus

   c) (i) 1 stores water and solutes and regulates water content of cell by osmosis,
          3 absorbs light energy needed for photosynthesis of food,
          9 acts as a site of the cell's biochemical reactions.

     (ii) 9.

3 See Figure A1.1.

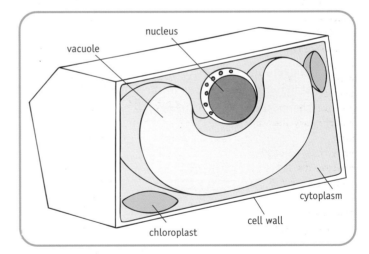

nucleus

vacuole

chloroplast

cell wall

cytoplasm

Figure A1.1

4 a) See Figure A1.2

   b) (i) 6.

     (ii) 4.

   c) 182 minutes.

   d) (i) 6.

     (ii) By repeating the experiment using a narrower, more detailed range of pH values such as 5.6, 5.8, 6.0, 6.2, 6.4, 6.6 and 6.8.

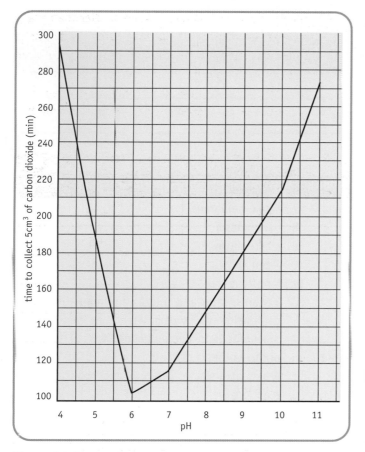

Figure A1.2

e) By maintaining the experiment at constant temperature (e.g. 20°C). This could be achieved by standing the apparatus in the water in a thermostatically controlled water bath.

5 a) S.
  b) P.
  c) V.
  d) S.
  e) V.

6 a) P = glucose, Q = live yeast cells, R = alcohol.
  b) Oxygen is still available for aerobic respiration so anaerobic respiration has not yet begun and no alcohol has been formed.
  c) It rises sharply as the population of yeast cells grows rapidly, it becomes less steep as population growth slows down and then it begins to drop as yeast cells start to die due to alcohol poisoning.
  d) It drops as yeast cells use glucose for respiration and growth. It levels off when the yeast cells die due to alcohol poisoning and stop using the glucose.
  e) (i) Alcohol.

(ii) The glucose supply did not run out so growth of yeast cells was not limited by lack of food.

7 a) Straw.
  b) Timber.
  c) Use of water.
  d) Straw.
  e) Sugar cane
  f) 9 times.

8 a) (i) B.
    (ii) A.
    (iii) W is not growing near B so it must be sensitive to B. W is growing in contact with A so it must be resistant to A.
    (iv) It makes an antibiotic which prevents W from growing by, for example, inhibiting the synthesis of W's cell wall.
  b) See Figure A1.3.

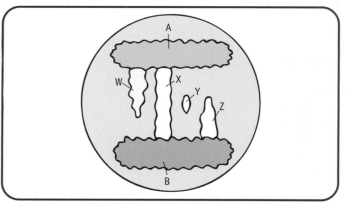

Figure A1.3

9 a) 3.
  b) P, Q and S.
  c) 3.
  d) S.
  e) None.
  f) 4.
  g) 6.
  h) 2.
  i) 4.

10 a) (i) At first their numbers increase equally then an inverse relationship develops and as the population of lactic acid bacteria increase in number, the others decrease.
    (ii) As the lactic acid bacteria continue to convert lactose to lactic acid, the environmental conditions become increasingly acidic. The low

pH that develops prevents growth of the other bacteria.

b) (i) and (ii) See Figure A1.4.

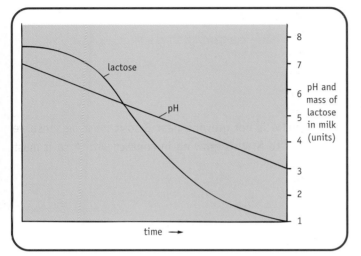

Figure A1.4

2 Diffusion and osmosis

Applying your knowledge

1 1) e
   2) f
   3) i
   4) h
   5) j
   6) c
   7) b
   8) d
   9) l
   10) g
   11) a
   12) k

2 a) 1 = Y, 2 = Z, 3 = X.
   b) (i)   Z.
      (ii)   Y.
   c) (i)   F.
      (ii)   T.
      (iii) T.
      (iv)   F.

3 a) (i)   Oxygen
      (ii)   Carbon dioxide.
   b) Starch.

4 a) 1 = C, 2 = A, 3 = B.
   b) Cylinder 1 is turgid (and bigger than cylinder 3) showing that it has taken water in by osmosis in

tube C. Cylinder 2 is flaccid (and smaller than the others) showing that it has lost water by osmosis to the hypertonic sugar solution in tube A. Cylinder 3 is neither turgid nor flaccid showing that it has neither gained nor lost water by osmosis when immersed in the isotonic sugar solution in tube B.

5 a) See Table A2.1.

change in length (mm)	percentage change in length
+8	+16
+3	+6
−2	−4
−7	−14
−12	−24

Table A2.1

b) See Figure A2.1.

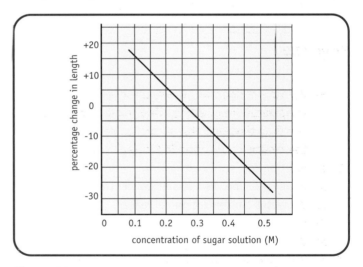

Figure A2.1

c) Hypotonic. The potato cylinder made a net gain in length at this concentration of sugar solution so water must have passed into it by osmosis from a HWC to a LWC.

d) Hypertonic. The potato cylinder suffered a net loss in length at 0.5M sugar solution showing that solution to be hypertonic to the cell sap. So 0.6M would also be hypertonic to the cell sap (but to an even greater extent).

e) 0.26.

f) So that a second variable factor is not introduced into the investigation.

g) Because the investigation is monitoring changes in length of the cylinders. Unlike mass, length is unaffected by surplus liquid adhering to the cylinders' surfaces.

6 a) A. HWC outside, LWC inside.
       B. HWC inside, LWC outside.

    b) (i) A up, B down.
       (ii) In A, water molecules move down a concentration gradient from the dilute sugar solution (HWC) outside the osmometer to the concentrated sugar solution (LWC) inside. This increases the volume of liquid inside the osmometer making the level rise.
            In B, water molecules move down a concentration gradient from the dilute sugar solution (HWC) inside the osmometer to the concentrated sugar solution (LWC) outside. This reduces the volume of liquid inside the osmometer making the level drop.

    c) It will increase.

7 a) (i) 5.
       (ii) 1.

    b) 1 = Y,
        2 = X,
        3 = Y,
        4 = Z,
        5 = X,
        6 = Y.

    c) Solution 5.

8 a) An inverse one. As the salt concentration increases, the water concentration decreases.

    b) (i) A direct one. As the salt concentration increases, so does the time taken for one pulsation.
       (ii) The higher the salt concentration of the bathing solution, the lower its water concentration and the lower the volume of water gained by osmosis by the animal immersed in it.

    c) (i) 1% salt solution.
       (ii) They worked at a faster rate.
       (iii) It would make them swell up and burst.

    d) (i) Percentage concentration of salt in bathing solution.
       (ii) Temperature and pH of salt solution.

    e) To increase the reliability of the results.

    f) So that they would stay in one place while feeding, making them easy to view under the microscope.

9 a) (i) 1 = cell wall, 2 = cell membrane.
       (ii) 1 is freely permeable, 2 is selectively permeable.

    b) (i) Z.
       (ii) W.
       (iii) Water concentration gradient.

    c) Y, X, W.

10 D.

3 Enzyme action

Applying your knowledge

1   1) f
    2) e
    3) j
    4) i
    5) d
    6) a
    7) k
    8) c
    9) g
    10) h
    11) b

2 a) Hydrogen peroxide.

    b) (i) A and C.
       (ii) The enzyme speeds up the rate of breakdown of hydrogen peroxide.
       (iii) Oxygen. It relit a glowing splint.

    c) B and D.

    d) When a molecule of enzyme is denatured the bonds holding the amino acids in its active site in a specific order become broken. This results in destruction of the active site and the enzyme can no longer fit on to and act on its substrate.

3 a) Enzyme-substrate complex.

    b) (i) Y, Z, X.
       (ii) X, Z, Y.

4   The <u>enzyme</u>, phosphorylase, combines with its <u>substrate</u>, <u>simple sugar</u>, at the <u>active site on the enzyme</u> molecule. The two fit together like a <u>lock and key</u> forming an <u>enzyme-substrate complex</u>. Once a bond has formed between the sugar molecules, the enzyme becomes free and repeats the process many times, adding another sugar molecule each time until the <u>product</u>, a <u>starch</u> molecule, has been formed.

5 a) B should contain **fresh** creamy milk (not sterilised) and **5** cm³ of bile salts (not 3 cm³).

    b) The control should differ from the experiment by only

the one factor being investigated. Here the effect of fresh lipase is being investigated, so, strictly speaking, boiled lipase is a better control than water because it makes tube B identical in every way to tube A except that the lipase in B has been denatured.

6 a) Type of potential substrate.
   b) Size of urease tablet, volume of potential substrate, length of litmus paper.
   c) To mix the urease with the potential substrate.
   d) (i) B.
      (ii) The red litmus paper turned blue indicating the release of ammonia gas formed by the action of urease on urea.
   e) Specificity of an enzyme to its substrate.

7 a) See Figure A3.1.

Figure A3.1

   b) (i) Optimum temperature means the temperature at which an enzyme works best.
      (ii) 40°C.
   c) (i) 2 times.
      (ii) As temperature increases from 20°C to 30°C, the rate of molecular movements and the frequency of collisions between enzyme and substrate molecules increase. A greater number of enzyme-substrate complexes are formed at the higher temperature resulting in a doubling of the reaction rate.
   d) 20 to 25°C.

   e) The enzyme molecules have become denatured and their active sites have been permanently destroyed.
   f) 0 g.

8 a) X = 0.5 – 4.5, Y = 5 – 9, Z = 8 – 12.
   b) (i) It is equal for each of these enzymes.
      (ii) There is no part of the pH working range that they all share.
   c) X = 2.5, Y = 7, Z = 10.
   d) (i) Y.
      (ii) X.
   e) A sudden drop in pH results in the breakdown of the bonds that hold together the chains of the amino acids in an enzyme molecule. The molecule becomes denatured.

9 a) Enzymes that control essential biochemical reactions in the body become denatured at this high temperature. The reactions slow down with fatal consequences.
   b) The low pH destroys the attacking microbe's enzymes by denaturing them. This saves the food from being attacked and going bad.
   c) Any fungal colonies on the cheese grow more slowly in the refrigerator since enzyme molecules are less effective at combining with and acting on their substrate molecules at low temperatures.

10 a) It would be digested to a soluble state.
   b) 3.
   c) 1.
   d) (i) 4.
      (ii) pH 3 at 30°C.
      (iii) Pepsin's optimum pH is around 2.5 and its optimum temperature is around 40°C. The values chosen for the answer to (ii) are closest to these optimum values.
   e) (i) 2.
      (ii) It contains 3 variable factors.
      (iii) See Figure A3.2.

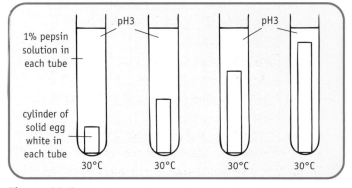

Figure A3.2

Answers

11 a) See Table A3.1.

b) See Figure A3.3.

c) (i) 9.

   (ii) 12

d) 9.

e) By repeating the experiment using a narrower but more detailed range of pH values such as 8.2, 8.4, 8.6, 8.8, 9.0, 9.2, 9.4, 9.6 and 9.8.

pH of hydrogen peroxide solution	time to collect 10cm³ of oxygen (s)	rate of breakdown of hydrogen peroxide (cm³/s)
6	50	0.2
7	40	**0.25**
8	20	**0.5**
9	**10**	1.0
10	25	**0.4**
11	**50**	0.2
12	100	**0.1**

Table A3.2

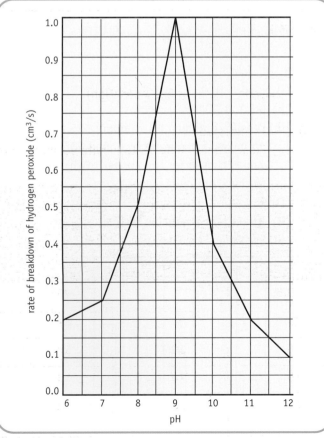

Figure A3.3

4 Respiration

Activity (Selecting and presenting data on the energy content of food)

1 a) Corn oil, lard, olive oil. 38 kJ/g on average.

b) (i) Protein = egg white. 19 kJ/g.

   (ii) Carbohydrate = sucrose. 19 kJ/g.

c) 1 : 2 : 1

2 3 times.

3 a) 2000 g.

b) 0.5 kg.

c) See Figure A4.1.

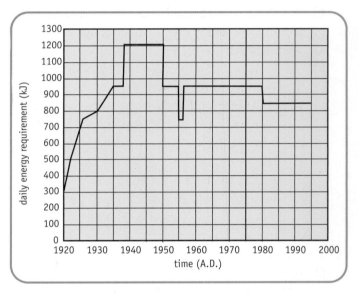

Figure A4.1

Applying your knowledge

1 1) o

    2) f

    3) g

    4) n

    5) k

    6) j

    7) b

    8) d

    9) l

    10) a

    11) h

    12) m

    13) i

    14) c

    15) e

2 a) A valid comparison between two set-ups can only be made if they differ by the one factor being investigated (in this case type of food) and by no other.

(i) A and B fail because they have the same food and differ from one another in two ways – needle or spoon and volume of water.

(ii) A and C fail because they have the same food and differ from one another in two ways – mass of food and volume of water.

(iii) Although A and E have different foods, they fail because they also differ from one another in two other ways – needle or spoon and mass of food.

b) C's volume of water would need to be reduced to 50 cm^3, its mass of food reduced to 1g and its mounted needle exchanged for a spoon.

c) (i) peanut = 9.24 kJ, glucose = 5.04 kJ.

(ii) Heat loss to the surroundings is greatly reduced by the burning food being enclosed in the food calorimeter and by the presence of a coiled chimney. Both of these promote maximum transfer of heat energy to the water.
Burning the food in air enriched with oxygen (instead of normal atmospheric air) ensures that the food is burned to ashes and that all of its energy has been released

3 a) 612 kJ.
 b) 48 g.
 c) Activity (iii) would expend most energy.
   (i) 65.1 × 20 = 1302 kJ.
   (ii) 53.2 × 25 = 1330 kJ.
   (iii) 28.6 × 50 = 1430 kJ.

4 a) Glycolysis.

b) (i) Q = 2, T = 18.
   (ii) R = ethanol, S = carbon dioxide.
 c) It will diffuse out of the yeast cell.
 d) Poisoning of the cell.
 e) (i) 2.
   (ii) 38.

5 a) See Table A4.1.
 b) See Figure A4.2.

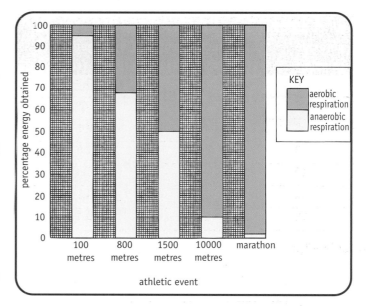

Figure A4.2

6 a) Anaerobic.
 b) Production of ATP (phosphorylation).
 c) ADP.
 d) (i) It increased.
   (ii) From about minute 75 to minute 90 shortage of

athletic event	volume of oxygen needed for event (l)	volume of oxygen consumed during event (l)	oxygen debt (l)	percentage of energy obtained for event from aerobic respiration	percentage of energy obtained for event from anaerobic respiration
100 metres	10	0.5	9.5	5	95
800 metres	25	8	17	32	68
1 500 metres	36	18	18	50	50
10 000 metres	150	135	15	90	10
marathon (42 186 metres)	700	686	14	98	2

Table A4.1

Pi had been holding up the processes of ATP regeneration and CO_2 production. Following the addition of Pi at minute 90, both of these processes resumed.

e) The supply of glucose needed for respiration was running out; the build-up of ethanol formed during anaerobic respiration was starting to poison the yeast cells.

7 a) See Figure A4.3.

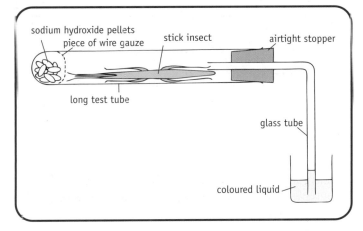

sodium hydroxide pellets
piece of wire gauze
stick insect
airtight stopper
long test tube
glass tube
coloured liquid

Figure A4.3

b) To absorb CO_2.
c) A rise in the level of the coloured liquid in the glass tube.
d) The control would contain a dead stick insect (of the same size, age and gender) or an equal mass of some inert material (such as plasticine or glass beads).

8 See figure A4.4.

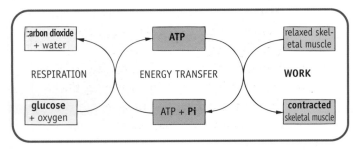

carbon dioxide + water

ATP

relaxed skeletal muscle

RESPIRATION

ENERGY TRANSFER

WORK

glucose + oxygen

ATP + Pi

contracted skeletal muscle

Figure A4.4

9 3), 5), 2), 4), 1).
10 a) By immersing them in a thermostatically controlled water bath.
b) To exclude light and prevent photosynthesis affecting the experiment.

c) It rises to take up the space previously occupied by the oxygen that has now been absorbed by the leaves for use in respiration.
d) The syringe is used to inject air until level A has been pushed back to the start.
e) 15°C = 0.7 ml, 25°C = 1.2 ml.
f) Increase in temperature brings about an increase in respiratory rate.
g) The experiment could be repeated several times and average results calculated.
h) 2 g of dead dandelion leaves would make a better control than an empty tube.

5 Photosynthesis

Applying your knowledge

1 1) d
2) e
3) g
4) p
5) n
6) a
7) j
8) k
9) m
10) o
11) l
12) h
13) b
14) f
15) c
16) i

2 a) Day 2, 12.00 – 15.00.
b) Photosynthesis.
c) (i) 05.00.
(ii) 21.00.
d) Day 3. The CO_2 curve shows the smallest dip suggesting least use for photosynthesis.
e) See Figure A5.1.

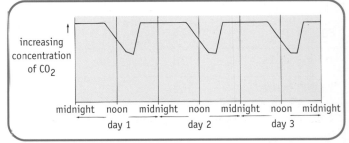

increasing concentration of CO_2

midnight   noon   midnight   noon   midnight   noon   midnight
day 1   day 2   day 3

Figure A5.1

3 a) 4.
  b) 2 times.
4 a) (i) Photosynthesis.
     (ii) Respiration.
  b) (i) Respiration.
     (ii) Photosynthesis.
  c) (i) Respiration.
     (ii) Respiration.
  d) 14.00. Up to this point, respiration had been the dominant process for several hours.
  e) The peak of the curve that represents increasing oxygen concentration would have been lower.
5 a) (i) 4 and 6.
     (ii) 3 and 4.
     (iii) 2 and 4.
  b) (i) The only difference between discs 4 and 6 is the presence or absence of light.
     (ii) The only difference between discs 3 and 4 is the presence or absence of chlorophyll.
     (iii) The only difference between discs 2 and 4 is the presence or absence of CO_2 (which is absorbed by sodium hydroxide).
6 a) (i) Sugar production as g/kg of dry plant.
     (ii) Time (taken by plant to produce known masses of sugar).
     (iii) Rate of oxygen production, rate of CO_2 uptake.
  b) 0.6%.
  c) See Figure A5.2.
  d) 2.5g/kg of dry plant.
  e) (i) Each increase in CO_2 concentration between 0 and 50 units was accompanied by a rise in the graph indicating an increase in sugar production (i.e. photosynthesis) and showing that CO_2 concentration was holding up the process until 50 units.
     (ii) Each increase in CO_2 concentration from 50 units onwards resulted in a continuous levelling-off of the graph indicating no further increase in sugar production despite the presence of plenty of CO_2. So CO_2 concentration was no longer limiting the process.
     (iii) Light intensity.
7 a) (i) Chloroplast.
     (ii) Green chlorophyll.
  b) (i) X = oxygen, Y = hydrogen, Z = glucose.
     (ii) Z.
     (iii) X.

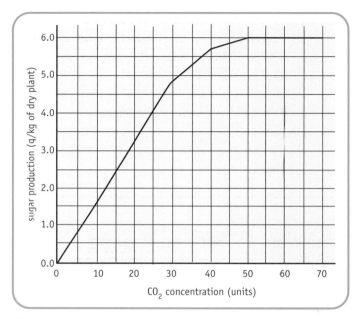

Figure A5.2

  c) (i) 1 = light-dependent stage (photolysis), 2 = temperature-dependent stage (carbon fixation).
     (ii) ATP.
8 a) Oxygen.
  b) 0.04 cm³/hour.
  c) (i) Exchange the coolant for a stream of water heated to the required temperature.
     (ii) To allow the plant to become acclimatised to the new temperature.
  d) (i) 55°C.
     (ii) The plant would probably be dead following the denaturation of its enzymes at this temperature.
  e) (i) To ensure that CO_2 is not a limiting factor when investigating the effect of varying temperature.
     (ii) Sodium (or potassium) hydrogen carbonate (bicarbonate).
  f) Keep the temperature constant at, say, 25°C and then vary the light intensity by varying the distance of the light source from the plant or by using a lamp with a dimmer switch.
9 a) (i) Light intensity.
     (ii) CO_2 concentration.
  b) The apparatus containing the water plant was immersed in a thermostatically controlled water bath.
  c) Light intensity.
  d) 5 units.
  e) (i) Light intensity.

(ii) Temperature.

10 a) Light intensity and carbon dioxide concentration.

b) (i) Law of the minimum.

(ii) It means that the growth rate of a plant is determined by the factor that is at its lowest level and is holding up the process.

c) (i) Photosynthesis during daylight hours.

(ii) Aerobic respiration during all 24 hours of the day and night.

(iii) It doubles the rate approximately.

(iv) At 20°C the plant respires so rapidly that it uses up much of its food reserves. At 10°C it respires much more slowly and the unused food is stored making it gain mass.

d) Increased light intensity also affects stomatal opening. Wider stomata may allow increased CO_2 uptake and increased loss of water vapour.

Answers to Section 2

6 Energy flow

Activity (Identifying an ecosystem's component parts and inter-relationships)

1 a) Common crab, limpet, brown seaweed, decay bacteria.

b) All of them.

2 a) Illuminated rock surface, shady recess, sand, sea water.

b) Illuminated rock surface and sea water.

c) A shady recess lacks sufficient light for photosynthesis; sand is too loose a material to provide the seaweed with a good source of attachment.

3 a) (i) The crab depends on the limpet as a source of food.

(ii) The limpet depends on the seaweed as a source of food.

(iii) All of the organisms depend on the decay bacteria to decompose dead material and allow essential elements such as nitrogen to be recycled round the ecosystem.

b) brown seaweed → limpet → common crab

4 Top and bottom parts of diagram to be combined.

5 Plants = plant plankton, green seaweed and red seaweed.

Animals = animal plankton, barnacle, cockle, dog whelk, mussel, oyster and periwinkle.

6 See Figure A6.1.

Applying your knowledge

1 1) d

2) i

3) e

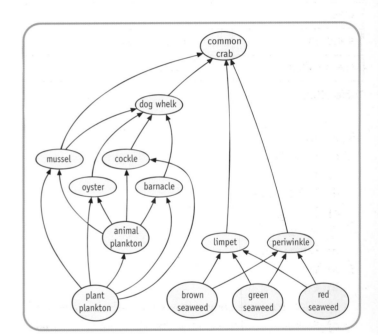

Figure A6.1

4) b

5) g

6) k

7) a

8) c

9) f

10) l

11) h

12) j

2 The diagram shows a soil ecosystem. The soil solution provides a habitat for the members of the nematode worm population. Air spaces in the soil provide habitats for the springtail population. Burrows in the soil are the habitats of the earthworm population. The producers in this ecosystem are a sycamore tree and a population of grass plants. All of these animals and plants together make up the <u>community</u> of this soil ecosystem.

3 See Table A6.1.

predator	prey
cat	mouse
diving beetle	frog tadpole
fox	rabbit
hawk	songbird
ladybird	greenfly
pike	perch
scorpion	locust hopper
sperm-whale	squid

Table A6.1

4 a) T.
  b) F, smaller.
  c) T.
  d) F, field mice.
  e) T.
  f) F, more.
  g) F, primary.
5 a) W.
  b) 4%.
  c) Heat lost from body and energy used for movement.
  d) Faeces and urine contain chemical substances that provide energy for decomposers.
  e) 9.6 kJ.
6 a) 1).
  b) Any course of action that involves keeping the hen alive would be less successful than course 1) because it would involve feeding wheat to the hen. This would lead to a loss of energy at two links in the food chain whereas the man eating all the wheat would lead to energy being lost at only one link.
7 a) (i) See Figure A6.2.
      (ii) The top one containing fish-eating birds.
      (iii) Algae.
  b) Sunlight.
  c) (i) 800 times.
      (ii) Between herbivorous fish and carnivorous fish.

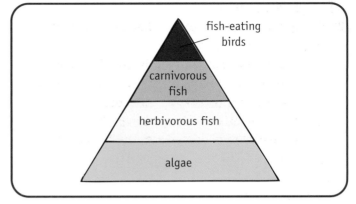

Figure A6.2

8 a) (i)   oak tree → wood mouse → owl
      (ii) grass → sheep → human
      (iii) algae → animal plankton → anchovy → tuna
      (iv) heather → mountain hare → eagle
  b) moorland = (iv), ocean = (iii), heavily grazed grassland = (ii), natural woodland = (i).
  c) A = (iii), B = (ii), C = (i), D = (iv).
9 a) (i)   100 kg.
      (ii) 10 kg.
  b) (i)   X = sheep, Y = salmon.
            Herbivores are less efficient than carnivores at both absorbing energy from food and converting the energy into body tissues. Therefore the two lower figures refer to the sheep (a herbivore) and the two higher figures refer to the salmon (a carnivore).
      (ii) It is used for movement or lost in waste materials.
  c) (i)   The animals are warmer indoors and therefore need to use less of their energy reserves to maintain body temperature. Instead of being used to generate heat, the energy is built into body tissues.
      (ii) The animals cannot move about and therefore use less energy. Instead of being used for movement energy reserves are built into body tissues.
  d) The trout farm would be more productive. Unlike birds (e.g. ostrich), fish (e.g. trout) are ectothermic ('cold-blooded') and do not use their food reserves to keep their bodies at a temperature higher than that of the environment. More energy is therefore available to be built up into body tissues.

7 Factors affecting the variety of species in an ecosystem

Applying your knowledge

1 1) i
  2) g
  3) l
  4) j
  5) h
  6) a
  7) k
  8) b
  9) f
  10) c
  11) d
  12) e

2 a) It increases the reliability of the results.
  b) 30%.
  c) (i) N.
    (ii) S.
  d) 750 lux.
  e) (i) S.
    (ii) N.
  f) (i) An inverse relationship. When light intensity is high, % *Pleurococcus* cover is low and *vice versa*.
    (ii) *Pleurococcus* is an alga that requires a damp habitat. Poorly lit habitats have more *Pleurococcus* since they are damper than well illuminated habitats that have been partly dried out by the heat of the sun.

3 Y was the square with 11 of the original species. It was left uncut and the dominant species competed with and choked out 9 of the weaker species. X was the square with all 20 species. Cropping kept the dominant species in check and the delicate ones managed to survive.

4 a) (i) 5.5.
    (ii) 5.8.
  b) Trout.
  c) Eel.
  d) (i) 8.
    (ii) 5.
    (iii) 3.
    (iv) 1.
    (v) 0.
  e) As the pH decreases so does the variety of species of fish present in the loch.

5 (i) and (ii) One of the effects of pollution on this ecosystem is that it reduces the variety of species. Before pollution there were about 20 different types of animal and plant; after pollution there were only about 3 different types. A second effect is the change in population numbers. Before pollution there were small populations of many different species; after pollution there were large populations of only a few species.

6 a) (i) 4 km.
    (ii) 20 km.
  b) (i) To increase the reliability of the results.
    (ii) It increases.
    (iii) Many species of lichen are sensitive to SO_2. The further away from the power station the lower the concentration of SO_2 in the air and the higher the number of lichen species that can survive.
  c) (i) 45.
    (ii) From NW.
    (iii) From SW.
  d) (i) The increase in biodiversity of lichen species along the SE line is much greater than that along the NE line.
    (ii) This is probably due to the fact that on 250 out of 365 days each year, SO_2 from the power station is blown along the NE line preventing many of the lichen species from surviving.

7 a) Soil mite and ladybird.
  b) Because spider mites survived and ate the crop.
  c) Because the natural predators of spider mites had been killed so the spider mites multiplied and severely damaged the crop.
  d) This example shows how the removal of the secondary consumers from a food web enables a primary consumer to undergo a population increase and destroy most of the producer.

8 a) An inverse relationship. The higher the concentration of glycerol solution, the lower the humidity of the air.
  b) When they are in the thin film of water and when they are inserted into the choice chamber.
  c) 2. 33% glycerol results in air of 90% humidity (i.e. 'damp'). 95% glycerol results in air of 10% humidity (i.e. 'dry').
  d) (i) Dry.
    (ii) Damp.

(iii) At first the animals' skins were too wet so the dry side was the favourable environment in which to slow down and gather in. However once their skins began to dry out, the damp side became the favourable environment in which to slow down and congregate in.

(iv) It ensures that the animals end up in a damp environment where they will not dry out and die.

9 a) (i) Light side.

(ii) They covered a longer distance for each 15-second time interval compared with distance covered each time in the dark.

b) Light side.

c) (i) Dark side.

(ii) They move more slowly in the dark side and change direction (by turning) less frequently.

(iii) If they gather in the dark they will not be seen by their predators.

10 a) Different numbers of seeds were carefully counted out for use.

b) Mass of cotton wool, volume of water added, size of carton.

c) (i) See Table A7.1.

number of seeds planted	number of healthy seedlings with green leaves	percentage number of healthy seedlings with green leaves
100	87	87
200	178	89
300	204	68
400	228	57
500	225	45

Table A7.1

(ii) To standardise the results so that they can be compared.

d) See Figure A7.1.

e) As competition increases, the percentage number of healthy seedlings with green leaves decreases.

f) Light.

g) If a plant's stomata are closed it cannot take in CO_2 so it cannot photosynthesise and make food. This results in stunted growth and, eventually, death.

h) See Figure A7.2.

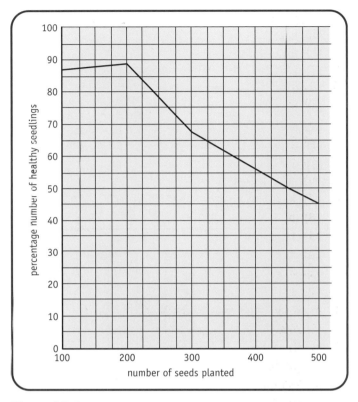

Figure A7.1

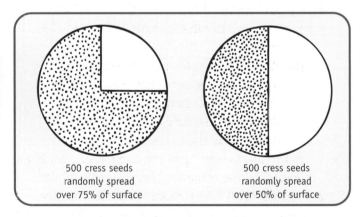

500 cress seeds randomly spread over 75% of surface

500 cress seeds randomly spread over 50% of surface

Figure A7.2

8 Factors affecting variation in a species

Applying you knowledge

1 1) e

2) f

3) g

4) l

5) r

6) a

7) q

8) m

9) k

10) n
11) h
12) p
13) j
14) c
15) i
16) d
17) b
18) o

2 c), a), e), b), d).

3 a) See Figure A8.1.

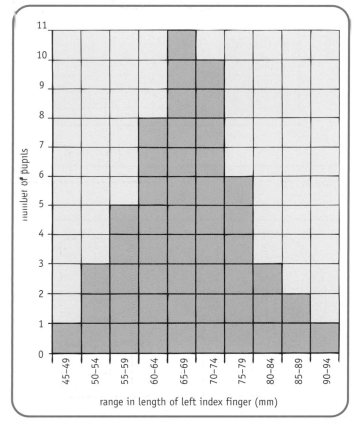

number of pupils

range in length of left index finger (mm)

Figure A8.1

b) 10.
c) (i) 65–69 mm.
   (ii) 11.
d) 12.

4 a) (i) A = head, = 5 tail.
   (ii) Digestion of egg membrane to allow sperm to enter the egg and fertilise it.
   b) 0.06.
   c) (i) X = 21 million/ml, Y = 19 million/ml, Z = 26 million/ml.
      (ii) Y.

d) Many sperm are needed so that there is a good chance that at least some of them will complete the long journey to the egg and that one will fertilise it.

5 a) (i) 3.
      (ii) 1.
      (iii) 2.
   b) It is too short a length of chromosome.

6 2), 4), 1), 3).

7 (i) 10.
   (ii) 20.
   (iii) 5.
   (iv) 5.

8 a) X = gamete production, Y = fertilisation, Z = mitosis.
   b) A = sperm, B = egg mother cell.
   c) Sperm and egg.
   d) Body cell, egg mother cell, zygote.

9 a) See Figure A8.2.
   b) See Figure A8.3.
   c) See Figure A8.4.
   d) See Figure A8.5.

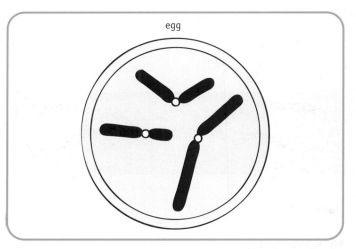

egg

Figure A8.2

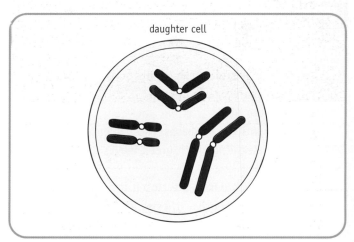

daughter cell

Figure A8.3

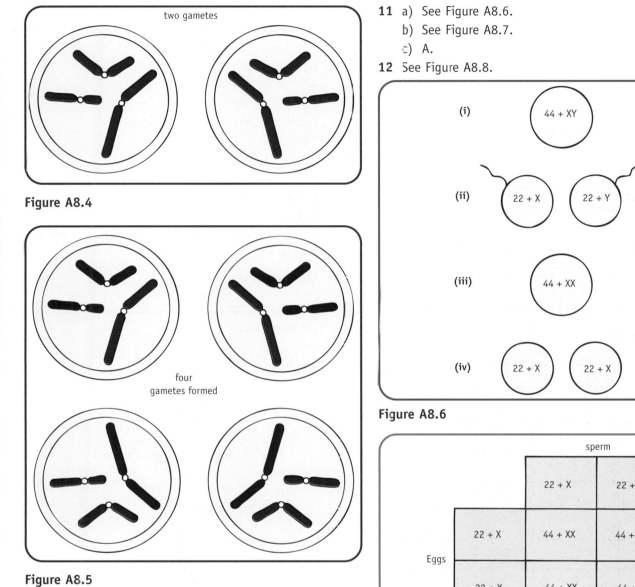

two gametes

Figure A8.4

four
gametes formed

Figure A8.5

10 See Table A8.1.

chromosome number		number of different combinations of homologous chromosomes that can arise in gametes following random assortment
in body cell	in gamete	
4	2	$2^2 = 4$
6	3	$2^3 = 8$
8	**4**	$2^4 = $ **16**
10	5	$2^5 = 32$
12	**6**	$2^6 = $ **64**
46	**23**	$2^{23} = 8\ 388\ 608$

Table A8.1

11 a) See Figure A8.6.
   b) See Figure A8.7.
   c) A.
12 See Figure A8.8.

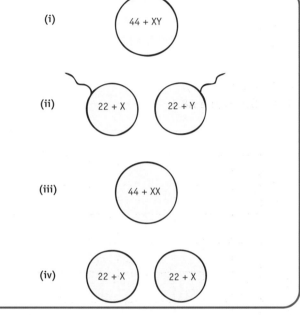

(i)   44 + XY

(ii)   22 + X   22 + Y

(iii)   44 + XX

(iv)   22 + X   22 + X

Figure A8.6

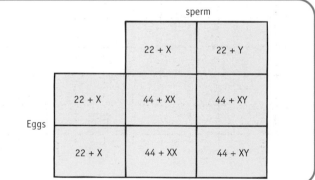

sperm

	22 + X	22 + Y
Eggs 22 + X	44 + XX	44 + XY
22 + X	44 + XX	44 + XY

Figure A8.7

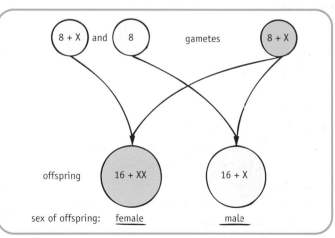

8 + X and 8   gametes   8 + X

offspring   16 + XX   16 + X

sex of offspring: <u>female</u>   <u>male</u>

Figure A8.8

9 Phenotype and genotype

Activity (Selecting and presenting information showing that characteristics are inherited from both parents)

1 -
2 a) Father.
   b) From her paternal grandmother and maternal grandfather via both of her parents.
   c) See Figure A9.1.

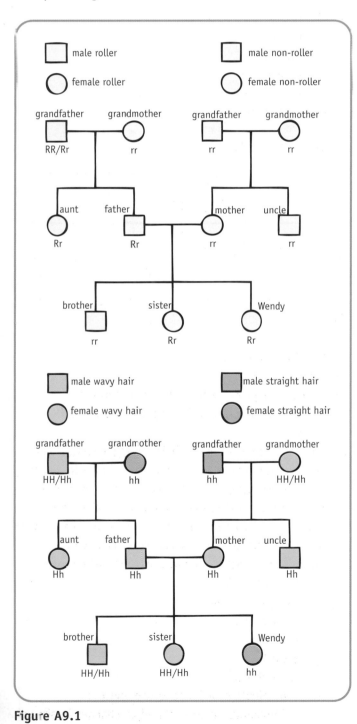

Figure A9.1

Applying your knowledge

1 1) k
   2) f
   3) d
   4) a
   5) h
   6) b
   7) c
   8) j
   9) l
   10) g
   11) m
   12) e
   13) i
2 a) See Figure A9.2.

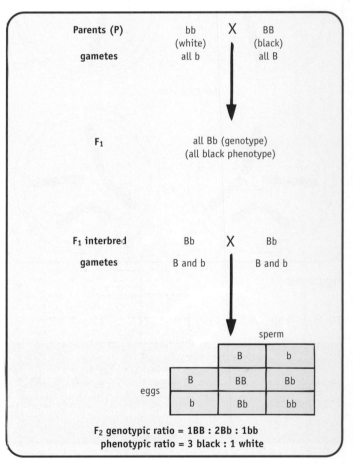

Figure A9.2

   b) BB and Bb.
3 a) Inflated is dominant because it has masked the appearance of constricted in the F1 generation.
   b) Ii.
   c) See Table A9.1

	genotypes of pollen	
	I	i
genotypes of ovules I	II	Ii
i	Ii	ii

Table A9.1

d) (i) 295.
(ii) 590.
(iii) 295.

4 A = RR, B = Rr, C = Rr, D = Rr, E = rr, F = rr.

5 a) See Figure A9.3.

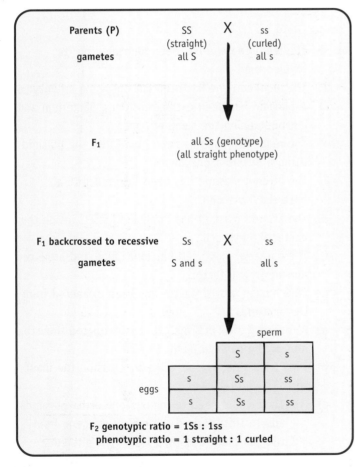

Figure A9.3

b) The allele for curled wing is recessive and masked by straight wing, the dominant allele.

c) (i) 1:1.
(ii) 84 straight to 84 curled.
(iii) Because fertilisation is a random process that involves an element of chance.

6 a) and b) See Figure A9.4.
c) Mandy has the white forelock.

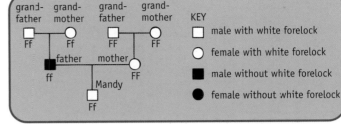

Figure A9.4

7 a) (i) nn.
(ii) nn.
(iii) Nn.
b) (i) NN or Nn.
(ii) NN or Nn.
c) No chance.
d) 1 in 4.

8 a) See Figure A9.5.

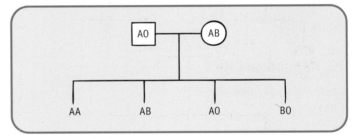

Figure A9.5

b) Blood group O.

9 See Figure A9.6.

10 a) (i) The plants will develop blue flowers.
(ii) The plants will develop pink flowers.
b) B.

11 Amongst a population of rabbits, <u>over-production</u> may occur resulting in more offspring than the environment can support. <u>Competition</u> for scarce resources follows. Since <u>variation</u> exists amongst the population, those rabbits that are better adapted to the environment survive by <u>natural selection</u>. The weaker ones die and by this means the population numbers are kept under control.

12 a) (i) B at 1400 m.
(ii) A at 3050 m.
b) (i) 3050 m.
(ii) 30 m.
c) Vertically.
d) Horizontally.
e) A combination of its genotype and the environmental factors related to changes in altitude.

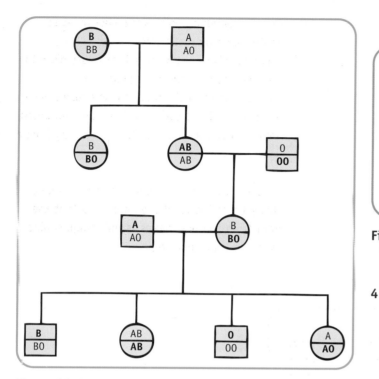

Figure A9.6

10 Applied genetics

Applying your knowledge

1 1) g
2) e
3) f
4) h
5) a
6) b
7) d
8) c

2 a) For both breeds of cattle the selective breeding programme has resulted in a continuous increase in milk yield and butter fat content. At the start, the Ayrshire cattle's milk yield was greater than that of the Jersey cattle but the Ayrshire cattle's butter fat content was less than that of the Jersey cattle. This difference was unchanged at the end of the 5-year period.

b) No. There comes a point in the selective breeding programme when the breed contains all of the desirable alleles and no further improvement can be achieved.

3 a) A = 7, B = 6, C = 6, D = 8, E = 7, F = 6, G = 8, H = 7, I = 7.

b) All of them.

c) D and G.

d) (i) See Figure A10.1.

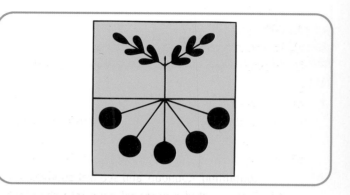

Figure A10.1

(ii) 9.

4 B The required gene has been identified on its chromosome.

E The required gene has been cut out using an enzyme.

D A plasmid has been extracted from a bacterium and cut open using the same enzyme.

F The required gene has been glued into the plasmid using a different enzyme.

H The altered plasmid has been inserted into a bacterial 'host' cell.

A The altered plasmid has duplicated itself inside the bacterial 'host' cell.

C The bacterial 'host' cell has multiplied and produced human growth factor.

G Pure human growth factor has been extracted from the bacteria.

5 a) A foodstuff rich in starch that is extracted from the root of the cassava plant.

b) They tried selective breeding by crossing the plant with its wild relatives.

c) (i) They would need to remove one of its plasmids, add to this plasmid the resistance gene from the pea plant and then insert the altered plasmid into *Agrobacterium*.

(ii) A new strain of greenfly might emerge in the future.

d) The case for: this industry would be based on a renewable resource (not a fossil fuel) and it could provide employment and profit for local people.
The case against: this industry might use land that was previously used to produce food and the multinational company might not share the profits with the local people.

6 a) See Figure A10.2.

b) X = genetically modified variety, Y = control.

c) (i) 7.

(ii) 70.

d) 90%.

e) The fruit would never ripen.

f) It extends the shelf life of the fruit in the shops.

g) -

7 a) (i) Weedkiller.

(ii) It blocks production of a protein needed by weeds for growth.

b) (i) Try growing the bacterium on nutrient agar containing 'Roundup' and see if it survives or not.

(ii) They inserted a bacterial gene into the plant.

(iii) They produce a protein different from the one blocked by 'Roundup' so their metabolism is unaffected.

c) (i) Use of 'Roundup' reduces the cost of weed control by more than 50%.

(ii) It is rapidly broken down by soil micro-organisms to carbon dioxide, ammonia and phosphate.

d) (i) They objected to the fact that the new soya beans were genetically modified and contained genetic material and a protein originally from a bacterium.

(ii) -

e) (i) They say that the new beans make up only a tiny part of the total soya crop and that the flour from the new beans is indistinguishable from the traditional type.

(ii) -

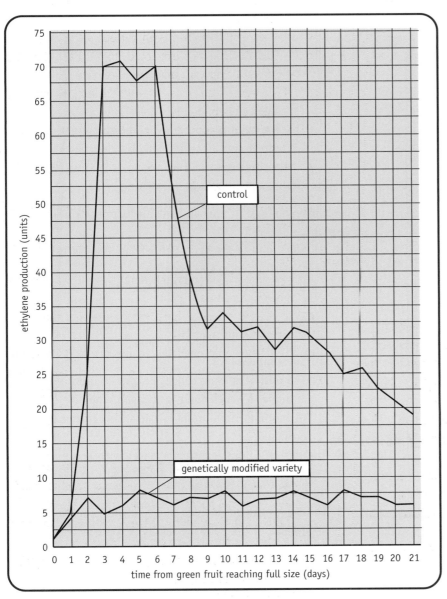

Figure A10.2

Answers to Section 3

11 Mammalian nutrition

Activity (Selecting and presenting information about foodstuffs)

1 a) Jam.
  b) Cornflakes.
  c) 7.
  d) 6.
  e) Almonds and pork chops.
  f) 4.5 times.
  g) 73.

2 See Figure A11.1.

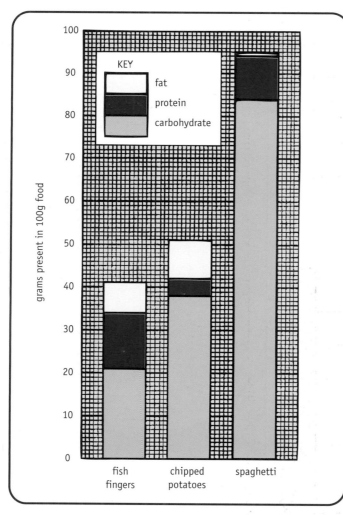

Figure A11.1

Activity (Selecting and presenting information to illustrate peristalsis)

1 -

2, 3 and 4 See Figure A11.2.

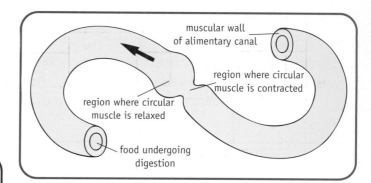

Figure A11.2

5 a) (i) Small intestine.
    (ii) The tube is too narrow to be the large intestine and it contains too many bends to be the oesophagus.
  b) Oesophagus and large intestine.

Activity (Selecting and presenting information about colonic cancer)

1 a) The colon is the main part of the large intestine and it is situated in the abdomen.
  b) Reabsorption of water from undigested waste material.

2 a) An abnormal, uncontrolled growth of cells.
  b) A lump of cells called a tumour.

3 a) See Figure A11.3.
  b) The rate of colonic cancer in every region of the UK is higher for men than for women. The rate is higher in Scotland than in any other region of the UK for both sexes.

4 The rate of colonic cancer increases with age for both sexes. The rate is higher for older men than older women.

5 The person should include plenty of fruit, vitamins, vegetables of the cabbage family and green tea in his/her diet. S/he should keep foods rich in saturated fats to a minimum, resist a diet low in fibre and not drink alcohol to excess. S/he should take regular exercise and avoid becoming grossly overweight. S/he

should resist smoking and try to keep stress down at a manageable level.

Applying your knowledge

1 1) f
   2) p
   3) r
   4) n
   5) h
   6) i
   7) k
   8) o
   9) q

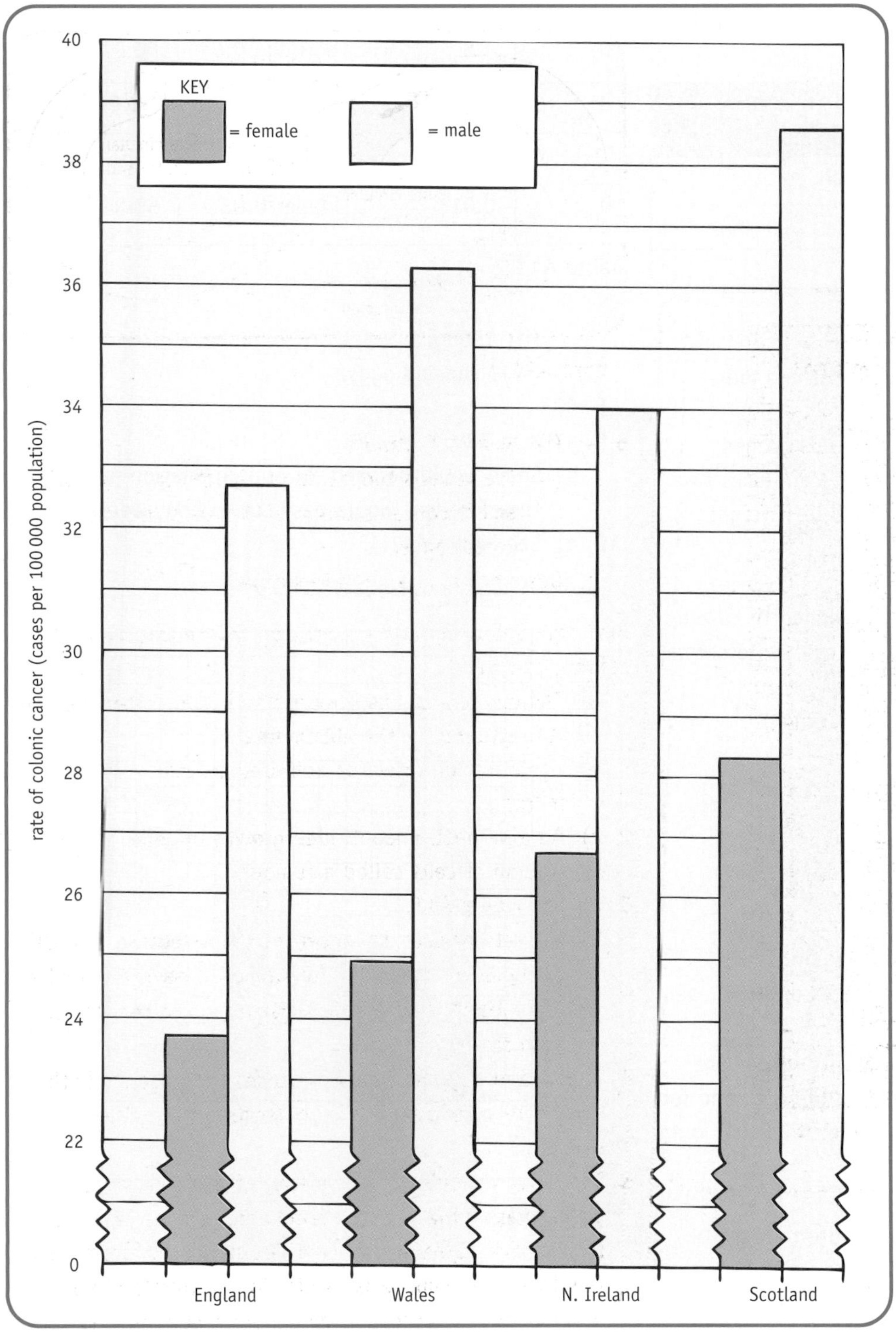

Figure A11.3

10) m
11) d
12) a
13) c
14) b
15) j
16) e
17) l
18) g

2 a) See Table A11.1.

type of consumption	class of food consumed (%)		
	protein	fat	carbohydrate
actual	20	40	40
recommended	15	30	55

Table A11.1

b) S/he should eat less protein, less fat and more carbohydrate.

3 a) Casein contains them all. Group 2 rats gained weight throughout the experiment. Zein lacks two essential amino acids. Group 1 rats lost weight throughout the experiment.

b) (i) Zein.

(ii) Their diet could have been changed to casein or to zein supplemented with the two essential amino acids that it lacks.

c) 20%.

4 a) 4.

b) 9.

c) From the store in their liver.

d) It causes heart rate to decrease.

e) (i) 280 beats/min.

(ii) 44%.

f) (i) The rats were fed vitamin B_1 again.

(ii) Heart rate returned rapidly to its original level.

5 a) C.

b) Vitamin C in solution loses its ability to prevent scurvy when the solution has been left to stand for a period of time or has been boiled.

6 See Table A11.2.

7 a) To make haemoglobin for blood.

b) Because he continues throughout life to lose 0.5–1.0 mg/day in dead skin cells, bile and urine.

c) (i) Because her body needs to take in iron for production of her own haemoglobin and that of the unborn baby.

vitamin	recommended daily allowance (mg)	mass of vitamin present in 100 g of named food (mg)	minimum mass of this food needed to supply recommended daily allowance (g)
A	1.0	carrots 2.0	**50.0**
B_1	1.5	porridge **0.5**	300.0
B_{12}	0.003	cheese 0.0015	**200.0**
C	**45**	apple 5.0	900.0
D	0.01	kipper **0.02**	50.0

Table A11.2

(ii) 1500g.

(iii) Meat and eggs.

8 90%.

9 a) X = 300, Y = 200.

b) Some would be used for bone formation and therefore less would pass out in body wastes.

10 a) See Figure A11.4.

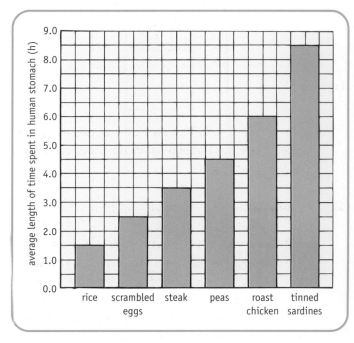

Figure A11.4

b) 2.4 times.

c) So that type of foodstuff is the only variable factor under investigation.

d) Because the rice moves quickly through their stomach leaving it empty and making them feel hungry.

11 a) A = contracted, B = relaxed.

b) Unlike food entering, food leaving will contain partly digested proteins and be at a lower pH (e.g. 3).

c) See Figure A11.5.

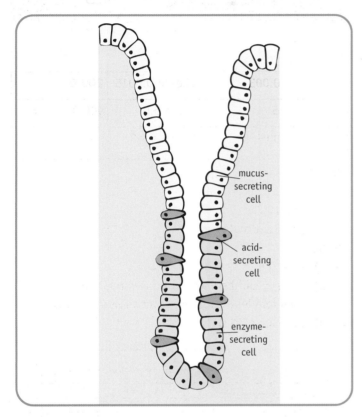

mucus-secreting cell

acid-secreting cell

enzyme-secreting cell

Figure A11.5

d) The person is expelling air from the stomach's air-filled space, up through sphincter 1 and out via the oesophagus.

12 a) insoluble protein $\xrightarrow{\text{pepsin}}$ soluble end products

b) (i) A and C (or B and D).

(ii) Pepsin works better at the warmer temperature.

(iii) The experiment would need to be carried out over a much wider and more detailed range of temperature.

c) (i) Surface area of egg white exposed to pepsin.

(ii) A and B (or C and D).

(iii) Pepsin works better when in contact with a large surface area of substrate.

d) (i) D.

(ii) Unlike the other tubes, no insoluble egg white was left since it had all been digested.

12 Control of the internal environment

Activity (Selecting and presenting data relating to volume and concentration of urine)

1 and 2 See Figure A12.1.

3 a) (i) 13.30.

(ii) See arrow in Figure A12.1.

b) 55 ml.

c) 2 hours.

d) 55 ml.

e) (i) An inverse relationship. As volume increases, concentration decreases.

(ii) Approximately the same quantity of urea is being produced and filtered out of the bloodstream during each half-hour interval. Sometimes it is dissolved in a large volume of water giving a low concentration of urea in urine; other times it is dissolved in a small volume of water giving a high concentration.

Activity (Selecting and presenting information on the role of ADH)

1 See Figure A12.2.

2 a) (i) When the blood's water concentration is low.

(ii) The kidney tubules and collecting ducts become more permeable so a large volume of water is reabsorbed into the bloodstream. By this means the blood's water concentration is returned to normal.

b) (i) When the blood's water concentration is high.

(ii) The kidney tubules and collecting ducts become less permeable so only a small volume of water is reabsorbed into the bloodstream. By this means the blood's water concentration is returned to set point.

c) (i) Increase.

(ii) Increase.

(iii) Decrease.

d) (i) Excessive production of urine.

(ii) S/he would reabsorb too large a volume of water and produce too small a volume of urine.

Applying your knowledge

1 1) k

2) j

3) m

4) p

5) q

6) f

7) r

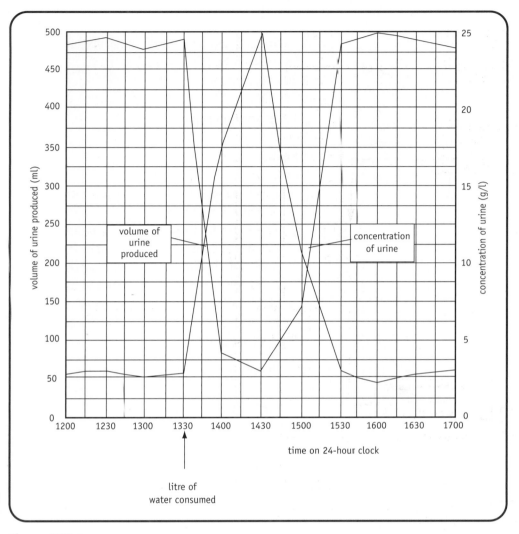

Figure A12.1

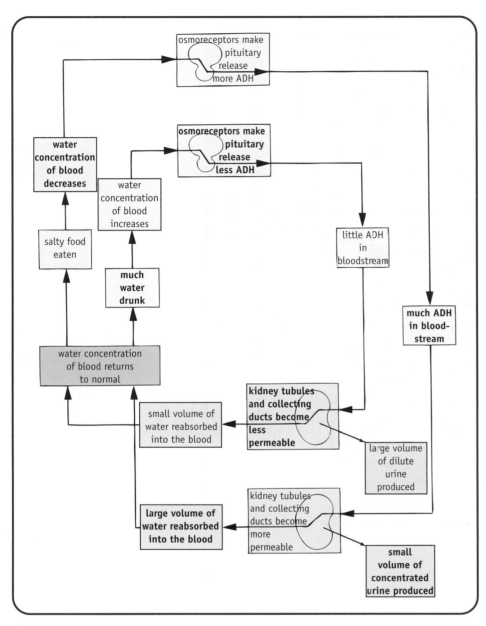

Figure A12.2

8) b
9) o
10) c
11) d
12) a
13) e
14) g
15) l
16) i
17) h
18) n

2 a) See Tables A12.1 and 2.

means by which water was gained	volume of water gained (ml)	percentage of total gain
in food	800	32
in drink	1350	54
metabolism	350	14
total	2500	100

Table A12.1

means by which water was lost	volume of water lost (ml)	percentage of total loss
by breathing	400	16
in sweat	550	22
in urine	1450	58
in faeces	100	4
total	2500	100

Table A12.2

b) See Figure A12.3.

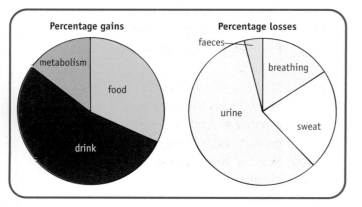

Figure A12.3

c) Urine.
3 a) 1 = renal artery, 2 = blood capillary from renal artery, 3 = blood capillary to renal vein, 4 = collecting duct, 5 = kidney, 6 = ureter, 7 = renal vein.
   b) Glomerulus, Bowman's capsule, blood capillary.
   c) (i) Blood.
      (ii) Urine.
   d) The liquid in 1 would contain more urea than the liquid in 7.
4 B.
5 a) To prevent a second variable factor being introduced into the investigation.
   b) Type of liquid (water or salt solution) consumed.
   c) 53 ml.
   d) (i) It increased.
      (ii) 2 hours.
   e) (i) No effect.
      (ii) Since the salt water consumed was isotonic to the blood, it did not alter the blood's water concentration. Therefore the osmoreceptors continued to respond as before and neither triggered an increase nor a decrease in ADH production.
6 a) Vessel A is a branch of the renal artery. It contains blood that is at high pressure caused by the beating of the heart; vessel B is a branch of the renal vein and contains blood at low pressure. Vessel A is wider than vessel B and this creates a 'bottle-neck' effect that further increases the pressure of blood entering the glomerulus.
   b) 1 = Bowman's capsule for filtration, 2 = kidney tubule for reabsorption of glucose (and water), 3 = collecting duct for reabsorption of water (and transport of urine).
   c) (i) It becomes more permeable.
      (ii) It allows more water to be reabsorbed into the bloodstream. This results in the formation of a smaller volume of more concentrated urine.
7 a) Glucose and amino acids.
   b) (i) Urea and salts.
      (ii) 70 times (urea) and 2 times (salts).
   c) 125 ml/min.
   d) (i) 1%.
      (ii) It was reabsorbed into the bloodstream.
8 See Table A12.3.

feature	freshwater bony fish	saltwater bony fish
relative number of glomeruli in kidney	**many**	**few**
relative size of glomeruli	**large**	**small**
rate of filtration of blood	**high**	**low**
state of fish's tissues compared with environment (hypo/hypertonic)	**hypertonic**	**hypotonic**
relative volume of urine produced	**large**	**small**

Table 12.3

13 Circulation and gas exchange

Applying your knowledge

1 1) i
  2) d
  3) j
  4) a
  5) h
  6) c
  7) f
  8) e
  9) b
  10) g

2 1) d
  2) j
  3) g
  4) f
  5) h
  6) b
  7) a
  8) e
  9) c
  10) i

3 1) g
  2) j
  3) i
  4) f
  5) h
  6) a
  7) e
  8) b
  9) d
  10) c

4 a) 12 units.
  b) B-C.
  c) (i) Q.
     (ii) Semi-lunar (SL) valve.
     (iii) R.

5 a) (i) Finland.
     (ii) Japan.
  b) (i) Scotland.
     (ii) Japan.
  c) 710.
  d) 0.2%.
  e) 80.
  f) 6 times

6 a) and b See Figure A13.1.

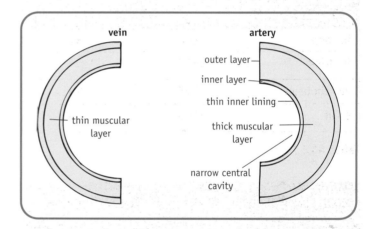

Figure A13.1

  c) (i) Unlike an artery, a vein has valves.
     (ii) They prevent backflow of blood.

7 a) See Figure A13.2.
  b) Rhythmical contractions of heart muscle forcing blood along arteries.
  c) P = 60 beats/min, Q = 70 beats/min.
  d) (i) Between minutes 3 and 4.
     (ii) Between minutes 8 and 9.
  e) (i) P = 5 minutes, Q = 7 minutes.
     (ii) P was fitter since her recovery time was shorter.

8 a) (i) It makes rate of blood flow increase.
     (ii) This supplies the muscle with the increased quantity of food and oxygen needed to generate extra energy for strenuous exercise.
  b) (i) Skin and heart muscle.
     (ii) Kidneys.
  c) (i) Brain.
     (ii) The brain is the body's control centre. To be

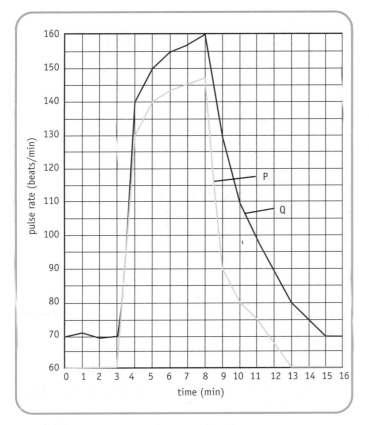

Figure A13.2

reliable and stay in perfect working order, it must remain unaffected by exercise.
d) (i) It would become redder.
   (ii) It would be affected by an increased rate of blood flow.

9 The molecule of urea present in the bloodstream would leave the liver by the <u>hepatic vein</u>, enter the <u>vena cava</u> and then be transported into the <u>right atrium</u> and then the <u>right ventricle</u> of the heart. From here it would pass by the <u>pulmonary artery</u> to the lungs then via the <u>pulmonary vein</u> back to the <u>left atrium</u> of the heart. It would be transported to the <u>left ventricle</u> of the heart and then be pumped out into the <u>aorta</u>. Finally it would enter the kidney by the <u>renal artery</u>.

10 a) They both increase.
   b) 7500 ml.
   c) 6250 ml.

11 a) (i)   C.
      (ii) A
      (iii) Blood at site C has a higher concentration of CO_2 than blood at site A.
   b) (i)   40 = C.
      (ii) 95 = B.
      (iii) 100 = A

c) X = pulmonary vein, Y = pulmonary artery.
d) Because they rid the body of CO_2 which is a waste product of metabolism.
e) 3), 5), 4), 2), 1).

14 Blood

Applying your knowledge
1   1) d
    2) g
    3) f
    4) h
    5) b
    6) i
    7) j
    8) l
    9) k
    10) c
    11) e
    12) a
2 a) Capillary.
   b) (i)   Red blood cell.
       (ii) Uptake and transport of oxygen.
       (iii) It is small and biconcave in shape giving it a relatively large surface area for oxygen absorption. Its cytoplasm is rich in haemoglobin, the oxygen-carrying pigment. It is flexible in shape enabling it to squeeze through narrow capillaries.
   c) (i)   Monocyte.
       (ii) It engulfs bacteria by phagocytosis.
3 a) (i)   98.
       (ii) 83.
       (iii) 25.
   b) (i)   58.
       (ii) Dissociation.
       (iii) Respiring tissues.
   c) (i)   94.
       (ii) 70.
       (iii) Increase in temperature brings about a reduction in haemoglobin's affinity for oxygen.
4   598:1.
5 a) See Figure A14.1.
   b) As partial pressure of oxygen increases, percentage saturation of haemoglobin with oxygen also increases.
   c) (i)   4.
       (ii) 3.

361

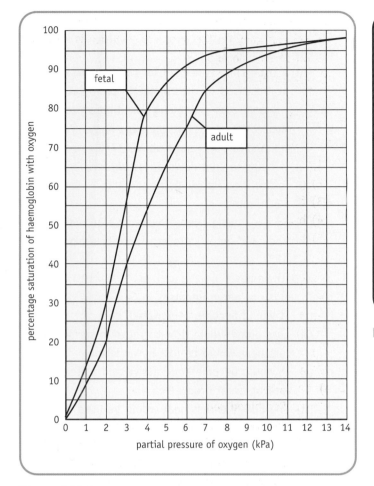

Figure A14.1

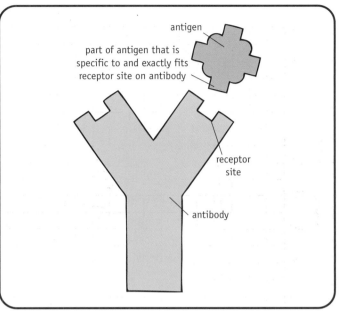

Figure A14.2

d) (i) 79.
   (ii) 91.

e) (i) Fetal haemoglobin has a higher affinity for oxygen than adult haemoglobin.
   (ii) Only by having a higher affinity for oxygen is fetal haemoglobin able to draw oxygen across the placenta from the mother's haemoglobin.

6 4), 2), 3), 1).

7 a) Lymphocyte.
   b) See Figure A14.2.

8 a) IgG.
   b) (i) IgM.
      (ii) IgE.
   c) During the secondary response, the production of antibodies is more rapid than during the primary response. In addition the concentration of antibodies reaches a higher level and lasts for longer in the secondary response.
   d) The graph shows that the response made to the second injection was faster than that made to the first one. This suggests that certain lymphocytes already 'knew what to do' from the previous time.

9 a) In response to the antigen (even in a damaged state) lymphocytes make antibodies which give immunity. In addition some of these lymphocytes act as memory cells allowing a rapid response to the antigen if invasion occurs in the future.
   b) Artificially.

15 Brain and nervous system

Applying your knowledge

1 1) g
   2) e
   3) n
   4) j
   5) i
   6) a
   7) o
   8) m
   9) k
   10) b
   11) f
   12) d
   13) l
   14) h
   15) c

2 a) 2.

b) A = cerebrum, B = medulla, C = cerebellum,
   D = spinal cord.
c) (i)   T.
   (ii) T.
   (iii) F, medulla.
   (iv) F, cerebrum.
   (v)   F, cerebellum.
   (vi) F, hypothalamus.
   (vii) T.
   (viii)T.

3 a) The leg is represented by a relatively small part of
     the cerebrum's sensory region because the number
     of receptor cells that it possesses in proportion to
     its size is small compared with many other parts of
     the body.
  b) D.
  c) B.

4 a) Lips and hands.
  b) Foot.
  c) (i) The pinna would be much bigger in relation to
          the rest of the rabbit's body.
     (ii) In the rabbit, the ear flaps are much more
          mobile than they are in humans.

5 a) 1 = (iv), 2 = (vi), 3 = (ii), 4 = (v), 5 = (iii),
     6 = (i).
  b) It makes the person quickly withdraw their foot
     from danger without wasting time thinking about
     it.

6 a) Stimulus = piece of dirt,
     receptor = surface of eye,
     effector = tear gland,
     response = secretion of 'water',
     protective value = dirt is washed out of eye.
  b) See Table A15.1.

completely involuntary	can be altered partly by voluntary means
churning of food in stomach	blinking
constriction of eye pupil	coughing
dilation of eye pupil	laughing
flushing of skin	muscular contraction making knee jerk
peristaltic contraction	sneezing

Table A15.1

7 a) and b) See Figure A15.1.
  c) (i)   It increases.
     (ii) It decreases.
     (iii) A little time is needed for the body to become
           acclimatised to the change in temperature.
  d) (i)   His body was employing vasodilation to
           increase heat loss because his body
           temperature has risen above 37°C.
     (ii) It helps to cool down an overheated body and
           return it to its set point at 37°C.
  e) (i)   42 minutes.
     (ii) By generating heat, it warms up an over-cooled
           body and helps to return it to its set point of
           37°C.

8 a) Hypothalamus.
  b) It sends out nerve impulses.
  c) (i)   It would become constricted.
     (ii) Less blood would flow to the skin surface so
           less heat would be lost by radiation.
  d) (i)   It would become more active and increase the
           rate of sweat production.
     (ii) The body's excess heat would be used to
           convert the water in sweat to water vapour and
           bring about a cooling effect.

9 a) (i) and (ii) See Figure A15.2.
     (iii) Blood vessels in the skin.
  b) Shivering which involves muscular contractions that
     generate heat energy.
  c) Put on more clothing and consume a hot drink.

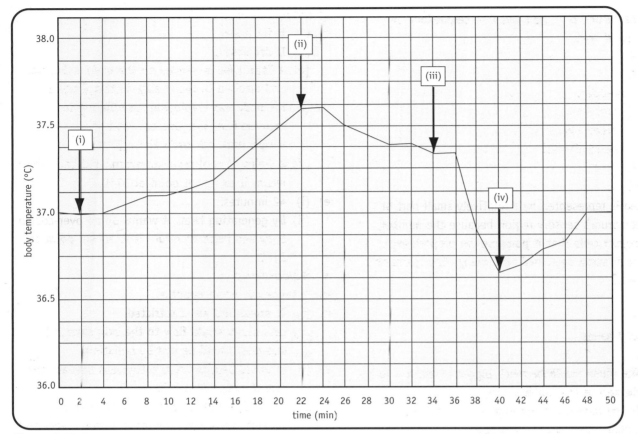

Figure A15.1

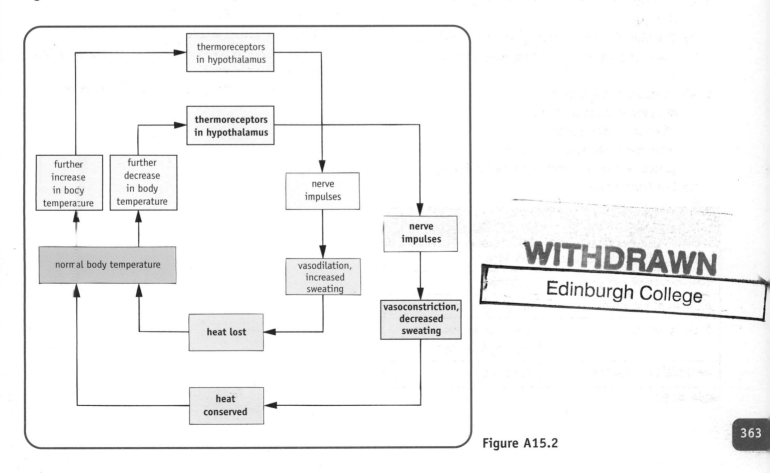

Figure A15.2

WITHDRAWN
Edinburgh College